T0353203

Introduction to Nuclear Science

Written to provide students who have limited backgrounds in the physical sciences and math with an accessible textbook on nuclear chemistry and physics, *Introduction to Nuclear Science, Fourth Edition* continues to provide a clear and complete introduction to nuclear chemistry and physics, from basic concepts to nuclear power and medical applications. Incorporating suggestions from adopting professors and collaborations with the U.S. Department of Energy-funded and American Chemical Society-sponsored Nuclear Chemistry Summer School, a new chapter on nuclear structure is now included.

Also new to this edition:

- A section covering mass excess calculations
- Isochron dating of rocks
- The section on statistics is completely re-written to better align with conventional instruction
- Expanded discussion of recent changes in the nuclear power industry and nuclear medicine

This book covers energetics, nuclear stability and structure, radioactive decay and reactions, interactions of radiation with matter, detection methods, and safety measures, including monitoring and regulations. This updated, expanded edition provides a much-needed textbook and resource for undergraduate students in science and engineering as well as those studying nuclear medicine and radiation therapy.

Introduction to Nuclear Science

Fourth Edition

Jeff C. Bryan
University of Wisconsin–La Crosse

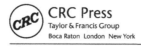

CRC Press
Taylor & Francis Group
Boca Raton London New York

CRC Press is an imprint of the
Taylor & Francis Group, an **informa** business

Designed cover image: Lori Hanson

Fourth edition published 2023
by CRC Press
6000 Broken Sound Parkway NW, Suite 300, Boca Raton, FL 33487-2742

and by CRC Press
4 Park Square, Milton Park, Abingdon, Oxon, OX14 4RN

CRC Press is an imprint of Taylor & Francis Group, LLC

First edition published by CRC Press 2008
Second edition published by CRC Press 2013
Third edition published by CRC Press 2018

ISBN: 9781032301808 (hbk)
ISBN: 9781032301839 (pbk)
ISBN: 9781003303787 (ebk)

DOI: 10.1201/9781003303787

Typeset in Times
by Newgen Publishing UK

Access the Support Material: www.routledge.com/9781032301808

Dedication

For Chris

Contents

Author Biography

Jeff C. Bryan was born in Minneapolis, Minnesota, in 1959, the second of four children. His arrival coincided with the purchase of a clothes dryer and was therefore somewhat joyous—no more freeze-dried diapers during the long Minnesota winters. After two winters of being bundled up and thrown out into the snow to play with his brother, the family moved to Sacramento, California. Dr. Bryan was raised in Northern California during the 1960s and 1970s on hot-dish, Jell-O™[1] salad, reticence, and hard work. The conflict of the free and open society, with the more reserved home life is the cause of many of Dr. Bryan's current personality disorders.

Dr. Bryan's father felt it was important that his children pay for their own way for their post-secondary education, thereby gaining an appreciation for advanced learning as well as being able to budget. Dr. Bryan decided to stay at home for his first two years of college, attending American River College in Sacramento. This was fortuitous, as he began his study of chemistry from excellent teachers in a small classroom and laboratory-intensive environment. Dr. Bryan later earned a Bachelor of Arts degree in chemistry from the University of California, Berkeley, and his doctorate in inorganic chemistry from the University of Washington, Seattle, under the supervision of Professor Jim Mayer. His thesis presented a new chemical reaction, the oxidative addition of multiple bonds to low-valent tungsten.

After one year of postdoctoral work with Professor Warren Roper at Auckland University, Dr. Bryan spent five years at Los Alamos National Laboratory, initially as a postdoctoral fellow, then as a staff member. At Los Alamos his mentor, Dr. Al Sattelberger, encouraged him to begin research into the chemistry of technetium, which turned out to be modestly successful. Dr. Bryan then accepted a staff position at Oak Ridge National Laboratory, working with Dr. Bruce Moyer's Chemical Separations Group as the group's crystallographer. During this time there was a great deal of work done in the group investigating the separation of ^{137}Cs from defense wastes.

In 2002, Dr. Bryan made a long-anticipated, and somewhat apprehensive, move to academia by accepting a faculty position at the University of Wisconsin–La Crosse. He was hired to teach nuclear and general chemistry courses, primarily on the basis of his 13 years of research experience in the National Laboratories. Apparently, the chemistry faculty was in desperate need to fill this position, as this background did not adequately prepare him for such a dramatic change. In spite of the odds, this arrangement has worked out well for all concerned. It has gone so well that Dr. Bryan was granted tenure in 2008 and was promoted to full professor in 2011. He also taught radiation physics for six semesters.

Dr. Bryan has over 100 publications in peer-reviewed journals, including one book chapter and an encyclopedia article. His research has been featured on the covers of one book, two journals, and two professional society newsletters. Dr. Bryan co-authored a nuclear chemistry laboratory manual with Dr. Sid Katz in 2011. In 2015, Dr. Bryan worked with Dr. Nivaldo Tro to revise the nuclear chemistry chapters in Dr. Tro's General Chemistry textbooks. They also worked as part of a larger group of authors to write *30-Second Chemistry* (a.k.a. *Know-it-All Chemistry*) in 2017.

Dr. Bryan married the love of his life, Chris, in 1983. They are still happily married and are very proud of their three children, Katrina, Lars, and Tom.

---

1   Jell-O is a delicious dessert and a registered trademark of the Kraft Foods Inc., Northfield, Illinois.

Preface to the Fourth Edition

While introducing myself to a high-school general chemistry class I co-instruct, I told the students about how my parents raised me to never be too sure of myself. Low self-esteem is pretty common in the north-central United States, so I think some of the students appreciated hearing this. I added that, in all modesty, I thought this made me a better teacher because I feel a constant need to improve.

For much of the past decade, I had settled into a pretty comfortable teaching routine of alternating semesters of general and nuclear chemistry. As I began to realize that I might be approaching retirement age, I wondered to myself if I shouldn't just ride this routine out until then. That thought did not sit well with me, but I could not figure out what I wanted to tackle next. Fortunately, I got a call from Dr. Lynn Francesconi asking if I might be interested in teaching at the U.S. Department of Energy-funded and American Chemical Society-sponsored Nuclear Chemistry Summer School (NCSS). I quickly realized this was the challenge I was looking for and agreed.

Almost all of the changes in this edition come from my experiences in the NCSS over the past three summers. My main goal was to make the book a useful tool for the NCSS, while maintaining its appeal as an introductory text. This was a bit of a challenge as the students that populate the NCSS are top-notch and the NCSS covers a bit more ground than what is normally covered by an introductory course. The main gap to fill was nuclear structure (now Chapter 6), but I also completely re-wrote the statistics section (2.7) to make it more consistent with common instruction in nuclear science. Having a chapter on nuclear structure also allowed a bit more in-depth discussion of nuclear reactions. I also added a section to the chapter covering energy and the nucleus (Section 3.4) describing mass excess and how to use it in nuclear decay.

I am immensely grateful to the other NCSS instructors who allowed me to sit in on their lectures and tolerated my many questions. In particular, I'd like to thank Trish Baisden, Cody Folden, Alice Mignerey, and Paul Karol. They all contributed significantly to my understanding of nuclear structure. I'd also like to thank Sylvia Jurisson, David Shuh, and Henry VanBrocklin who all contributed to deepening my knowledge of nuclear science in their respective areas of expertise. Finally, I'd like to thank Lori Hanson for this edition's cover design, especially the radiation symbol which is reminiscent of a Viking shield.

Like previous new editions, the applications chapters have also all been updated to maintain a reasonably contemporary view of nuclear reactors and medicine. I also corrected an embarrassing number of errors present in the third edition. If you spot any in this edition, they are my responsibility but please let me know.

Jeff C. Bryan
La Crosse, Wisconsin

Preface to the Third Edition

Those of you that made the transition from the first to second editions of this book can let out a sigh of relief. The changes this time are much less extensive. This book is only about 5% bigger than the second edition, and the table of contents remains unchanged. My primary goals for the third edition were three-fold: updating contemporary information, using more standard conventions, and adding a few more questions at the end of each chapter.

Bringing the book up-to-date mostly involved changes in the nuclear power industry. When I wrote the first edition, there was a lot of excitement over what was assumed to be a renaissance for construction of new nuclear reactors. The second edition came out shortly after the Fukushima meltdowns, and there was a great deal of uncertainty. Now (mid-2017), there seems to be some divergence. China, Russia, India, and some Middle Eastern countries are continuing to move aggressively forward with nuclear power plant construction, while most Western democracies are stepping away. I can't help but wonder if the rather pervasive, and largely irrational, fear of anthropogenic sources of ionizing radiation is at least partly to blame for so many countries turning away from nuclear.

When I first wrote this book, I used units and symbols that were familiar to me. I favored units such as curies and rems because I was accustomed to seeing them in my work at the U.S. national laboratories. I also used k to represent the decay constant instead of λ because many general chemistry textbooks did the same, reinforcing the relationship between radioactive decay and first order chemical kinetics while avoiding potential confusion with the symbol for wavelength. I did not anticipate this book enjoying any level of popularity outside of the United States. I've been pleasantly surprised. A couple of the anonymous reviewers responded to Taylor & Francis' request for suggestions by encouraging me to use more conventional units. So has my colleague, Shelly Lesher, a faculty member in the physics department at the University of Wisconsin–La Crosse. While I did not abandon U.S. units, I tried to provide more parity to international units. For any other k fans out there, I'm sorry, but this book was rather singular among nuclear science textbooks in eschewing λ as the decay constant.

When adding questions at the end of each chapter, I looked for holes in coverage. Remarkably lacking were vocabulary questions. I try to emphasize the importance of speaking the language of nuclear science to my students, yet I wasn't holding them to it with the end of chapter questions. I hope that more students will now correctly differentiate between "nuclide" and "isotope."

While I corrected a number of the errors present in the second edition, I probably missed a few, and likely created some more. I accept full responsibility for all of them. Please feel free to let me know when you find them.

Jeff C. Bryan
La Crosse, Wisconsin

Preface to the Second Edition

I never expected to be writing this. When I wrote the first edition of this textbook, I thought I had really gotten away with something. I wrote the book for students in my nuclear chemistry and radiation physics classes—almost all of whom are nuclear medicine or radiation therapy majors with a limited background in physical sciences and math. My main motivation was to provide them with an accessible textbook that would be a useful tool in their studies. I honestly didn't expect that anyone outside of my students (besides my mom) would be interested in buying it. Much to my surprise, it exceeded the publisher's sales expectations, and a second edition became possible.

I was happy to have this opportunity, as I wanted to add a chapter on dosimetry as well as flesh out a few sections in the existing text. It was also a chance to find out what readers would like to see in a new edition. I asked a student, Lucas Bartlett, to survey teachers who used the text in their classes, as well as those who had examined the text but were not yet using it with their students. I am grateful to all who responded, especially Jason Donev, Thomas Semkow, Brett McCollum, Craig McLachlan, Lawrence Duffy, Pierre St. Raymond, Dot Sulock, and Magdalen Normandeau. I wasn't able to incorporate all of their suggestions, but they will likely see their influence in some additions listed here:

- The mathematics in some of the examples in the first few chapters are explained in more detail
- An appendix containing nuclear data for all nuclides mentioned in the text (plus some others) is added
- Discussion of applications are expanded and, generally, moved to earlier chapters
- Coverage of nuclear reactors is significantly expanded, with a new chapter looking at more reactor types, their safety systems, and recent accidents
- The number of end-of-chapter questions was increased for every chapter

Along with the new chapters and appendix, new sections, or subsections, were added on nuclear forensics, radiology, gamma cameras, and decay through proton or neutron emission. In my hands, the manuscript is about 60% larger than the first edition, so it is definitely bigger. Hopefully, you will also find that it is better and more flexible.

In trying to write a textbook that would be more generally useful, I was especially conscious of how much the first edition centered on nuclear science in the United States. I have tried to broaden this perspective somewhat by pointing out how things are sometimes done differently outside of the U.S. I was also concerned that those teaching classes more focussed on applications of nuclear science be better served by the second edition. By moving the applications chapters earlier in the text, it gives those instructors an opportunity to get their classes into discussions of applications sooner.

As was true with the first edition, I had a lot of help with this one. I'd like to thank all the students in my nuclear chemistry and radiation physics classes over the past few years for their suggestions and attention to detail. I'd especially like to thank those that worked directly on this edition with me: Mikaela Barrett, Travis Warmka, Kelsey Schilawski, Lucas Bartlett, Dan Hanson, and Mike Hoppe. I'd also like to thank my colleagues at the University of Wisconsin—La Crosse for their suggestions and review of portions of the manuscript, especially Shelly Lesher, Sherwin Toribio, Melissa Weege, Nishele Lenards, Bruce Osterby, and Aileen Staffaroni. I would also like to thank Mary Ellen Jafari of Gundersen-Lutheran Hospital for her comments on Chapter 8. Finally, I need to thank Douglas Boreham for getting me to think more critically of the LNT and ALARA (you'll have to read the book if these acronyms are unfamiliar!).

While I was able to correct a number of the errors in the first edition, the extensive nature of the revisions for this edition have undoubtably created some more. I accept full responsibility for all of them. If you spot one, I'd appreciate hearing about it, as we may be able to correct it before the next printing.

I have enjoyed writing this book. I hope you enjoy reading it.

Jeff C. Bryan
La Crosse, Wisconsin

Preface to the First Edition

Like many textbooks for specialized classes, this one was born out of frustration with the applicability of existing offerings. This text was initially designed for the Nuclear Chemistry course I have taught for the past five years. This class is almost entirely populated by nuclear medicine technology majors who need a solid background in fundamental nuclear science, but have not studied math beyond college algebra and statistics. There are many excellent Nuclear Chemistry textbooks available to teach science or engineering seniors or graduate students, but very few for those lacking a more extensive science and math background. Ehmann and Vance's *Radiochemistry and Nuclear Chemistry* served as my primary text for this class for the past four years. It is very well written and nicely descriptive, but it is really intended for radiochemists, not medical professionals. It is also slipping out of date, with no future edition planned. I also learned that some of my students were passing my class without ever cracking the primary text. Clearly, a more appropriate text was necessary.

Writing a Nuclear Chemistry textbook for nuclear medicine majors seemed too limiting, so this text also includes material relevant to other medical professionals using ionizing radiation for diagnosis and treatment of diseases. In particular, chapters were added discussing radiation therapy and X-ray production. Additionally, it seemed that an introductory nuclear science text should address important contemporary topics such as nuclear power, weapons, and food/mail irradiation. All too often, those speaking passionately to these issues are inadequately versed, and resort to emotional appeals while ignoring some of the science. Therefore this text is also meant to serve as a general primer in all things nuclear for the general public. Very little science and math background is assumed, only some knowledge of algebra and general chemistry.

The more extensive chapters on fission nuclear reactors are also designed to help those that will soon be entering the nuclear workforce. The nuclear power industry in United States has been static for almost 30 years, but now seems on the verge of renaissance. Nuclear power now appears to a more environmentally benign source of electricity, and a significant number of new plants may be constructed in the near future. Workers at these plants will need a fundamental understanding of nuclear science as well as a basic idea of how the plant works. This text can serve has a primer for these workers.

Finally, this text can also be of use to scientists making a career start or move to the growing nuclear industry or to the National Laboratory system. I began 13 years of work at Los Alamos and Oak Ridge National Laboratories in 1988. At that point I was thoroughly trained as a synthetic inorganic chemist, and was largely ignorant of nuclear science. Some of my motivation in writing this text is to provide scientists in similar situations an easier way better understand their new surroundings.

The term nuclear science is used rather deliberately in the title and throughout the text. It is meant to encompass the physics, mathematics, chemistry, and biology related to nuclear transformations and all forms of ionizing radiation. Fundamentally, nuclear processes are within the realm of physics, but it can also be argued that all of the disciplines mentioned above play an important role in understanding the uses of these processes and the resulting ionizing radiation in today's world.

When this text was initially proposed, some of the reviewers rightly questioned whether my background was appropriate. After all, much of it could be characterized as synthetic inorganic chemistry and crystallography – there's not a lot of hard-core nuclear work there. In my defense, I would point out that I did pay attention when my National Laboratory colleagues "went nuclear." After 13 years at Los Alamos and Oak Ridge, some of it rubs off. I would also argue that a book that wishes to bring nuclear science to a broader audience must be written by someone who originally comes from outside the field and can still see it from an external perspective. My successful experiences teaching nuclear science to those that have a limited math and science background suggests that I might be a good choice to write this book. Ultimately, it will be the readers who decide if I have been successful in this effort.

This book draws heavily on those that have proceeded it, especially Ehmann and Vance's *Radiochemistry and Nuclear Chemistry*, and Choppin, Rydberg, and Liljenzin's *Radiochemistry and Nuclear Chemistry*. It is not meant as a replacement to either, as they are both much more detailed, rather it is a contemporary resource for those wishing to learn a little bit of nuclear science. For those that wish to explore nuclear science beyond this text, a complete listing can be found in the bibliography.

I had often considered writing this textbook over the past five years of teaching nuclear chemistry and physics at the University of Wisconsin – La Crosse. It did not become a reality until Lance Wobus of Taylor & Francis asked to meet with me at a meeting of the American Chemical Society about a year and a half ago. I am very grateful to Lance for seeking me out, and for his constant support during the writing process.

This is easily the most ambitious writing project I have undertaken, and it would not have been possible without the assistance and encouragement of many others. First, I must acknowledge my predecessor in my current position at the University of Wisconsin – La Crosse, Willie Nieckarz. Based on his background in the nuclear navy, he pioneered our nuclear science programs. My nuclear chemistry course remains strongly based on his materials. I also acknowledge my students who have never been reluctant to question me beyond the limits of my knowledge, and for their critical reading of earlier drafts of this text. I would especially like to single out former students Stephanie Rice, Rachel Borgen, and Jim Ironside, for their extensive suggestions, and help preparing the solutions manual and figures. I must also thank my colleagues in the Departments of Chemistry and Physics at the University of Wisconsin – La Crosse, without their support and encouragement, this book would not have been written. I would also like to acknowledge my former colleagues and teachers who took their time to introduce me to nuclear science. Without them, I would not have been hired into the wonderful job I now enjoy. Finally, I need to thank my family. Not only have they endured my many absences while preparing this text, but my father (William Bryan Sr., formerly an engineer who worked on the nuclear rocket program NERVA) and son (Lars Bryan, currently a sophomore studying mathematics at the University of Wisconsin–Madison) have read much of the text and provided numerous helpful suggestions. It is rare that three family generations have the opportunity to work together on a project like this.

Jeff C. Bryan
La Crosse, Wisconsin

1 Introduction

1.1 RADIATION

Radiation is scary. We can't see, hear, smell, or touch it, and we know that it can do terrible things to us. In large enough doses, it can kill us. Significant doses received over a long period of time can increase our odds of getting cancer. It is easy to fear. Radiation is also well studied, and this knowledge can be used to dispel some of the fear it so easily engenders. This chapter will focus on the fundamentals of physics that can help us answer questions like "what is radiation," "where does it come from," and "what does it want with us?"

Much of this chapter will be devoted to definitions. In a sense this chapter is a foreign language tutorial where we'll learn to speak some nuclear science. Let's start with *radiation*. **Radiation** is energy that moves. More scientifically, it is energy propagated through a medium. There are all kinds of radiation all around us, all of the time, such as radio waves, microwaves, and infrared and visible light. Radio stations use radio waves to transmit audio signals to our homes and cars. We also use radio waves to transmit signals to and from our growing number of wireless devices, such as cell phones, networked computers, and remote controls. Remote controls can also use infrared light to transmit signals to televisions, game consoles, and other video or audio equipment. We use microwaves to transmit energy to water molecules in our food. Visible light is also a form of radiation. Green light generated by the sun travels through space and our atmosphere, bounces off a plant, travels through some more air to our eyes, which tells us the plant is green.

All of the examples of radiation in the preceding paragraph use photons to transmit their energy. Photons are commonly thought of as the individual packets of visible light, but are also handy at transmitting all kinds of energy as shown in Figure 1.1. Collectively, these types of radiation are known as electromagnetic (EM) radiation.

The various forms of EM radiation interact with matter in different ways. Radio waves don't interact much with most matter but produce electrical currents in wires and circuits. By varying aspects of the waves, complex signals such as those containing audio and video can be transmitted from one electronic device to another. Microwave radiation causes molecular agitation. Specifically, a microwave oven causes water molecules to rotate faster, thereby increasing their temperature. This cleverly allows the food to be warmed without directly heating up everything else in the oven (because the microwaves do not interact with plastic, glass, or most dishware). Infrared radiation causes molecular vibration and bending, again increasing the temperature of the matter it hits. Warming lamps in restaurants typically use infrared lamps to keep prepared meals warm while they are waiting to be distributed to customers. Visible light causes some electron excitations within atoms and molecules. This means an electron is temporarily promoted to a higher energy level. Chemical receptors in our eyes use these temporary changes to send electrical signals to our brain,

DOI: 10.1201/9781003303787-1

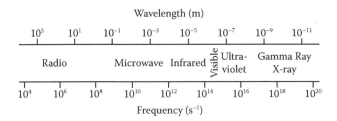

FIGURE 1.1   The electromagnetic spectrum.

allowing us to see. Ultraviolet radiation also causes electron excitations but does so to a broader number of chemical compounds and is more likely to result in a lasting chemical change. Sufficient exposure to our skin can cause enough chemical change to take place to cause a burn. Gamma and X-rays are powerful enough to cause ionizations in any atom or molecule and are therefore rather destructive in how they interact with matter. An ionization occurs when an electron is completely separated from its atom or molecule.

Like most things in the subatomic universe, photons are a little strange. Not only do they behave like individual packets of energy, but they also exhibit wave behavior. For example, the type of energy they propagate depends on the wavelength (λ) and frequency (ν) of the photon. **Wavelength** is the distance between two equivalent points on a wave (like two peaks), and **frequency** is how often the same part (like peaks) of the wave goes past. If the waves are traveling the same speed, then frequency is also how many waves are packed into the same distance. Wavelength and frequency for any wave are illustrated in Figure 1.2. The top wave has a longer wavelength (more distance between peaks), but a smaller frequency (fewer waves in the same distance). In fact, there are twice as many waves in the bottom wave than the top, meaning its frequency must be two times larger. Notice also that wavelength of the top wave is twice as long as that of the bottom, i.e. two bottom waves fit in the same space as one of the top waves. This pattern is also apparent in Figure 1.1 where there is a regular decrease in wavelength and a regular increase in frequency when moving from left to right.

It would seem that there is a simple mathematical relationship between wavelength and frequency, and there is. If the wavelength of a particular photon is multiplied by its frequency, the result is always the same: the speed of light ($c = 2.998 \times 10^8$ m/s). This relationship is quite handy for calculating wavelength from frequency or frequency from wavelength.

$$c = \lambda\nu \tag{1.1}$$

Example 1.1   Green light has a wavelength of 525 nm, what is its frequency?

Solving Equation 1.1 for frequency and substituting in the wavelength (525 nm) and the speed of light gives the equation, and answer, below. Note that the wavelength units need to be converted to meters to cancel properly. Notice also that frequency can be expressed in units of per second (s^{-1}) or Hertz (Hz), and the two are numerically equivalent.

FIGURE 1.2   Two waves.

$$v = \frac{c}{\lambda} = \frac{2.998 \times 10^8 \frac{m}{s}}{525 \text{ nm} \times \frac{m}{10^9 \text{ nm}}} = 5.71 \times 10^{14} \text{ s}^{-1} \text{ or } 5.71 \times 10^{14} \text{ Hz}$$

The amount of energy each photon is packing is directly related to the frequency.

$$E = h\nu \tag{1.2}$$

where h is Planck's constant (6.626×10^{-34} J·s).

Example 1.2 A therapeutic X-ray machine delivers a beam of photons with a maximum frequency of 9.67×10^{20} s^{-1}. Calculate the energy of one of these photons.

$$E = h\nu = (6.626 \times 10^{-34} \text{ J} \cdot \text{s}) \times (9.67 \times 10^{20} \text{s}^{-1}) = 6.41 \times 10^{-13} \text{ J}$$

Combining Equations 1.1 and 1.2 shows that energy and wavelength are inversely related.

$$E = hc/\lambda$$

Therefore, electromagnetic radiation with high frequency and short wavelength, such as X-rays and gamma rays, are also the highest energy photons known, while radio waves are relatively low in energy. This is why Figure 1.1 is called the electromagnetic *spectrum*. It is an arrangement of electromagnetic radiation according to energy (increasing from left to right). We can also begin to appreciate why certain forms (X- and gamma rays) of electromagnetic radiation are feared. Being more energetic has its consequences.

1.2 ATOMIC STRUCTURE

Before we get into the more energetic forms of radiation, we need to know a bit about how atoms are put together, since most of the radiation we'll be concerned with in this book has its origin in the nuclei of atoms (**radioactivity**). Atoms are made up of a small, positively charged nucleus surrounded by a relatively large space (orbitals) occupied by tiny, fast-moving, negatively charged **electrons**. The nucleus is 10,000 to 100,000 times smaller than the atom, which is defined by the approximate boundaries of the electron orbitals. This means that if the nucleus were the size of a marble, the atom would be as big as a football stadium! Atoms and all the objects that they make up, like this book, bananas, and people, are therefore mostly empty space. Despite its relatively small size, the nucleus contains over 99.9% of the mass of the atom, which means that almost all of the mass of an atom occupies very little space within the atom. How could this be? It is because the particles that make up the nucleus are about 2000 times more massive than the electrons.

In this book, we'll use the **Bohr model** to describe the electronic structure of the atom. As shown in Figure 1.3, the Bohr model shows the electrons orbiting the nucleus in circles. Readers should know that this is not terribly accurate. In reality, electrons occupy three-dimensional spaces called orbitals. These spaces are not exactly delineated since there is always a finite (however small!) probability of finding the electron just a little further from the nucleus. The Bohr model works just fine for us, because the circular orbits nicely represent the average distance between the electron and nucleus. Note also that the nucleus is not drawn to scale in Figure 1.3—if it were, it would be so small, we wouldn't be able to see it.

FIGURE 1.3 The Bohr atomic model.

The nucleus is made up of positively charged **protons** and neutral **neutrons**. Both have a mass of approximately 1 unified atomic mass unit or u (1 u = 1.66×10^{-24} g), while the mass of an electron is a mere 5.5×10^{-4} u.[1] The number of protons in a nucleus is equal to the atom's atomic number and determines what element it is. For example, an atom with eight protons in its nucleus is an oxygen atom, the eighth element listed in the periodic table. It is an oxygen atom regardless of the number of neutrons present. The number of neutrons will simply affect the mass of the atom and determine which **isotope** of oxygen an atom is.

If both protons and neutrons vary, we'll need a more general term than isotope. The term **nuclide** refers to any isotope of any element, i.e., an atom with any number of protons and neutrons in its nucleus. We can now define isotope a bit more efficiently as nuclides with the same number of protons, but different numbers of neutrons. Any nuclide can be represented using a nuclide symbol:

$$\ce{^{A}_{Z}X_{N}}$$

where X is the elemental symbol from the periodic table, A is the **mass number** (number of protons plus neutrons), Z is the **atomic number** (number of protons), and N is the **neutron number** (number of neutrons). Since the mass number is equal to the number of protons plus the number of neutrons, then:

$$A = Z + N \tag{1.3}$$

It is fairly rare that the neutron number is included on a nuclide symbol. To find the number of neutrons, one only needs to subtract the number of protons from the mass number. Also, the atomic number is sometimes left off a nuclide symbol, since the elemental symbol also gives us that value. When spoken aloud, the nuclide symbol given as "element-A." For example, ^{99}Tc would be said as "technetium-99." Technetium-99 is another way to represent a nuclide symbol in writing. Finally, ^{99}Tc can also be represented as "Tc-99." These other methods of representing a nuclide in print are likely holdovers from days when it was difficult to place superscripts into manuscripts.

If isotopes have a constant number of **p**rotons, then isoto**n**es have the same number of **n**eutrons. More formally, **isotones** are nuclides with the same number of neutrons, but different numbers of protons. Finally, **isobars** are nuclides with the same mass number, but differing numbers of protons and neutrons. As examples, consider the following nuclides:

$$\ce{^{16}_{8}O_{8}} \quad \ce{^{18}_{8}O_{10}} \quad \ce{^{15}_{7}N_{8}} \quad \ce{^{18}_{9}F_{9}}$$

---

1   Some readers may be familiar with the "atomic mass unit" (amu). It is an older, albeit numerically equivalent, way of expressing atomic masses. 1 u = 1 amu.

Of these, only ^{16}O and ^{18}O are isotopes, ^{16}O and ^{15}N are isotones, and ^{18}O and ^{18}F are isobars. Isotope, isotone, and isobar are all more narrowly defined groups of nuclides, i.e., nuclide is the most general term, and the "iso" terms are more specific.

Collectively, neutrons and protons are known as **nucleons**. Can nucleons be further subdivided into even smaller particles? Yes, both are partially made up of **quarks**. There are six types, or flavors, of quarks: up, down, top, bottom, charmed, and strange. We are only concerned with up quarks, which have a +2/3 charge, and down quarks, which have a –1/3 charge. Protons have two up quarks, and one down quark, for a net charge of +1. Neutrons are made up of one up quark and two down quarks, for a net charge of 0. It is interesting to note that the difference between a proton and a neutron is a single quark. We can imagine the interconversion between them as simple as flipping a quark, from down to up or up to down. Viewed this way, it would seem relatively easy for protons and neutrons to convert to each other.

So, what holds the nucleus together? At first glance it should appear impossible to assemble a number of protons so closely. All those positive charges in a relatively small space should push themselves apart. The force pushing them apart is electrostatic: like charges repel each other, and opposite charges attract. This is also known as **Coulomb repulsion** (or attraction), and it acts over fairly long distances. Since we know that nuclei place protons into close quarters, there must be another force holding them together. This force is known as the nuclear **strong force**, which has a very short range (~10^{-15} m). The average nucleus is ~10^{-14} m in diameter, so this is a force that has a very limited range, even in a really small space like a nucleus. In fact, any given nucleon only feels a significant strong force attraction to its immediate neighbors, i.e., the nucleons that are right next to it. While the strong force acts over a very short distance, it is much more powerful (~100 × !) than the Coulomb repulsion threatening to rip the nucleus apart. It could be said that the ability to destroy a nucleus (through Coulomb repulsion) is insignificant next to the power of the (strong) force. After all, it is the strong force that binds the nucleons together in the nucleus.

1.3   NUCLEAR TRANSFORMATIONS

Now that we have an idea of what nuclei are made of, why they hold together, and how to represent them as symbols, let's start messing with them. Fundamentally, there are two types of nuclear transformations, decay and reaction. This is wonderful simplicity when compared to the world of chemistry where the numbers and types of chemical reactions are constantly proliferating. However, it is an oddity of nuclear science that decay processes are not referred to as "reactions." Nuclear reactions, which we'll study in more detail in Chapter 10, are only those transformations where a particle (like a neutron or a ^{4}He nucleus) or a photon interacts with a nucleus, resulting in the formation of another nuclide and some other stuff. Let's set that aside for a moment, so we don't get ahead of ourselves.

This first main transformation type we'll consider is nuclear decay. **Decay** starts with an unstable nuclide (the **parent**) that typically spits out a particle and/or a photon while turning itself into something more stable (the **daughter**, or **progeny**). An example would be ^{222}Rn decaying to ^{218}Po by ejecting two neutrons and two protons all at once as a single particle. This combination of nucleons is a ^{4}He nucleus and is also known as an **alpha particle**.

$$^{222}_{86}\text{Rn} \rightarrow {}^{218}_{84}\text{Po} + {}^{4}_{2}\text{He}$$

This is an example of **alpha decay**, or, it could be said that ^{222}Rn undergoes alpha decay. Notice that the atomic and mass numbers are the same on both sides of the arrow. ^{222}Rn starts out with 86 protons and 222 total nucleons. Together, ^{218}Po and an alpha particle have 86 protons and 222 total nucleons (check this for yourself!). This is how most nuclear equations are "balanced," by simply comparing the mass numbers (superscripts) and atomic numbers (subscripts). They should always add up to the same values on either side of the arrow.

There are two other basic types of nuclear decay, **beta** and **gamma**. Additionally, there are three different kinds of beta decay. The first is often referred to simply as "**beta decay**" but is sometimes called "beta minus" or "negatron" decay. This form of beta decay involves the ejection of an electron from the nucleus—a statement that shouldn't make much sense! Electrons are supposed to spend their time in the space *around* the nucleus, so how can one be emitted *by* the nucleus? Let's take a look at an example and see if we can understand its origin. ^{14}C decays to ^{14}N by spitting out an electron, also known as a **beta particle**.

$$^{14}_{6}\text{C} \rightarrow \, ^{14}_{7}\text{N} + \, ^{0}_{-1}\text{e}$$

The beta particle is symbolized here as $^{0}_{-1}\text{e}$, but is sometimes represented as $^{0}_{-1}\beta$ or β^-. The mass number for an electron should be equal to zero because it contains no protons or neutrons. The "atomic number" of -1 simply accounts for its negative charge. It also balances our nuclear decay equation quite nicely.

How does this electron come from the nucleus? Look at what happens to the numbers of protons and neutrons in this decay. The carbon nuclide has six protons and eight neutrons, while the nitrogen nuclide has seven protons and seven neutrons. A neutron has turned itself into a proton in this decay (flipped a quark), as shown in the decay equation of a free neutron.

$$^{1}_{0}\text{n} \rightarrow \, ^{1}_{1}\text{p} + \, ^{0}_{-1}\text{e}$$

When that positive charge is created, a negative charge must also be created, for balance. We can therefore think of a neutron converting to a proton and an electron within the ^{14}C nucleus. The proton remains a part of the nucleus while the electron is sent packing. Note that a neutron is represented in nuclear equations as $^{1}_{0}\text{n}$. As a nuclide, it has no protons, but it is a single nucleon; therefore, it has a mass number of one and an atomic number of zero. Similarly, a proton is a nucleon with an atomic number of one and can be symbolized as $^{1}_{1}\text{p}$ or as $^{1}_{1}\text{H}$ (a hydrogen-1 nucleus) in nuclear equations.

The second kind of beta decay is known as **positron** or "beta plus" decay. In this type of decay, a proton is converted into a neutron and a positron. A positron is a particle with the same mass as an electron, but with a positive charge. An example is the decay of ^{18}F to ^{18}O.

$$^{18}_{9}\text{F} \rightarrow \, ^{18}_{8}\text{O} + \, ^{0}_{+1}\text{e}$$

Positrons can also be symbolized as $^{0}_{+1}\beta$ or β^+. Positrons are a form of anti-matter (the same stuff that fuels the starships in *Star Trek*!). Every particle that we know about in our universe, such as electrons, has a corresponding anti-particle. In this case, the difference between particle and anti-particle is simply charge.

The third, and last, type of beta decay is **electron capture**. Just like positron decay, electron capture involves the conversion of a proton into a neutron, but it's done a little differently. As its name implies, one of the electrons surrounding the nucleus is drawn into the nucleus, transforming a proton to a neutron. An example is the decay of ^{67}Ga to ^{67}Zn.

$$^{67}_{31}\text{Ga} + \, ^{0}_{-1}\text{e} \rightarrow \, ^{67}_{30}\text{Zn}$$

The words "orbital electron" are often added below the electron in the decay equation to indicate this particle does not originate in the nucleus. In other words, we should not think of it as a beta particle, nor should it ever be represented with the Greek letter beta (β). Electron capture is unique among all forms of nuclear decay in that it has more than the unstable nuclide on the left-hand side of the arrow. We can excuse this poor behavior because the electron is not a nuclear particle.

Notice that this is the first time since atomic structure was described earlier in this chapter that the atom's (orbital) electrons have been mentioned. Generally speaking, nuclear scientists are not concerned with the electrons or even the chemical form of the atom. We are instead focused on the nucleus and nuclear changes. So why all the stuff about the Bohr model a few pages back? Patience, young padawan.

Our final major type of nuclear decay is **gamma decay**. Gamma decay is unique in that no particles[2] are emitted or absorbed, and both the numbers of protons and neutrons remain constant. It is simply a rearrangement of the nucleons in a nucleus. Initially, the nucleons are arranged poorly, in terms of energy. That is to say one or more nucleons occupy a higher energy state than they need to, and the nucleus is said to be in an **excited state**. Rearrangement allows the nucleons that are in unusually high-energy states to drop down into lower ones. The lowest energy configuration would be called the **ground state**. Rearrangement from a higher energy state to a lower one requires the release of energy, which can be done in the form of a high-energy photon, or gamma ray. An example is the decay of ^{99}Tc* to ^{99}Tc.

$$^{99}_{43}\text{Tc}^* \rightarrow {}^{99}_{43}\text{Tc} + \gamma$$

The asterisk indicates that the nuclide to the left of the arrow is in an excited state. Two nuclides that differ only in energy state (such as ^{99}Tc* and ^{99}Tc) are called **isomers**.[3] The gamma photon can also be represented as $^{0}_{0}\gamma$, but the zeroes are usually left off. We all know that photons have no protons, neutrons, or electrical charge.

Some excited (isomeric) states exist for reasonable amounts of time. If it hangs around for more than a second or two, it is often referred to as a **metastable** (or simply "meta") state, and a lower-case m is added after the mass number in the nuclide symbol, instead of the asterisk following the element symbol. For example, ^{99}Tc has a metastate, which is symbolized as ^{99m}Tc.

Nuclear reactions are the other main kind of nuclear transformation we'll study. They involve the collision of a particle or photon with a nucleus (or two nuclei crashing into each other) with the formation of a product nuclide and either a particle or photon. An example would be a neutron colliding with a ^{98}Mo nucleus to form ^{99}Mo and a gamma photon.

$$^{98}_{42}\text{Mo} + {}^{1}_{0}\text{n} \rightarrow {}^{99}_{42}\text{Mo} + \gamma$$

In this reaction, the neutron is the **projectile**, the ^{98}Mo is the **target**, and the ^{99}Mo is the **product**. The easy way to distinguish between decay and nuclear reactions is that reactions always have at least two nuclides, or a nuclide with a particle or photons on either side of the arrow. Decay equations generally only have a nuclide on the left side (except electron capture). Like all of the nuclear

---

2   Some would correctly argue that photons *are* particles. This textbook is trying to keep things simple and is restricting particles to things that have mass.

3   As noted in Section 5.7, many nuclear science textbooks define isomers as only those excited nuclear states with measurable lifetimes. Because of the arbitrarily limiting nature of this definition, and a desire to define our "iso" terms consistently, isomers are defined more broadly here.

equations in this section, this one is balanced, both in terms of atomic and mass numbers. Like decay, there are special kinds of nuclear reactions, but we'll worry about them later.

1.4 NUCLEAR STABILITY

A **radioactive** nuclide is one that spontaneously undergoes nuclear decay. Why do some nuclides decay but others do not? The question is really one of nuclear stability, and the answer depends on numbers. Perhaps the most important number related to nuclear stability is the neutron-to-proton ratio (N/Z). If we plot the number of protons (Z) vs. the number of neutrons (N), and make a mark for every stable nuclide known, the result will look like Figure 1.4. The region where stable nuclides are almost always found is called the **belt of stability**. Notice that for low Z nuclides (near the top of the periodic table, or the bottom left of Figure 1.4), the belt is narrow and stable nuclides have ratios very close to 1:1. It is difficult to see on this figure, but there are only two stable nuclides with an N/Z ratio less than 1: ^{1}H and ^{3}He. As atomic number increases, the belt widens, and the belt's N/Z ratio also increases.

How can we use this information? If a nuclide's N/Z ratio is too high or too low, it will not lie within the belt, and odds are very good that it will not be stable. Lying outside of the belt means that the nuclide has too many protons or too many neutrons to be stable. Therefore, it will likely undergo some type of beta decay to an isobar. A large number of known (unstable) nuclides lie off the belt, so examination of the N/Z ratio is an important method to determine nuclear instability.

The belt is not completely filled with dots, telling us that there are a number of unstable nuclides with good N/Z ratios. Why? Again, we need to look at some numbers. If we take all the known *stable* nuclides and count how many have an even number of protons (Z) and an even number of neutrons (N), there are about 159. Statistically speaking there should be equal numbers of stable nuclides with an even Z and odd N, odd Z and even N, or odd Z and odd N.

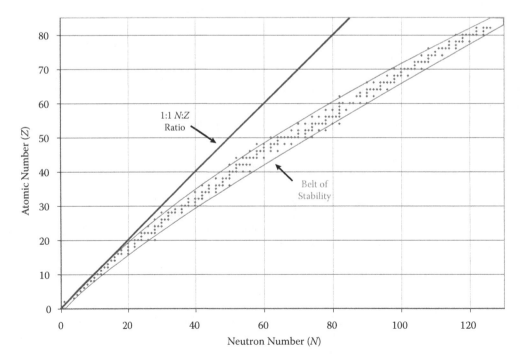

FIGURE 1.4   The belt of stability—dots indicate stable nuclides.

TABLE 1.1
Stable Nuclides

Z	N	#Stable
Even	Even	159
Even	Odd	53
Odd	Even	50
Odd	Odd	4

Curiously enough, this is not true. As can be seen in Table 1.1, if either Z or N is odd and the other is even, then there are only about 50 stable nuclides. If *both* Z and N are odd, there are only four! These four odd balls are $^2_1H_1$, $^6_3Li_3$, $^{10}_5B_5$, and $^{14}_7N_7$. It seems that odd numbers for both Z and N generally means a nuclide is unstable. Why do these exceptions exist? It's related to the fact that they all have a low number of protons (low Z) and that the belt is narrow at low Z. A more detailed explanation will have to wait until Chapters 5 and 6. For now, we know that even is good and odd is bad, but why? Apparently, protons and neutrons like to pair up with others of their kind, and, when they do, they are more stable. Therefore, an odd/odd nuclide can become even/even by converting a proton to a neutron or a neutron to a proton. So, some type of beta decay looks likely for odd/odd nuclides.

Why does Figure 1.4 stop at $Z = 82$? Beyond element 82 (Pb) there are no stable nuclides. They are simply too big for the very short-ranged strong force to hold the nucleus together against the tremendous Coulomb repulsion of all those protons. Nuclides this large *tend* to undergo alpha decay, as this lowers both Z and N.

Finally, close examination of Figure 1.4 reveals that some dots seem to lie just outside of the belt of stability. Apparently, there are some *stable* nuclides with *bad* N/Z ratios. All the rationalizations (bad N/Z, odd/odd, too big) we've seen so far are different ways to understand *instability*. How can we rationalize unusual *stability*? We'll need more information first. Are there other instances where unusual stability is observed? Yes, certain elements seem to have an unusually large number of stable isotopes. Most elements only have one, two or three stable isotopes, but lead has four, nickel has five, calcium has six, and tin has ten! There is something special about certain numbers of protons or neutrons. Detailed analysis reveals that the numbers 2, 8, 20, 28, 50, 82, and 126 seem to impart special stability to nuclides that contain these numbers of protons and/or neutrons. Because of this characteristic, they are sometimes referred to as **magic numbers**.

Far from having supernatural qualities, these numbers have a great analogy in chemistry. Students of chemistry (and some others) know that chemical stability is imparted on those elements occupying the far-right column of the periodic table, the so-called noble gases. This stability comes from exactly filling their valence shells with electrons. In chemistry, these numbers would be 2, 8, 18, and 32. They are not referred to as "magic," but they could be. Applying this analogy to numbers of nucleons, we could say that nuclei with magic numbers of nucleons have complete, or filled nuclear shells, and therefore exhibit extra stability. Therefore nucleons, just like electrons, are filling some sort of pattern of energy levels. When a level is filled, the nucleus gives a sigh, as it suddenly feels more stable.

Table 1.2 lists all stable nuclides containing magic numbers of nucleons. Note that a few have magic numbers of both protons *and* neutrons. These nuclides are referred to as "doubly magic" and can be expected to be extra stable compared to others with similar (but not quite the same) numbers of nucleons.

Not all stable nuclides lying outside the belt have magic numbers of protons and/or neutrons, nor are all nuclides with magic numbers of nucleons stable. In general, if a nuclide is known to be stable,

TABLE 1.2
Stable Nuclides with Magic Numbers of Nucleons

2	8	20	28	50	82	126
Neutrons						
^{4}He	^{15}N	^{36}S	^{48}Ca	^{86}Kr	^{136}Xe	^{208}Pb
	^{16}O	^{37}Cl	^{50}Ti	^{88}Sr	^{138}Ba	
		^{38}Ar	^{51}V	^{89}Y	^{139}La	
		^{39}K	^{52}Cr	^{90}Zr	^{140}Ce	
		^{40}Ca	^{54}Fe	^{92}Mo	^{141}Pr	
					^{142}Nd	
					^{144}Sm	
Protons						
^{3}He	^{16}O	^{40}Ca	^{58}Ni	^{112}Sn	^{204}Pb	
^{4}He	^{17}O	^{42}Ca	^{60}Ni	^{114}Sn	^{206}Pb	
	^{18}O	^{43}Ca	^{61}Ni	^{115}Sn	^{207}Pb	
		^{44}Ca	^{62}Ni	^{116}Sn	^{208}Pb	
		^{46}Ca	^{64}Ni	^{117}Sn		
		^{48}Ca		^{118}Sn		
				^{119}Sn		
				^{120}Sn		
				^{122}Sn		
				^{124}Sn		

lies outside the belt of stability, *and* has magic numbers, *then* the numbers can be used to rationalize its apparently unusual stability.

Let's look at some examples and use this newfound knowledge about numbers and nuclear stability to help rationalize why specific nuclides are stable or unstable.

Example 1.3 Why is ^{58}Ni stable?

For questions like this, it's best to start by looking at the numbers of protons and neutrons in the nucleus (are they odd or even?) and the N/Z ratio. Nickel is element number 28, so ^{58}Ni has 28 protons and 30 neutrons (58–28 = 30). It is an even/even nuclide, which increases the odds of stability (but does not determine it!). Its N/Z ratio is approximately 1.07.

$$\frac{N}{Z} = \frac{30}{28} \cong 1.07$$

Close examination of Figure 1.4 suggests this ratio is a little low. In fact, it appears that a dot representing ^{58}Ni lies just outside (above) the belt of stability. If this nuclide is outside the belt, how can its stability be rationalized? It is because of the extra stability imparted by having a filled nuclear shell (magic number) of protons ($Z = 28$).

Example 1.4 Why are ^{138}Te, ^{120}Xe, ^{241}Am, and ^{46}Sc unstable?

Table 1.3 summarizes the relevant numerical information for these four nuclides. Of these four nuclides, only ^{46}Sc has odd numbers of both protons and neutrons. Since it is not one of the previously

TABLE 1.3
Z, N, and N/Z for Specific Unstable Nuclides

Nuclide	Z	N	N/Z
^{138}Te	52	86	1.65
^{120}Xe	54	66	1.22
^{241}Am	95	146	1.54
^{46}Sc	21	25	1.19

mentioned exceptions, we can say that it is unstable because it is odd/odd. Note that, ironically, its N/Z ratio is actually pretty good, placing it near the center of the belt of stability.

The first two nuclides listed in Table 1.3 are even/even but are unstable because they have poor N/Z ratios. ^{138}Te has a high N/Z ratio and is below the belt of stability, while ^{120}Xe has a low ratio and is above the belt.

We can only guess about the quality of the N/Z ratio for ^{241}Am, because Figure 1.4 doesn't go that high. ^{207}Pb is more or less in the middle of the belt in the upper right corner of Figure 1.4 and has an N/Z ratio of 1.52. Since this ratio (within the belt) increases as atomic number (Z) increases, a value of 1.54 for $Z = 95$ is probably okay. With all of our hard work examining the N/Z ratio for this nuclide wasted, we suddenly realize that this nuclide is unstable simply because it is too big.

1.5   IONIZING RADIATION

The first section of this chapter discussed radiation as energy propagated through a medium but stayed focused on electromagnetic radiation (photons). We have since learned about radioactive decay and know that this happens to achieve greater nuclear stability—therefore energy must be released as part of all nuclear decay processes.

Ionizing radiation is when a particle or a photon has enough energy to turn an unsuspecting atom into an ion (Figure 1.5). In other words, the radiation needs to have enough energy to knock an electron or a proton loose from an atom. The former process is much more probable, so much so that we'll ignore the latter.

As we've already seen, the only types of electromagnetic radiation with sufficient energy to ionize matter are X-rays and gamma rays. All alpha and beta particles produced in nuclear decay also have enough energy to ionize matter. Therefore X-rays, gamma rays, and alpha and beta particles can all be considered forms of ionizing radiation.

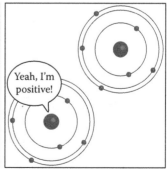

FIGURE 1.5   Ionizing radiation.

Unfortunately, the more general term "radiation" is commonly, but incorrectly, used to refer to the energetic photons and particles emitted as part of nuclear processes. This text will strive to use the terms "radiation" and "ionizing radiation" properly. The reader should know that misuse of "radiation" is very common, and even those that know better will sometimes slip up.

Even though the three major forms of decay (alpha, beta, and gamma) can all create ions when their emitted particles or photons interact with matter, their relative abilities to penetrate matter differ greatly. Alpha particles have a high probability of interacting with matter. They carry a 2+ charge, and, relative to electrons and photons, are very large. Alpha particles can be effectively stopped by thin pieces of paper, the outer layers of your skin, or by just a few centimeters of air. Beta particles are electrons, so they have a lot less mass and only half the charge of alpha particles, but still have a reasonably high probability of interacting with matter. They can generally be stopped by a centimeter (or two) of water or plastic. Energetic photons such as X-rays and gamma rays have no charge or mass, and therefore have a relatively low probability of interacting with matter. The best shielding material for them is high-density matter, such as lead.

1.6   A BIOLOGICAL THREAT?

One of the reasons humans fear ionizing radiation is that we know that, in large enough doses, it can hurt us. From what we've learned so far it shouldn't be hard to figure out how this might happen. Ionizations inside our bodies and possibly in our DNA can cause cells to die, or to replicate poorly. With enough of a dose, we start to observe measurable effects such as lower blood cell counts. As the dose increases, we know that the health effects gradually get worse until death becomes pretty much certain.

What about below the level where measurable adverse health effects are known? Since the health effects cannot be measured there is some uncertainty as to what effect low doses might have. Regulatory agencies have decided to be prudent and assume that the roughly linear relationship between risk and dose at higher exposure levels should be extrapolated through low doses. This is known as the "linear non-threshold (LNT) model," and is illustrated by the straight, dashed black line in Figure 1.6. This means that any additional exposure *could* lead to negative health effects. If this is true, then it is a good idea to minimize dose due to exposure to ionizing radiation, when possible. This is the basis of the **ALARA** (*As Low As Reasonably Achievable*) policy followed by those who work with radioactive materials. If there's more than one reasonable way to do something, pick the path that gives a lower dose.

While used by regulators and radiation workers, it should be noted that the LNT is not supported by science. Our bodies can repair or replace many of the cells that are damaged or killed due to exposure to ionizing radiation. We are all exposed to naturally occurring ionizing radiation every minute of every day. Some are exposed to more than others, such as those that live at higher altitudes

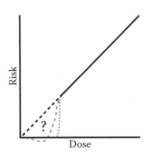

FIGURE 1.6   Risk vs. dose.

or in areas with relatively high concentrations of naturally occurring radioactive nuclides. There are no measurable negative health effects due to the extra exposure on the large populations of humans that live in these areas, even though background radiation doses vary by as much as a factor of 100. This suggests that our bodies can safely cope with low levels of ionizing radiation, and that there might be a threshold dose, below which no health effects can be anticipated. The *dotted* gray curve in Figure 1.6 represents the threshold model.

A few others have gone a bit further than those that suggest there is a threshold dose. These scientists suggest there may be a health benefit to low doses of ionizing radiation (hormesis), and there is some evidence to indicate this might be the case. This is represented by the *dashed* gray curve in Figure 1.6. Hormesis acts like an immunization. A flu shot stimulates the immune system such that, when the body is exposed to the virus, it can fend it off. Exposure to a little extra ionizing radiation can stimulate cell repair mechanisms, making the body healthier, and better able to effectively deal with future exposures.

The bottom line is that none of these hypotheses have enough science behind them to gain widespread acceptance. For the time being, those of us working with ionizing radiation will continue to follow prudence in the form of ALARA. However, we must also recognize that application of ALARA outside of the radiation workplace can have drastic consequences. For example, how extensively must areas contaminated by the Fukushima reactors be cleaned before we allow people to return? ALARA and the LNT suggest that we decontaminate until it is no longer more radioactive than before the accident. This may be ridiculously expensive and unnecessary, if we instead assume a threshold dose or hormesis.

1.7   NATURAL AND ANTHROPOGENIC RADIATION

If we are being exposed to ionizing radiation all the time, where is it coming from? **Naturally occurring** ionizing radiation comes from several sources. The universe is a very radioactive place, constantly bombarding the Earth with cosmic radiation. The Earth's magnetic field and atmosphere shield us from most of it, but some makes it all the way to the surface. People living at higher altitudes receive two to three times the dose from cosmic radiation than those living near sea level, because there's less shielding from the atmosphere. This also means those that spend a lot of time flying (watch out Supergirl!) are getting a significantly higher dose than those of us that stay on the ground.

What about radioactive materials here on planet Earth? Well, there's a lot, and almost all of it has always been here. There are three main sources of radioactive materials. First, some of the lighter elements present in the upper atmosphere (like N, O, and H) can undergo nuclear reactions that are a result of cosmic ray interactions. ^{3}H and ^{14}C are constantly produced in this way. These nuclides find their way back to the planet surface and are incorporated into molecules that make up living things since their chemistries remain the same as other (stable) hydrogen and carbon atoms. Thus, all living things are naturally radioactive.

Another source comes from radioactive nuclides that take a very long time to decay. If that time is longer than the age of the solar system (4.6×10^9 years), then they're still around us. Altogether, there are about 20 of them. One of the most notable is ^{40}K. Since all isotopes of an element have the same chemical behavior, and because we humans need potassium in our bodies, we all have some ^{40}K in our bodies. Anything that contains potassium (bananas, Gatorade®, your friends) is naturally radioactive.

Almost all 20 of these naturally occurring, long-lived radioactive nuclides decay to stable nuclides. Three that don't are ^{232}Th, ^{235}U, and ^{238}U. They all eventually decay to stable isotopes of lead, but along the way, they produce a number of other naturally occurring radioactive nuclides, including ^{222}Rn. This is our final source of naturally occurring radioactive nuclides. The paths they take to Pb are called the naturally occurring **decay series** (or **chains**). To get there, a number of

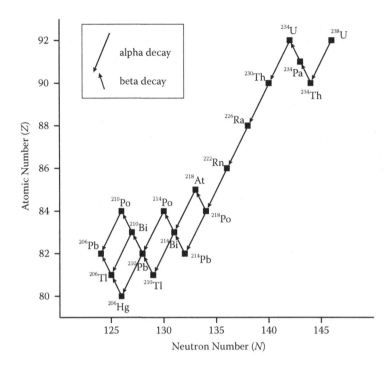

FIGURE 1.7 The ^{238}U decay series.

alpha and beta (β^-) decays need to take place, as illustrated in Figure 1.7 for the series originating with ^{238}U. Notice that an alpha decay lowers both the proton and neutron numbers by 2, while a beta decay lowers N by 1, but increases Z by 1. At certain nuclides (like ^{218}Po), the path splits into two. This means that more than one decay pathway exists for that particular nuclide, i.e., it can decay via alpha or beta.

What about all the radioactive materials released by humans (anthropogenic sources)? Despite our best efforts, we've allowed a fair bit of radioactive material to get into the environment. The following are some of the more significant sources:

1. The development, manufacture, testing, and use of nuclear weapons.
2. Accidental releases from nuclear power plants such as Fukushima (2011), Chernobyl (1986), and Three Mile Island (1979).
3. Regulated releases from nuclear power plants.
4. Irresponsible disposal of radioactive waste.
5. The manufacture, use, and disposal of radioactive nuclides for research or medical applications.

Sounds bad, but all five sources combined account for less than 0.1% of the average annual human dose in the United States. They are combined under "Other" in Figure 1.8, which illustrates the major sources for average annual doses from ionizing radiation in the United States. The dose rates are given in millisieverts per year. The millisievert is a commonly used unit for dose and is explained in Chapter 13. Notice also that the unit abbreviation for years is "a," which comes from the Latin word for year, *annum*.

The first four bars shown in Figure 1.8 are from natural sources. By far, we get the most from radon. That's because it is a gas. Once formed in a decay series, it can percolate through the ground

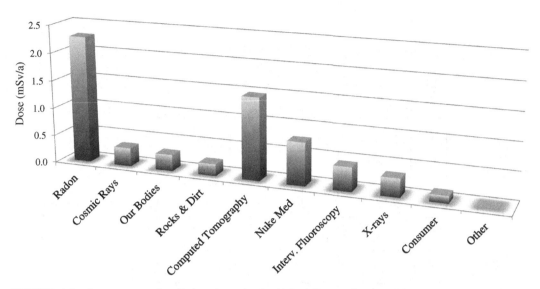

FIGURE 1.8 Average annual radiation doses in the United States. *Source:* NCRP Report #160, *Ionizing Radiation Exposure of the Population of the United States*, 2009.

and get into the water we drink or into the air we breathe. It gives us a large dose because we ingest it, and it can undergo alpha decay while inside of us. Because of its high probability of interacting with matter, the alpha particles cannot escape our body and end up doing a lot of local damage. Don't stop breathing the air or drinking the water, your friends and relatives might miss you! Besides, radon has always been present on the planet, and humans have evolved with some ability to repair the damage it causes. At most concentrations we humans normally encounter, there is no adverse health effect, in fact, there seems to be a hormetic effect at moderate concentrations. At very high concentrations, long-term exposure to radon gas can cause lung cancer. Since a good chunk of this dose comes from the air we breathe at home, it is a good idea to test your home for radon, especially if your home is in an area known to have high radon levels. If it is found in high concentration,[4] there are ways to lower it. As we'll soon learn, the best results come from a longer test—choose one that is in your home for a few months or longer.

You might have expected cosmic rays to account for a larger percentage of our annual dose. Much of the cosmic radiation we're exposed to is in the form of high-energy photons and particles, most of which have a very low probability of interacting with the matter we are made of. That's the difference between exposure and dose, our greatest exposure to naturally occurring ionizing radiation is cosmic, but our greatest dose comes from radon because radon is inside us and emits ionizing radiation with a very high probability of interacting with matter, while most cosmic radiation has a relatively low probability of interacting.

The final six sources shown in Figure 1.8 are anthropogenic. Notice that the first four are all medical diagnostic procedures and make up almost the entire U.S. dose from human-produced ionizing radiation. Of course, not everyone has a CT scan or nuclear medicine procedure every year, so total dose for individuals will vary somewhat. These numbers come from the dose for all sources of ionizing radiation each year divided by the total population. The total average dose also varies from

---

4 How high is too high? Governments typically set a recommended threshold for potential remediation, but they are set much lower than they should be based on scientific evidence. For example, the U.S. sets its threshold at 4 pCi/L, yet 56 pCi/L is the observed threshold for negative health effects.

Z \ N	0	1	2	3	4	5	6	7	8	9	10	11	12	13	14	15	16	17	18
8					O-12 1.77	O-13 17.77	O-14 5.1439	O-15 2.7542	O-16 99.757	O-17 0.038	O-18 0.205	O-19 4.822	O-20 3.815	O-21 8.11	O-22 6.5	O-23 11.3	O-24 11.5	O-25 15.9	O-26 16.9
7				N-10 2.64	N-11 2.290	N-12 17.338	N-13 2.2205	N-14 99.636	N-15 0.364	N-16 10.421	N-17 8.68	N-18 13.90	N-19 12.53	N-20 18.0	N-21 17.2	N-22 22.8	N-23 23.8	N-24 1.2	
6			C-8 2.142	C-9 16.495	C-10 3.6480	C-11 1.982	C-12 98.93	C-13 1.07	C-14 0.156476	C-15 9.772	C-16 8.010	C-17 13.17	C-18 11.81	C-19 16.6	C-20 15.8		C-22 21		
5			B-7 2.200	B-8 17.980	B-9 0.277	B-10 19.9	B-11 80.1	B-12 13.369	B-13 13.437	B-14 20.64	B-15 19.10	B-16 0.040	B-17 22.7		B-19 26.9				
4			Be-6 1.372	Be-7 0.8619	Be-8 0.092	Be-9 100	Be-10 0.556	Be-11 11.51	Be-12 11.71	Be-13 0.50	Be-14 16.3								
3		Li-4 3.100	Li-5 1.970	Li-6 7.59	Li-7 92.41	Li-8 16.0052	Li-9 13.607	Li-10 0.42	Li-11 20.62										
2		He-3 0.000134	He-4 99.999863	He-5 0.890	He-6 3.508	He-7 0.440	He-8 10.65	He-9 1.15	He-10 15.8										
1	H-1 99.9885	H-2 0.0115	H-3 0.18591	H-4 2.910	H-5 25	H-6 24.3													

$N \longrightarrow$

$\longleftarrow Z$

FIGURE 1.9   A chart of the nuclides for the first eight elements.

country to country. Adding up all the bars in Figure 1.8 tells us that the average American picks up roughly 6.2 mSv each year, yet the average Canadian sees only about 2.6 mSv per year. The difference between the two numbers seems to largely be an American affinity for diagnostic medical procedures.

1.8   THE CHART OF THE NUCLIDES

The chart is a valuable tool for anyone studying nuclear science. We've already seen a couple of extremely condensed versions of it in Figures 1.4 and 1.7. A more conventional, albeit limited, representation is shown in Figure 1.9. Basically, it is a way to provide information on all known nuclides, stable (white) and unstable (gray). It is laid out with neutron number (N) increasing from left to right (the x-axis) and atomic number (Z) increasing from bottom to top (the y-axis). Thereby, each square represents a unique N and Z combination, i.e., a unique nuclide. Percent natural abundance is given at the bottom of the square for each stable nuclide, and total energy (MeV, see Chapter 3) given off during decay is given for all the unstable nuclides.

Printed versions are prepared by scientists at the Knolls Atomic Power Laboratory (KAPL) and are available for purchase through www.nuclidechart.com/. The printed charts also come with extensive text explaining how to properly interpret the information provided for each nuclide. There are also web-based versions of the chart, just ask your favorite AI assistant. Many of the questions following each chapter in this textbook will assume that the reader has access to the kinds of information available in the chart. For the readers' convenience, Appendix A also lists limited information about select nuclides. This appendix should be sufficiently complete to perform the problems posed in this text.

QUESTIONS

1.1 Which of the following statements are true?
   a. All radiation is harmful.
   b. All ionizing radiation is anthropogenic.
   c. Atoms cannot be changed from one element to another.
   d. You are exposed to ionizing radiation every minute.
   e. The human body is naturally radioactive.
   f. The human body is capable of detecting radioactivity at any level.

1.2 Which of the following statements are false?
   a. An electron is an example of a nucleon.
   b. There are three different kinds of beta decay.
   c. A nuclide having a magic number of protons and/or neutrons will always be stable.
   d. Anthropogenic radiation is more harmful than naturally occurring radiation.
   e. A nuclide with an N/Z ratio of exactly 1 will always be stable.
   f. The nucleus accounts for almost all of the mass of an atom.

1.3 List some of the apprehensions you have or have had concerning ionizing radiation.

1.4 What is radiation? What is ionizing radiation and where does it come from?

1.5 What is the wavelength of a photon with a frequency of 3.86×10^{18} s^{-1}?

1.6 What is the energy of a photon with a wavelength of 6.87×10^{-5} m? What type of radiation is this?

1.7 Millimeter wave imaging systems are used for personnel screening. They emit electromagnetic radiation with wavelengths in the range of 1 to 10 mm, which penetrate clothing but are reflected by denser matter such as the human body. Calculate the frequency range for these photons. Are they ionizing? The radiation used by these scanners is typically characterized as "high-frequency radio waves." Is this correct? What might motivate the use of this moniker?

1.8 What is the wavelength of a photon with an energy of 7.84×10^{-18} J? Is this ionizing radiation? Briefly explain your answer.

1.9 What do gamma rays and UV-light have in common? What distinguishes them?

1.10 Is the phrase "nature abhors a vacuum" accurate based on our knowledge of atomic structure?

1.11 If you built a scale model of an atom, and used a baseball to represent the nucleus, how big would the atom have to be?

1.12 Define the following: nuclide, nucleon, beta particle, positron decay, the electromagnetic spectrum, isotope, isobar, and ALARA.

1.13 Define the following: belt of stability, metastable state, alpha particle, isotone, electron capture, and isomer.

1.14 Which of the following are isotopes? Isotones? Isobars?

$$^{14}\text{C} \qquad ^{16}\text{O} \qquad ^{16}\text{N} \qquad ^{12}\text{C} \qquad ^{13}\text{B} \qquad ^{19}\text{F}$$

1.15 How many electrons could fit in the same space occupied by a proton?

1.16 Complete and balance the following. Identify each as nuclear decay or reaction. If it is decay, give the type of decay.

$$^{81}_{37}\text{Rb} \rightarrow {}^{81}_{36}\text{Kr} + \underline{\quad} \qquad {}^{14}_{7}\text{N} + \underline{\quad} \rightarrow {}^{1}_{1}\text{H} + {}^{14}_{6}\text{C} \qquad \underline{\quad} \rightarrow {}^{137}_{56}\text{Ba} + {}^{0}_{-1}\text{e}$$

$$^{239}_{94}\text{Pu} \rightarrow {}^{4}_{2}\text{He} + \underline{\quad} \qquad {}^{201}_{80}\text{Hg} + {}^{2}_{1}\text{H} \rightarrow 2\,{}^{1}_{0}\text{n} + \underline{\quad} \qquad {}^{208}_{83}\text{Bi} + {}^{0}_{-1}\text{e} \rightarrow \underline{\quad}$$

1.17 Give three reasons why a nuclide might be stable.

1.18 Why are ^{32}P, ^{64}Ge, ^{20}O, ^{100}Ru*, and ^{212}Po unstable nuclides? Write out a possible decay equation for each.

1.19 Are the following nuclides stable or unstable? Why? ^{11}C, ^{190}W, ^{114}In, ^{232}Th, ^{76}Se.

1.20 Complete and balance the following nuclear decays. Is the product nuclide of each of the decays stable? If not, why not?

$$^{225}_{90}\text{Th} \rightarrow {}^{221}_{88}\text{Ra} + \underline{\quad} \qquad\qquad {}^{184}_{73}\text{Ta} \rightarrow {}^{184}_{74}\text{W} + \underline{\quad}$$

$$^{133m}_{58}\text{Ce} \rightarrow {}^{133}_{58}\text{Ce} + \underline{\quad} \qquad\qquad {}^{126}_{53}\text{I} + \underline{\quad} \rightarrow {}^{126}_{52}\text{Te}$$

1.21 ^{40}Ca shouldn't be a stable nuclide but is. Explain this apparent contradiction.

1.22 One of the naturally occurring decay series begins with ^{232}Th and ends with ^{208}Pb. What is the minimum number of alpha and beta decays required for this series?

1.23 A fourth naturally occurring radioactive decay series starting with ^{237}Np may have existed in the past. Using a chart of the nuclides, graph this decay series (similar to Figure 1.7) and explain why it does not exist today.

1.24 Using Figure 1.8, estimate the percentage of the average annual U.S. dose due to natural sources.

1.25 List at least two anthropogenic sources of radioactive nuclides now found in the environment. Do a little research outside of this textbook and explain how these nuclides were (are) released.

1.26 Why are so many more neutron-rich nuclides shown in Figure 1.9 than proton-rich nuclides?

1.27 What is the most common nuclide in air?

1.28 Otto Hahn and Lise Meitner reported the discovery of an isotope of a new element in 1918. They described it as follows:

    a.   The previously hypothetical mother substance of actinium has been found and prepared in alkali acids in radioactively pure form. It is a homologue of tantalum.

    b.   It emits alpha rays with a penetration of 3.14 cm.

    c.   The half-life is a minimum of 1200 years and a maximum of 180,000 years.

What nuclide did they isolate? Briefly explain.

2 The Mathematics of Radioactive Decay

2.1 ATOMIC MASSES AND AVERAGE ATOMIC MASSES

Before looking at the mathematics of decay, it is important to remember the difference between atomic mass and average atomic mass. **Atomic mass** is the mass of a single atom and is numerically equivalent to the mass of one mole of atoms. Atomic mass is expressed in unified atomic mass units (u). When we are talking about a mole of atoms, the units are grams per mole (g/mol). For example, the mass of a single ^{12}C atom is 12.00 u, while the mass of a whole mole (6.022×10^{23}) of ^{12}C atoms is 12.00 g. The unified atomic mass unit is more convenient when referring to the mass of single atoms and is more commonly used in nuclear science. To convert one to the other: one unified atomic mass unit is equal to 1.66054×10^{-24} g.

The masses listed in the periodic table are not always the mass of a single atom of each element; rather it lists the element's *average* atomic mass. **Average atomic mass** is a weighted average of the masses of each naturally occurring isotope for a particular element. "Weighted" means that if one isotope has a greater natural abundance for a particular element, it contributes more significantly to the average atomic mass. The only time the periodic table gives the mass of a single atom of an element is when there is only one naturally occurring isotope for that element. The best way to see where the average atomic masses in the periodic table come from is through a sample calculation.

Example 2.1 Calculate the average atomic mass of copper.

Copper exists in nature as a mixture of two isotopes, ^{63}Cu and ^{65}Cu. The percent natural abundances are 69.15% and 30.85%, respectively. This means that if we find 100 copper atoms in nature, we'll likely have ~69 atoms of ^{63}Cu and ~31 atoms of ^{65}Cu. Their masses are 62.9296 u and 64.9278 u, respectively. These data are available in Appendix A for all elements.

To determine the average atomic mass, calculate the sum of the products of the atomic masses with the *fractional* abundances. In this way, the average mass of copper atoms is weighted by the amount of each isotopic mass present.

$$(62.9296 \text{ u} \times 0.6915) + (64.9278 \text{ u} \times 0.3085) = 63.55 \text{ u}$$

Notice that the average atomic mass is closer to 63 u than it is to 65 u. This makes sense because there's more of the mass ~63 u isotope than there is of the mass ~65 u isotope.

DOI: 10.1201/9781003303787-2

2.2 THE NATURE OF DECAY

Since humans can't sense radioactive decay, we need some other way to measure it. There are a variety of radiation detectors, as discussed in Chapter 8. For now, we just need to know that it can be done, albeit not always with high efficiency. Radioactive decay is **isotropic**, which means that particles and/or photons are emitted equally in all directions by a sample of a radioactive nuclide. Detectors can only "see" radiation that interacts with it, typically a small fraction of all radiation being emitted at the time, i.e., those that happen to enter the detector instead of flying off in some other direction. Data recorded by detectors are given units of count rate, such as counts per minute (cpm). If it were possible to detect *all* of the decays from a source over time, then the decay rate, or **activity**, would be measured. Decay rate is often given units of decays per minute (dpm) or decays per second (dps). Those who work with radioactive materials often use count rate and decay rate interchangeably. This is understandable, since they are proportional to each other. Readers of this book would do well to know the difference.

Over time, as a sample of a radioactive nuclide decays, it becomes less radioactive. We could make measurements of this change with a detector and a clock. Once we've collected these data, it could be plotted as shown in Figure 2.1a. Mathematically, radioactive decay is an exponential process, meaning that for equal intervals of time, the count rate will decrease by an equal percentage. This also means that if instead of plotting count rate vs. time, we plot the logarithm of the count rate vs. time, we'll get a straight line, as shown in Figure 2.1b.

Two different forms of the same mathematical equation can describe the change in radioactivity of a source:

$$\ln \frac{A_1}{A_2} = \lambda t \tag{2.1}$$

$$A_2 = A_1 e^{-\lambda t} \tag{2.2}$$

where A represents activity, or count rate so long as it is applied consistently in one equation. A_1 is the activity at the beginning of the time elapsed (t), and A_2 is the activity at the end of that time. ln is the natural logarithm function and e is natural antilogarithm. λ is the **decay constant** and has units of reciprocal time (time^{-1}). The decay constant is an indication of how quickly a source will decay,

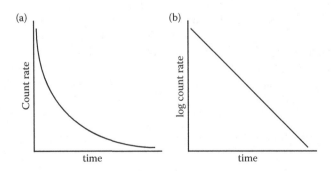

FIGURE 2.1   (a) The change in count rate over time for radioactive decay. (b) The change in the logarithm of the count rate with time.

and is sometimes represented by the letter k. The decay constant (λ) is also equal to the natural logarithm of two divided by half-life ($t_{1/2}$).

$$\lambda = \frac{\ln 2}{t_{1/2}} \qquad (2.3)$$

Half-life is defined as the amount of time it takes for the decay rate (or count rate) to reach half of its original value. If a nuclide starts out with an activity of 1000 dpm, it will be 500 dpm after one half-life, 250 dpm after two, and so on. Half-life is commonly used in nuclear science because it is a bit more conceptually tractable than the decay constant. A nuclide with a very long half-life will decay slowly and be radioactive for a very long time. In contrast, a nuclide with a short half-life will decay away quickly.

Example 2.2 A radionuclide has a half-life of 6.01 hours. If its activity is currently 5430 dpm, what will its activity be 4.00 hours from now?

$$A_2 = A_1 e^{-\lambda t} = A_1 e^{-\frac{\ln 2}{t_{1/2}}t} = 5430 \text{ dpm} \times e^{-\frac{\ln 2}{6.01 \text{ h}} \times 4.00 \text{ h}} = 5430 \text{ dpm} \times e^{-0.461} = 3423 \text{ dpm}$$

Less than one half-life has gone by, so we expect the answer to be greater than half of the original. Notice also that the time units need to match within the exponent (both are hours in this example), and the activity units for A_1 and A_2 must also match. If your instructor is a stickler for significant figures, then the answer should probably be rounded to 3420 dpm. Generally speaking, three significant figures are the best that can be expected from most detector measurements.

Example 2.3 A radioactive nuclide had an activity of 1.38×10^5 dpm exactly 60 days ago, but now has an activity of 6.05×10^4 dpm. What is its half-life?

$$\ln \frac{A_1}{A_2} = \lambda t$$

$$\ln \left(\frac{1.38 \times 10^5 \text{ dpm}}{6.05 \times 10^4 \text{ dpm}} \right) = \frac{\ln 2}{t_{1/2}} (60 \text{ d})$$

$$0.825 = \frac{41.6 \text{ d}}{t_{1/2}}$$

$$t_{1/2} = 50.4 \text{ d}$$

In this example, approximately one half-life has passed, and the activities differ by (roughly) a factor of two. Notice that units for half-life ($t_{1/2}$) come from the elapsed time (t).

Activity is also commonly expressed in units of bequerels (Bq) or curies (Ci). The becquerel (Bq) is named for Henri Becquerel, the French physicist who first observed radioactivity. It is the *Systéme International d'Units* (SI) unit and is equal to 1 dps. This is a tiny value, so megabecquerels (MBq) or kilobecquerels (kBq) are often used. Remember that the kilo prefix means 10^3 and the mega prefix multiplies by 10^6. The curie is named for Marie Curie, the Polish scientist who performed much of the early groundbreaking work with radioactive materials. One curie is defined as exactly 3.7×10^{10}

dps, which is the rate of decay of exactly 1 g of ^{226}Ra. This is a rather large number, so millicuries (mCi = 10^{-3} Ci) or microcuries (μCi, 1 μCi = 10^{-6} Ci = 2.22 × 10^6 dpm) are often used for smaller sources.

Example 2.4 A ^{133}Ba source was calibrated to emit 36.9 kBq of radiation. If its activity today is measured at 5.55 × 10^5 dpm, how long ago was it calibrated?

First, convert the initial activity units from kBq to dpm.

$$36.9\,\text{kBq} \times \frac{1000\,\text{Bq}}{\text{kBq}} \times \left(\frac{1\,\text{dps}}{1\,\text{Bq}}\right) = 3.69 \times 10^4\,\text{dps}$$

$$3.69 \times 10^4\,\text{decays/s} \times 60\,\text{s/min} = 2.21 \times 10^6\,\text{dpm}$$

Now that the activity units are all on the same page, we can use Equation 2.1 to solve for time.

$$\ln\frac{2.21 \times 10^6\,\text{dpm}}{5.55 \times 10^5\,\text{dpm}} = \frac{\ln 2}{10.54\,\text{a}}t$$

$$1.384 = 0.06576\,\text{a}^{-1} \times t$$

$$t = 21.04\,\text{a}$$

In this example, it's necessary to convert one of the two given activities to match the units of the other. The choice to convert A_1 to dpm was arbitrary. A_2 could've been converted to kBq instead. It's also necessary to look up the half-life for ^{133}Ba (10.54 years, Appendix A). Finally, notice that roughly two half-lives have passed, and the activity is roughly 1/4 of its original value.

As mentioned at the beginning of the chapter, detectors typically only "see" a fraction of all decays. The percentage of all decays observed is called the **percent efficiency**.

$$\% \text{ efficiency} = \frac{\text{cpm}}{\text{dpm}} \times 100\% \qquad (2.4)$$

Example 2.5 A ^{210}Pb source was calibrated at 0.0202 μCi 31.0 years ago. Under your detector it records 2.16 × 10^3 cpm. What is your detector's percent efficiency?

First calculate the calibrated activity in units of decays per minute:

$$0.0202\,\mu\text{Ci} \times \frac{2.22 \times 10^6\,\text{dpm}}{\mu\text{Ci}} = 4.48 \times 10^4\,\text{dpm}$$

After looking up the half-life of ^{210}Pb in Appendix A, calculate what the *activity* is today. In other words, how much has the source decayed since it was calibrated?

$$A_2 = 4.48 \times 10^4 e^{-\left(\frac{\ln 2}{22.3a}\right)31.0a} = 1.71 \times 10^4\,\text{dpm}$$

Finally, use the count data to determine the percent efficiency:

$$\% \text{ efficiency} = \frac{\text{cpm}}{\text{dpm}} \times 100\% = \frac{2.16 \times 10^3 \text{cpm}}{1.71 \times 10^4 \text{dpm}} \times 100\% = 12.6\%$$

In addition to the limited size of the detector, there are a number of experimental conditions that can affect the percent efficiency and, therefore, the observed count rate. These are discussed in Chapter 8.

So long as the conditions of the experiment remain constant, the count rate will be proportional to activity and can be used in Equations 2.1 and 2.2. There's another important proportionality that needs to be considered. If there's more radioactive material, there should be more activity. The mathematical relationship is remarkably simple.

$$A = \lambda N \tag{2.5}$$

where A is activity (it cannot be count rate in this equation), λ is the decay constant, and N is the number of atoms of the radioactive nuclide. The more you've got, the toastier it is. This is a simple, yet powerful little equation that relates the amount of material with its activity. If the number of atoms is known, it can be converted to moles of atoms using Avogadro's number (6.022×10^{23} atoms/mole), and then into mass using the atomic mass (grams/mole). Likewise, if the mass is known, the number of atoms can be calculated, as well as activity.

Example 2.6 What is the mass of 45.5 kBq of ^{137}Cs?

First convert activity to dpm:

$$45.5 \text{ kBq} \times \frac{1000 \text{ Bq}}{\text{kBq}} \times \left(\frac{1 \text{ dps}}{1 \text{ Bq}}\right) \times \frac{60 \text{ s}}{\text{min}} = 2.73 \times 10^6 \text{ dpm}$$

Next, calculate the number of ^{137}Cs atoms (after looking up the half-life for ^{137}Cs); but first we'll have to rearrange Equation 2.5, and then substitute $\ln 2/t_{1/2}$ for λ:

$$A = \lambda N \quad N = \frac{A}{\lambda} = \frac{A \cdot t_{1/2}}{\ln 2} = \frac{\left(2.73 \times 10^6 \text{ dpm}\right)\left(30.07 \text{ a}\right)}{\ln 2}$$

But wait! Notice the problem with the time units? Decays per minute doesn't cancel out when multiplied by years. To fix this, it's handy to know that there are approximately 5.259×10^5 minutes in a year.

$$N = \frac{\left(2.73 \times 10^6 \text{ dpm}\right) \times \left(30.07 \text{ a}\right) \times \left(\dfrac{5.259 \times 10^5 \text{ min}}{\text{a}}\right)}{\ln 2} = 6.23 \times 10^{13} \text{ atoms}$$

Convert to mass:

$$\left(6.23\times10^{13}\,\text{atoms}\right)\times\frac{\text{mol}}{6.022\times10^{23}\,\text{atoms}}\times\frac{137\,\text{g}}{\text{mol}}=1.42\times10^{-8}\,\text{g}$$

Wow! That's a very small mass for something spitting out radiation over two million times each minute! A little radioactive material can often go a long way. Notice also that 137 g/mole was used as the atomic mass. In a pinch, the mass number (A) of a particular nuclide can be used as an approximate atomic mass. We could've easily used a more precise value from Appendix A, but an approximate value is adequate for this problem.

Because of the direct proportionality of activity (A) and number of atoms (N), Equations 2.1 and 2.2 can be rewritten as:

$$\ln\frac{N_1}{N_2}=\lambda t \tag{2.6}$$

$$N_2=N_1 e^{-\lambda t} \tag{2.7}$$

This is particularly handy if we know the mass, or number of atoms of a radioactive nuclide.

2.3 SPECIFIC ACTIVITY

Activity can easily be determined using a detector with a known efficiency. First measure the count rate, then use Equation 2.4 to solve for activity. Mass is also easily measured using an appropriate balance. So why is the equation $A=\lambda N$ such a big deal? Even if atoms of a particular radioactive nuclide are painstakingly purified, it is not possible to exactly determine the mass of the nuclides that have not yet decayed. In other words, there will always be some contribution to the mass by the product nuclide(s). Additionally, radioactive nuclides are often incorporated into molecules or other non-radioactive materials. If mass is determined, it is the mass of all atoms present, not just the radioactive ones. Therefore, determining mass (and thereby N) of a radioactive nuclide is not as simple as throwing it onto a balance.

Occasionally, it is important to know how radioactive something is per gram of sample, regardless of what other stuff might be present. **Specific activity** is the activity per unit mass of a substance, whether it is atoms of a single nuclide, a compound, or some mixture of compounds. The SI units for specific activity are Bq/kg, but, in the United States, variations on Ci/g are often used. Different activity and mass units can be combined. For our first example, let's keep it simple and calculate the specific activity of a pure nuclide.

Example 2.7 Calculate the specific activity of ^{133}Xe in units of dpm/g.

Notice that the problem doesn't give us a mass! The specific activity will be the same no matter how much ^{133}Xe there is. When we get to the last step, it'd be convenient if we had assumed there's exactly 1 gram of ^{133}Xe. Since the mass doesn't matter, why not go with something that will make the math easier later on? Assuming exactly 1 g, let's calculate N.

$$1\,\text{g}\times\frac{\text{mol}}{133\,\text{g}}\times\frac{6.022\times10^{23}\,\text{atoms}}{\text{mol}}=4.53\times10^{21}\,\text{atoms}$$

Now calculate activity:

$$A = \lambda N = \frac{\ln 2}{5.243\ \text{d}} \times \left(4.53 \times 10^{21}\ \text{atoms}\right) = 5.99 \times 10^{20}\ \text{decays per day}$$

Note that, because the half-life has units of days, the units for activity are decays per *day*. Calculating specific activity from here involves conversion of activity to units of decays per minute, then dividing by the total mass of the sample, which we cleverly chose as exactly 1 g.

$$\frac{\dfrac{5.99 \times 10^{20}\ \text{decays}}{\text{day}} \times \dfrac{\text{d}}{24\ \text{h}} \times \dfrac{\text{h}}{60\ \text{min}}}{1\ \text{g}} = 4.16 \times 10^{17}\ \frac{\text{dpm}}{\text{g}}$$

Example 2.8 Calculate the specific activity (dpm/g) of a gas mixture containing 78 g N_2, 21 g O_2, and 2.01 g of ^{133}Xe.

There are no naturally occurring radioactive isotopes of nitrogen or of oxygen; therefore, ^{133}Xe is the only hot component of this mixture. The activity of any mass of ^{133}Xe can be calculated using its specific activity (conveniently determined in the previous example).

$$\frac{4.16 \times 10^{17}\ \text{dpm}}{\text{g}} \times 2.01\ \text{g} = 8.36 \times 10^{17}\ \text{dpm}$$

Okay, maybe that was trivial, but this *is* only Chapter 2 … the final step is to divide by the total mass of the sample.

$$\frac{8.36 \times 10^{17}\ \text{dpm}}{78\ \text{g} + 21\ \text{g} + 2.01\ \text{g}} = 8.27 \times 10^{15}\ \frac{\text{dpm}}{\text{g}}$$

Notice that specific activity decreases as a radioactive nuclide becomes more "dilute" (Examples 2.7 and 2.8) within a sample.

2.4 DATING

The fact that radioactive decay can easily be predicted through mathematics opens up some interesting practical applications. One application is **radioactive dating**—the determination of the age of an object by measuring how much radioactive material is left in it. The most common form of dating is carbon dating. Small amounts of radioactive ^{14}C are continuously produced in the upper atmosphere through the bombardment of cosmic rays (Section 9.5). ^{14}C has a half-life of 5715 years, so it is also continually decaying. The rates of production have long since balanced out with the rate of decay, meaning that there has been a small but constant amount of ^{14}C on the planet for almost all its history. Carbon is an essential element for all living things; therefore, a small fraction of their carbon is ^{14}C. Once an organism dies, it no longer takes in ^{14}C from the biosphere, and the ^{14}C in the dead thing slowly begins to decay. Over time the ratio of ^{14}C to non-radioactive carbon (mostly ^{12}C) gradually decreases.

The specific activity of carbon is a convenient way to express the relative amount of ^{14}C present in all carbon. Prior to 1950, the specific activity of carbon on Earth was roughly 14 dpm/g. It varied somewhat before that time because cosmic ray bombardment is not constant but fluctuates through fairly regular long-running cycles. The specific activity of carbon has varied in recent decades primarily because of the tremendous amount of fossil fuels being burned. Fossil fuels are made up of carbon from living things that died quite a long time ago. Since the time they died, all of the ^{14}C that was once part of them has decayed away. Burning fossil fuels therefore pumps the atmosphere full of non-radioactive isotopes of carbon, lowering the specific activity of all carbon. It causes other problems too, but they are topics for other books.

Carbon dating works well for dating stuff that has been dead less than ~60,000 years (~10 half-lives). If it's been dead longer than that, it may be difficult to accurately measure the amount of ^{14}C present, because it's almost all decayed away. Carbon dating was famously used to date the Shroud of Turin, which is believed by some to be the burial shroud of Jesus.

Example 2.9 The specific activity of carbon from the Shroud of Turin is 12.8 dpm/g today. When were the plants killed to make this cloth?

The equation we'd like to use here is 2.1, but we'll need to calculate the decay constant from the half-life of ^{14}C first, using Equation 2.3.

$$\lambda = \frac{\ln 2}{t_{1/2}} = \frac{\ln 2}{5715 \text{ a}} = 1.213 \times 10^{-4} \text{a}^{-1}$$

First, we rearrange Equation 2.1, solving for t, and then substituting in the appropriate values.

$$\ln \frac{A_1}{A_2} = \lambda t \quad t = \frac{\ln \dfrac{A_1}{A_2}}{\lambda} = \frac{\ln \left(\dfrac{14 \text{ dpm/g}}{12.8 \text{ dpm/g}} \right)}{1.213 \times 10^{-4} \text{a}^{-1}} = \frac{0.090}{1.213 \times 10^{-4} \text{a}^{-1}} = 740 \text{ a}$$

The cloth was made about 740 years ago, or approximately 1300. Note that the answer is only expressed to two significant figures. As mentioned earlier, the specific activity of carbon has fluctuated in the past, so the answer is an approximate value. Generally speaking, ^{14}C-dating calculations done at this level of sophistication are only good to about two significant figures. Notice also that we can use specific activity values for A_1 and A_2 because they are both for pure carbon; therefore, the gram units in their denominators are equivalent (grams of pure carbon for both) and cancel out.

Other nuclides can be used for similar dating methods. For example, the age of rocks containing ^{238}U can be determined by measuring the ^{238}U:^{206}Pb ratio, assuming that all of the ^{206}Pb now present in the rock comes from the ^{238}U decay series (Figure 1.7). The same can be done when ^{87}Rb is found in a rock.

Example 2.10: A meteorite contains 0.530 g of ^{87}Rb and 0.042 g of ^{87}Sr. If 5 mg of the ^{87}Sr was present when the rock was formed, how old is the rock?

Over time, ^{87}Rb undergoes beta decay to ^{87}Sr.

$$^{87}_{37}\text{Rb} \rightarrow ^{87}_{38}\text{Sr} + ^{0}_{-1}\text{e}$$

How much ^{87}Rb was present in the rock when it was formed? It should be the sum of what's there now, and what has decayed to ^{87}Sr since the rock was formed.

$$0.530 \text{ g } ^{87}\text{Rb} + \left(\left(0.042 \text{ g } ^{87}\text{Sr} - 0.005 \text{ g } ^{87}\text{Sr}\right) \times \frac{\text{mol } ^{87}\text{Sr}}{87 \text{ g } ^{87}\text{Sr}} \times \frac{1 \text{ mol } ^{87}\text{Rb}}{1 \text{ mol } ^{87}\text{Sr}} \times \frac{87 \text{ g } ^{87}\text{Rb}}{\text{mol } ^{87}\text{Rb}} \right)$$
$$= 0.567 \text{ g } ^{87}\text{Rb}$$

Notice that the conversion from grams of ^{87}Sr to grams of ^{87}Rb is really unnecessary since they have the same mass number and one ^{87}Sr is produced for every ^{87}Rb that decays. This is often the case in dating problems, but not always.

Now the math is the same as in the ^{14}C example, except here we use the gram ratio instead of the activity ratio (since $A \propto N$, and N is proportional to mass so long as the nuclide is the same).

$$t = \frac{\ln \dfrac{N_1}{N_2}}{\dfrac{\ln 2}{t_{1/2}}} = \frac{\ln \left(\dfrac{0.567 \text{ g}}{0.530 \text{ g}} \right)}{\dfrac{\ln 2}{4.9 \times 10^{10} \text{ a}}} = 4.8 \times 10^9 \text{ a}$$

An apparent drawback when dating rocks is knowing how much daughter was present when the rock formed. Example 2.10 conveniently supplied that value, but how can we know? The amount of daughter present today (N_{2D}) is the sum of the amount of daughter originally present (N_{1D}) plus the amount of parent that has decayed ($N_{1P} - N_{2P}$).

$$N_{2D} = N_{1D} + (N_{1P} - N_{2P})$$

That's a lot of variables! We can eliminate one by applying Equation 2.7 to the parent amounts. After some algebraic gymnastics we have.

$$N_{2D} = \left(e^{\lambda t} - 1 \right) N_{2P} + N_{1D}$$

Sweet! We can measure the amount of parent and daughter currently present in the rock (N_{2D} and N_{2P}, respectively) from several samples from the same rock and plot them against each other. The intercept is then the amount of daughter originally present and the slope can give us how much time has passed since the rock was formed. This dating method is referred to as **isochron dating** because it requires multiple samples that were formed at the same time. As a practical matter, it is often helpful if the N values are all "normalized" vs. the amount of a stable isotope of the daughter—one whose amount remains constant throughout the lifetime of the rock. In other words, it doesn't decay, nor is it the daughter of some other decay process. For the ^{87}Rb/^{87}Sr pair, ^{86}Sr is a good choice.

Example 2.11: Three samples from a rock exhibit the nuclide ratios listed in Table 2.1. How long ago was this rock formed?

First plot the two columns of data (Figure 2.2). If the slope is equal to m, then the elapsed time (age of the rock) is

$$t = \frac{\ln(m+1)}{\lambda} = \frac{\left(\ln(0.00134 + 1) \right) \times \left(4.97 \times 10^{10} \text{ a} \right)}{\ln 2} = 9.60 \times 10^7 \text{ a}$$

TABLE 2.1
Isochron Data. First Plot the Two Columns of Data (Figure 2.2)

Sample	$^{87}Sr/^{86}Sr$	$^{87}Rb/^{86}Sr$
1	0.70651	0.131
2	0.70724	0.659
3	0.70745	0.841

FIGURE 2.2   Isochron plot for Example 2.11.

TABLE 2.2
Branch Ratios for ^{64}Cu Decay

Decay Mode	Branch Ratio	Nuclide Produced
β^-	38.5%	^{64}Zn
β^+	17.6%	^{64}Ni
EC	43.9%	^{64}Ni

2.5   BRANCHED DECAY

Radioactive decay is often a messy process. As we've already seen in the naturally occurring decay series (Figure 1.7), some nuclides make matters worse by decaying by more than one pathway. ^{64}Cu is a classic example of a nuclide that has more than one decay mode. It can decay by β^- to ^{64}Zn, or by β^+ or electron capture to ^{64}Ni. The half-life for ^{64}Cu is given as 12.701 h, but this accounts for all decays. Information on the relative percentages of each form of decay, or **branch ratio**, is given in Table 2.2. These data are from the National Nuclear Data Center at Brookhaven National Laboratory (www.nndc.bnl.gov/nudat3/). Table 2.2 tells us that ^{64}Cu decays via β^- to ^{64}Zn only 38.5% of the time, and the remaining 61.5% of the time it decays via positron emission or electron capture to ^{64}Ni.

When does branch ratio matter? Any time a decay rate (activity) is calculated from a count rate.

Example 2.12: ^{198}Au emits a 412 keV gamma photon in 95.6% of its decays. If a count rate of 52,997 cpm is observed for this photon using a detector with an efficiency of 6.31%, what is the nuclide's decay rate (dps)?

First calculate the observed activity in decays per second using Equation 2.4, the percent efficiency.

$$\text{dpm} = \frac{52{,}997 \text{ cpm}}{6.31\%} \times 100\% = 8.40 \times 10^5 \text{ dpm} \qquad \frac{8.40 \times 10^5 \text{ decays}}{\text{min}} \times \frac{\text{min}}{60 \text{ s}} = 1.40 \times 10^4 \text{ dps}$$

Next, account for the branch ratio to determine the total activity. We know the decay rate is only for the 412 keV gamma, so the decay rate for all decays must be higher, so the math is the same as converting count rate to decay rate.

$$\frac{1.40 \times 10^4 \text{ dps}}{95.6\%} \times 100\% = 1.46 \times 10^4 \text{ dps}$$

We can experimentally determine half-life a couple different ways. We can observe the change in count rate over time and use

$$t_{1/2} = \frac{\ln 2}{\ln \dfrac{A_1}{A_2}} \times t$$

which is a rearrangement of Equation 2.1. Using this method does not require application of the branch ratio, i.e., the change in activity (or count rate) with time does not depend on the branch ratio.

The other way to find half-life is to determine the activity and mass of the nuclide (convert to N) and use

$$t_{1/2} = \frac{\ln 2 \times N}{A}$$

which is a rearrangement of $A = \lambda N$ (Equation 2.5). As you might guess the first method is more useful when the nuclide has a short half-life (a change in activity can be observed in a reasonable amount of time) and the second method works best for nuclides with long half-lives (measure mass and activity—take your time, the decay rate's not changing any time soon). If this latter method is used with a nuclide that undergoes branched decay, the branch ratio must be taken into account to properly calculate activity. Activity results from all decays. If only one mode is observed during an experiment, branch ratio and efficiency must be used to calculate activity. It's a bit like trying to figure out how quickly a bucket of water will drain through a hole based on the amount of water in the bucket and the flow rate out of the hole. The problem is that the bucket has another hole that you can't see. The bucket will always drain faster than you think it should, based on the flow rate from the hole you can see.

Example 2.13: A sample of copper contains 9.2×10^7 atoms of ^{64}Cu, and registers 2320 cpm on a detector that can only detect beta particles (β^-). If this detector has an efficiency of 7.20%, what is the half-life of ^{64}Cu?

Following Example 2.12, convert to the observed decay rate, then divide by the branch ratio (from Table 2.2).

$$\text{dpm} = \frac{2320 \text{ cpm}}{7.2\%} \times 100\% = 3.2 \times 10^4 \text{ dpm}$$

$$\frac{3.2 \times 10^4 \text{ dpm}}{38.5\%} \times 100\% = 8.4 \times 10^4 \text{ dpm}$$

Now plug into $A = \lambda N$.

$$8.4 \times 10^4 \, \text{dpm} = \lambda \times (9.2 \times 10^7 \, \text{atoms})$$
$$\lambda = 9.1 \times 10^{-4} \, \text{min}^{-1}$$

Because the units for activity were decays per minute, the units on the decay constant are reciprocal minutes. Now convert decay constant to half-life.

$$t_{1/2} = \frac{\ln 2}{\lambda} = \frac{\ln 2}{9.1 \times 10^{-4} \, \text{min}^{-1}} = 7.6 \times 10^2 \, \text{min}$$

It looks like this half-life would be more conveniently expressed in hours:

$$t_{1/2} = 7.6 \times 10^2 \, \text{min} \times \frac{1 \, \text{h}}{60 \, \text{min}} = 13 \, \text{h}$$

With only two sig figs to work with, this is pretty close to the actual value of 12.701 h.

If we had left out the branch ratio in these calculations, the half-life would be 33 hours—quite a way off! A half-life calculated using data from one branch of the decay is called a **partial half-life**. A partial half-life is always longer than the true or total half-life. If we know the fraction of the decays that go by each branch, we can calculate the partial half-life by simply dividing the total half-life by the fraction. Since the true half-life for ^{64}Cu is 12.701 h, the electron capture (EC, the other decay pathway for ^{64}Cu) partial half-life is:

$$\frac{12.701 \, \text{h}}{0.439} = 28.9 \, \text{h}$$

This makes sense! Since only about half of the decays are EC, the partial half-life should appear to be about twice as long.

Example 2.14: ^{192}Ir decays via β^- emission 95.2% of the time. If its β^- partial half-life is observed to be 77.5 d, what is its true half-life?

$$77.5 \, \text{d} \times 0.952 = 73.8 \, \text{d}$$

2.6 EQUILIBRIA

The naturally occurring decay series that begins with ^{238}U starts with an alpha decay forming ^{234}Th (Figure 1.7). ^{234}Th is also unstable toward decay, spitting out a beta particle, forming ^{234}Pa. These two decay processes can be combined into a single equation.

$$^{238}_{92}\text{U} \xrightarrow{\alpha} \, ^{234}_{90}\text{Th} \xrightarrow{\beta^-} \, ^{234}_{91}\text{Pa}$$

As we learned back in Section 1.3, the original nuclide in any decay is called the parent nuclide and the product nuclide is called the daughter. In the case of a decay series, all nuclides formed after the parent decays are called daughters. Here, ^{238}U is the parent and ^{234}Th and ^{234}Pa are daughters. A decay series could also be generically represented by:

$$A \xrightarrow{t_{1/2(A)}} B \xrightarrow{t_{1/2(B)}} C \ldots$$

Since daughter B is constantly being formed by decay of parent A, and is constantly decaying to daughter C, how can we figure out how much B exists at any one moment in time? More generally, how can we determine the total activity of the sample or the amount of any one nuclide at a particular point in time? It seems almost too complex to comprehend, but we can look at three simple scenarios where the math becomes more accessible.

2.6.1 SECULAR EQUILIBRIUM

If the half-life of the parent is very long relative to the half-life of the daughter, and the parent is isolated at time = 0, then the following equations describe the system *at any time* after the parent is isolated:

$$A_B = A_A \left(1 - e^{-\lambda_B t}\right) \tag{2.8}$$

$$N_B = \frac{\lambda_A}{\lambda_B} N_A \left(1 - e^{-\lambda_B t}\right) \tag{2.9}$$

The subscripts A and B refer to parent and daughter respectively. Therefore, A_B is the activity of the daughter and λ_A is the decay constant for the parent. These two equations are mathematically identical. To derive the second from the first, substitute $\lambda_X N_X$ (X = A, B) for the corresponding activities.

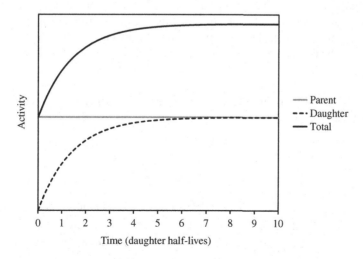

FIGURE 2.3 Secular equilibrium.

If the activities of both parent and daughter are plotted after the isolation of pure parent, something like Figure 2.3 will result. Because the parent half-life is so incredibly long, its activity is constant over the period of observation. Daughter activity starts at zero because the parent was chemically isolated. As daughter is produced, its activity gradually increases. However, as the amount of daughter builds up, its rate of decay gradually becomes more significant. When the rate of formation of B is equal to the rate of decay of B, it has reached **secular equilibrium**, much like the ^{14}C in our biosphere discussed in Section 2.4. Secular equilibrium is reached roughly seven daughter half-lives after the parent is isolated. At this point the activity of the daughter is equal to the activity of the parent, and the activity equation given above (Eq. 2.8) can be simplified to $A_B = A_A$. This is true for all radioactive daughters (not just B) of a long-lived parent A. If instead, C is a stable nuclide (i.e., A and B are the only radioactive species present), the total activity after secular equilibrium is reached is twice that of either the parent or the daughter, as illustrated in Figure 2.3.

Because $A_B = A_A$ at any point in time *after* secular equilibrium is reached, we can substitute in λN for the activities, and generate more equations like the following.

$$\lambda_A N_A = \lambda_B N_B \tag{2.10}$$

$$\frac{N_B}{N_A} = \frac{\lambda_A}{\lambda_B} = \frac{t_{1/2(B)}}{t_{1/2(A)}} \tag{2.11}$$

These equations can only be applied after secular equilibrium is reached. Secular equilibrium is very handy for solving all kinds of problems. It is especially useful with the three naturally occurring radioactive decay series. As you may recall, these decay series are a significant source of many of the radionuclides found in nature. Since they all start with a long-lived nuclide, and have been around for a few billion years, secular equilibrium applies!

1. Since the activities of all nuclides in secular equilibrium are equal, the mass ratios (N_B/N_A) can easily be determined.
2. We have another method for determining the half-life of a long-lived nuclide. As we saw in Section 2.5, it is difficult to determine a long half-life by sitting around waiting to observe a change in activity over time. If we know the half-life of one of the daughters, and its mass ratio to the parent, we can calculate the half-life of the long-lived parent using Equation 2.11.
3. The mass of the parent can be determined from the activity of any daughter.
4. The mass of the parent can be used to calculate the activity of a daughter.

Example 2.15: 10.2 kg of uranium is isolated from some ore, which also contains 3.33×10^{-3} g of ^{231}Pa. What is the half-life of ^{235}U?

$$t_{1/2(U\text{-}235)} = \frac{N_{U\text{-}235}}{N_{Pa\text{-}231}} t_{1/2(Pa\text{-}231)}$$

$$= \frac{10{,}200\text{ g} \times \frac{\text{mol U}}{238.0\text{ g}} \times \frac{6.022 \times 10^{23}\text{atoms}}{\text{mol}} \times \frac{0.7204\text{ atoms }^{235}\text{U}}{100\text{ atoms U}}}{\left(3.33 \times 10^{-3}\text{g}\right) \times \frac{\text{mol }^{231}\text{Pa}}{231\text{ g}} \times \frac{6.022 \times 10^{23}\text{atoms}}{\text{mol}}} \times \left(3.28 \times 10^4\text{ a}\right)$$

$$= 7.02 \times 10^8\text{ a}$$

It is important to realize that uranium isolated from ore will contain all naturally occurring isotopes of uranium in their normal proportions. To do this problem, the number of ^{235}U atoms within 10.2 kg (10,200 g) of uranium must be calculated. ^{235}U makes up only 0.7204% of all uranium. Notice that the conversion from moles to atoms (Avogadro's number) is unnecessary, in both the numerator and the denominator, as it cancels out.

Example 2.16: What is the activity of ^{222}Rn in an ore sample containing 3.00 g of uranium?

In this example, the correct decay series must first be determined. ^{222}Rn is only observed in the ^{238}U decay series; therefore, it is only produced by ^{238}U, which is 99.2742% abundant.

$$A_{\text{Rn-222}} = A_{\text{U-238}} = \lambda_{\text{U-238}} N_{\text{U-238}}$$

$$= \left(\frac{\ln 2}{4.468 \times 10^9 \text{a}} \right) \left(3.00 \text{ g U} \times \frac{\text{mol U}}{238.0 \text{ g U}} \times \frac{6.022 \times 10^{23} \text{atoms U}}{\text{mol U}} \times \frac{99.2742 \text{ atoms }^{238}\text{U}}{100 \text{ atoms U}} \right)$$

$$= 1.17 \times 10^{12} \text{ decays per year}$$

$$= 2.22 \times 10^6 \text{ dpm}$$

The above discussion of secular equilibria assumes that the parent does not undergo branched decay and that no branched decay happens between parent and a daughter that doesn't also go through that daughter. Close examination of the decay series for the nuclides involved in Examples 2.15 and 2.16 should bring a sigh of relief. No branches involving or between these nuclides. If the parent (A) undergoes branched decay, then the activity of the daughter (A_B) is equal to the activity of the parent (A_A) multiplied by the (fractional) branch ratio (BR):

$$A_B = A_A \times BR \tag{2.12}$$

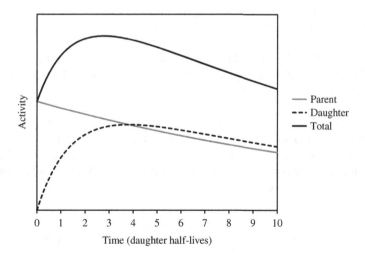

FIGURE 2.4 Transient equilibrium.

2.6.2 Transient Equilibrium

Transient equilibrium is like secular equilibrium in that the half-life of the parent is longer than the half-life of the daughter. The difference is the time of observation. A nuclear equilibrium is considered transient if significant parental decay occurs during the time of observation (usually more than 7–10 daughter half-lives). When transient equilibrium occurs, the parent and daughter half-lives tend to be closer together in value.

If the parent is isolated at time = 0, then as the daughter is produced, and begins to decay, the parent is also decaying. Parental activity will decrease in an exponential fashion, as shown in Figure 2.4. Keep in mind that the parent's half-life is longer than the daughter's, so a dramatic decrease in parent activity is not observed in Figure 2.4 (i.e., it is only slightly curved). Daughter activity grows at first, peaks, then decreases, eventually paralleling the decrease in parent activity. Notice that daughter activity eventually exceeds parent activity. It is after ~7 daughter half-lives that the relative activities of parent and daughter stabilize, establishing transient equilibrium. Again, assuming parent (A) and daughter (B) are the only significant radioactive species present, total activity starts at the same point as parent activity, increases as the amount of daughter increases, and then parallels the decrease in parent and daughter activity. *After* equilibrium is achieved, the following equations can be applied.

$$\frac{A_A}{A_B} = 1 - \frac{t_{1/2(B)}}{t_{1/2(A)}} \tag{2.13}$$

$$\frac{N_B}{N_A} = \frac{t_{1/2(B)}}{t_{1/2(A)} - t_{1/2(B)}} \tag{2.14}$$

Again, these two equations are mathematically related through the awesome power of $A = \lambda N$. The choice of which to use is largely determined by the problem. Does it involve activities or numbers of atoms? The former is much more likely.

Example 2.17: A sample of ^{47}Ca is isolated from other radionuclides. Twenty-eight days later, the activity due to ^{47}Ca is 2.19×10^4 dpm. What activity would you expect for ^{47}Sc, also 28 days after the original sample of ^{47}Ca was isolated?

The half-life for ^{47}Ca is 4.536 d and for ^{47}Sc it is 3.349 d. This sample has attained transient equilibrium. The equilibrium is transient because a lot of the isolated ^{47}Ca will have decayed after 28 days, i.e., it's been sitting around for 28 d/4.536 d = 6.2 parent half-lives. Twenty-eight days is also long enough for equilibrium to be established because 28 d/3.349 d = 8.4 daughter half-lives. Rearrange Equation 2.13 to solve for A_B, and plug in the numbers.

$$A_{Sc-47} = \frac{A_{Ca-47}}{1 - \frac{t_{1/2(Sc-47)}}{t_{1/2(Ca-47)}}} = \frac{2.19 \times 10^4 \, \text{dpm}}{1 - \frac{3.349 \, \text{d}}{4.536 \, \text{d}}} = 8.37 \times 10^4 \, \text{dpm}$$

For all transient equilibria, daughter activity is maximized at time t_{max}, which can be calculated from the parent and daughter half-lives using Equation 2.15.

$$t_{max} = \left[\frac{1.44 t_{1/2(A)} t_{1/2(B)}}{\left(t_{1/2(A)} - t_{1/2(B)} \right)} \right] \times \ln \frac{t_{1/2(A)}}{t_{1/2(B)}} \tag{2.15}$$

Example 2.18: In the previous example, how long did it take the daughter to reach maximum activity?

$$t_{max} = \left[\frac{1.44 t_{1/2(Ca-47)} t_{1/2(Sc-47)}}{\left(t_{1/2(Ca-47)} - t_{1/2(Sc-47)} \right)} \right] \times \ln \frac{t_{1/2(Ca-47)}}{t_{1/2(Sc-47)}} = \left[\frac{1.44 \times 4.536\,d \times 3.349\,d}{\left(4.536\,d - 3.349\,d \right)} \right] \times \ln \left(\frac{4.536\,d}{3.349\,d} \right)$$
$$= 5.59\,d$$

2.6.3 No Equilibrium

When the half-life of the parent is shorter than the daughter, equilibrium is not reached. If pure parent is again separated at time = 0, the parent activity will decrease rapidly, as shown in Figure 2.5. Daughter activity will increase until the parent is nearing depletion, then it will gradually decrease. When the parent is gone (> 7 parent half-lives), the daughter's decrease in activity will be based entirely on its own half-life ($t_{1/2(B)}$). Since no equilibrium is reached, no special equations are necessary to describe the change in daughter activity after the parent is gone. The activity equations given at the beginning of the chapter (Equations 2.1 and 2.2) are sufficient. However, Equation 2.15, used to calculate maximum daughter activity for transient equilibria, also applies here.

Nuclear equilibria are sometimes confusing to students. Perhaps it's because the term "equilibrium" is not the best one to describe what is actually happening. We say the nuclear equilibria are

FIGURE 2.5 Equilibrium is not established when $t_{1/2A} < t_{1/2B}$.

reached when the relative rate of formation of daughter equals its relative rate of decay (A → B → C...). For most students, the more familiar use of the word "equilibrium" likely comes from chemistry. In a chemical reaction, equilibrium is reached when the rate of formation of products equals the rate of formation of reactants. In other words, reactants are combining to form products at the same rate that products are combining to form reactants (A ⇌ B). To extend this analogy to a radioactive decay would mean that the daughter would have to "decay" back into the parent, which is not possible. "Steady-state" would be a better term to describe nuclear equilibria. Steady-state is like pouring water into a bucket that's already full—the rate water is being added is equal to the rate it is spilling out, but we always have the same amount of water in the bucket. Nuclear equilibria are all about the defined mathematical relationships between the amount of parent and daughter. Parent is constantly converting to daughter and daughter is constantly decaying but the ratio between parent and daughter is constant at equilibrium.

Two other points of confusion are trying to decide if a system under investigation is at equilibrium, and what kind of equilibrium it is. In this text, the passage of seven daughter half-lives is consistently used as the point where equilibrium is first reached. Other texts state that equilibrium is only attained after ten daughter half-lives have elapsed. If the point in time is within the controversial $7–10 \times t_{1/2(B)}$ period, then it is largely a matter of the level of precision required. For most work, seven half-lives will be sufficient, and that's why it is used as a point of demarcation in this book. However, if very precise measurements are being made, then waiting until ten half-lives have passed before declaring the system to be at equilibrium may be necessary.

The confusion in determining whether secular or transient equilibrium is appropriate to a particular problem is understandable. The difference can be ambiguous. The main distinction is that the half-life of the parent needs to be long compared to the time of observation for the equilibrium to be secular. In other words, the parent cannot decay significantly (< 5–10%) during the time of observation. Again, the exact percentage of decay that is tolerable will depend on the level of precision required by the problem.

2.7 STATISTICS

Radioactive decay is a random event. We can't say when an unstable nucleus will decay, only that it will probably decay within seven or so half-lives. If we've got a whole lot of unstable nuclides, the number of them that decay in a certain amount of time will vary, even if the total number of these nuclides does not change significantly. Table 2.3 gives count data for a ^{204}Tl source counted 20 times for one minute each. With a half-life of 3.8 years, it did not decay appreciably during the 20 minutes these data were collected. The lowest count is 2071 counts, and the highest 2217 counts—a difference of 146 counts! So, what is the true number of counts for this sample?

As you might guess, the best we can do is average our data and hope it is close to the true value. How much error is in this estimate? Statistics can give us the answers!

The probability that we will get a particular count can be approximated by a Gaussian (or normal) distribution. A Gaussian is a curve, symmetric about the mean ($\bar{x}$) value as shown in Figure 2.6. The mean value is determined by adding all the individual values (x) together and dividing by the number of runs (N).

$$\bar{x} = \frac{\Sigma x}{N} \tag{2.16}$$

The mean value for the 20 data in Table 2.3 is 2142 counts.

The vertical axis of the Gaussian plot corresponds to the probability that a value along the horizontal axis will be observed. The curve is highest at the mean and tapers away in both directions, so we are more likely to get a count close to the mean rather than far from it. For us, the mean

TABLE 2.3
Count Data

Run	Counts
1	2123
2	2189
3	2120
4	2167
5	2125
6	2217
7	2192
8	2098
9	2154
10	2130
11	2112
12	2151
13	2191
14	2124
15	2071
16	2085
17	2179
18	2134
19	2096
20	2187

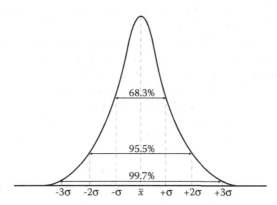

FIGURE 2.6 A Gaussian distribution.

value serves as our best guess as to the true value. In other words, we'll assume that the mean *is* the true value.

Why do decay data, like those in Table 2.3, approximately fit a Gaussian distribution? First, radioactive decay is a perfectly random event and there are only two possible outcomes (a nuclide either decays or it doesn't). As a result, a set of counts, like those given in Table 2.3, follow the binomial distribution law. Unfortunately for us, binomial distributions are mathematically complex, and become even more unwieldy when sample sizes are large. Because nuclear scientists typically work with such large numbers of nuclides, sample sizes are pretty much through the roof. A similar, but mathematically simpler distribution is the Poisson, which handles the large numbers of nuclides more easily. Finally, since the number of nuclides that undergo decay each minute is sufficiently large (greater than ~100 cpm), we can further approximate the distribution as a Gaussian.

So, radioactive decay data fit a binomial distribution which is approximated by a Poisson distribution which can be further simplified by a Gaussian.

While approximate, assuming a Gaussian means that the uncertainty associated with a count (the standard deviation or σ) can be approximated as the square root of the value.

$$\sigma \approx \sqrt{x} \quad \text{or} \quad \sigma \approx \sqrt{\bar{x}} \tag{2.17}$$

This equation applies to any single count or an average of a set of counts. This calculation of standard deviation is likely a little different than the one on your calculator or in a spreadsheet. Remember that we are looking at perfectly random data that fit a Gaussian, and that it is an approximation. The standard deviation of the average count from Table 2.3 is $\sqrt{2142} = 46$ counts. Nuclear scientists often use uncertainties expressed to 1σ, which means that 68.3% of the time an additional count will fall within the range of $\bar{x} \pm 1\sigma$, or 2142±46 counts. For some, the mean plus or minus two standard deviations provides a bit more comfort as it tells us there's a 95.5% chance that an additional count will fall within that range. If we wanted to see if our sample data (Table 2.3) fit a Gaussian, we could state that the count mean is 2142±92 counts to a confidence level of 95.5% ($\bar{x} \pm 2\sigma$). The 2σ range of these counts is therefore 2050 counts to 2235 counts. In other words, if we collected data for another one-minute count, 95.5% of the time, it will fall between 2050 counts and 2235 counts. Note that none of the 20 counts in Table 2.3 exceeds these limits. With 20 data, we expect that 4.5% would fall outside of $\bar{x} \pm 2\sigma$. Multiplying 20 by 0.045 gives 0.90. In other words, approximately one count should have fallen outside of the two standard deviation range. Should we be concerned that one did not? No. With such a small sample size, the fit to a Gaussian will not be perfect. Zero, one, or two data falling outside of 2σ are likely reasonable. We could also examine our data to see if approximately 68.3% of the data fall within $\bar{x} \pm 1\sigma$ (2142±46 counts). Careful examination shows that 15 out of the 20 data (75%) fall within one standard deviation of the mean value. Again, it is an approximate fit because of the small sample size.

The mean and standard deviation allow us to estimate the true value for, and characterize the precision of, a single datum or set of data. A relatively large standard deviation means poor precision and a relatively small σ means good precision. Notice that the magnitude of the standard deviation depends only on the magnitude of the count (Eq. 2.17). For example, let's say we collect another count from a toastier source and observe 4284 counts, twice what we observed from the data in Table 2.3. This second datum then has a 2σ value of 131 counts ($2\sqrt{4284} = 131$). This is larger than the 95.5% confidence level we observed (92 counts) when the average count was 2142 counts. Which is more precise?

A good way to compare the precision for two or more data is to calculate the relative standard deviation (RSD). RSD is determined by dividing the standard deviation by the measured value.

$$\text{RSD} = \frac{\sigma}{x} \approx \frac{\sqrt{x}}{x} = x^{-\frac{1}{2}} \quad \text{or} \quad \text{RSD} = \frac{\sigma}{\bar{x}} \approx \frac{\sqrt{\bar{x}}}{\bar{x}} = \bar{x}^{-\frac{1}{2}} \tag{2.18}$$

Since the standard deviation of a count is approximately equal to the square root of the count, we can also determine RSD by calculating the reciprocal square root of the count. Like calculation of standard deviation (Equation 2.16), we can use the same equations for a mean count (Dracula?) value.

The RSD for the mean count from Table 2.3 is 0.022, while the RSD for 4284 counts is 0.015. Relatively speaking, the higher count has less uncertainty and is therefore more precise. This is generally true; higher counts (and higher count rates) have lower relative uncertainty.

Notice that calculation of the RSD is basically the same as if we were going to calculate percent error. The only change we'd have to make to Equations 2.18 is to multiply by 100%. Therefore, we could say that 2142 counts has an error (or uncertainty) of 2.2% to 1σ and 4282 counts has an error of 1.5%. Like standard deviation, these percentages depend solely on the magnitude of the count. The more counts, the lower the percent uncertainty.

Up until now, we've been talking about the uncertainty associated with a count or the average count. In nuclear science we're more likely to be dealing with count rates. How can we determine uncertainty for count rates? The standard deviation of the count rate (σ_R) is determined by dividing the standard deviation of the count (from the first of the two Equations 2.17) by the time (t) it took to collect the data (in minutes).

$$\sigma_R = \frac{\sigma}{t} \approx \sqrt{\frac{R}{t}} \tag{2.19}$$

Mathematically, this is approximately equivalent to taking the square root of the count rate (R), in units of counts per minute (cpm), divided by the time in minutes (t).

If we repeat the above experiment (that generated data for Table 2.3) exactly as it was originally performed and obtain a single 20-minute count of 43,000 counts, the standard deviation of the count is 207 counts.

$$\sigma \approx \sqrt{x} = \sqrt{43000} = 207 \text{counts}$$

In other words, if we took another 20-minute count, there's a 68.3% chance we'd see a count in the range of 43,000 ± 207 counts. The count rate from this datum is 43,000 counts/20 minutes = 2150 cpm. The standard deviation of the count rate (σ_R) is 10 cpm.

$$\sigma_R = \frac{207 \text{ counts}}{20 \text{ min}} = \sqrt{\frac{2150 \text{ cpm}}{20 \text{ min}}} = 10 \text{ cpm}$$

If we took another 20-minute count, there's a 68.3% chance we'd see a count *rate* in the range of 2150 ± 10 cpm. Notice that the standard deviation of the count rate is quite a bit smaller than the standard deviation of the count in this case. The longer we collect count data, the narrower the precision on the count rate.

If we had a very large sample of data, we could calculate the true mean (usually represented as μ). The standard deviation (σ) would then be exactly equal to the square root of the true mean ($\sqrt{\mu}$). In a limited number of count runs, like the 20 in our first data set, we can calculate a good value of σ by using the average of our counts ($\bar{x}$) instead of the true mean μ. But we should keep in mind that this is an approximation because the calculated mean ($\bar{x}$) is only an approximation of the true mean (μ).

Can we compare the count rate we just obtained with the mean value from Table 2.3? You bet! The values given in Table 2.3 are also count rates because they were collected for one minute each. In total, it took 20 minutes to collect the 20 data in Table 2.3, so the standard deviation of the count rate is 10 cpm.

$$\sigma_R = \sqrt{\frac{2142 \text{ cpm}}{20 \text{ min}}} = 10 \text{ cpm}$$

Notice in this case that we cannot calculate the standard deviation of the count rate (σ_R) by dividing standard deviation by the time in minutes (σ/t). This part of Equation 2.19 does not work with mean values.

Notice also that the standard deviation of the count rate is the same (10 cpm) for both sets of data. Because they were collected for the same total amount of time, the count rate obtained in a series of 20 one-minute counts should have the same uncertainty as one obtained from a single 20-minute count.

Numerically, the two count rates we've obtained in our two sets of data are different, 2150 cpm vs. 2142 cpm. Is this difference statistically significant? To answer this question, we need to compare the ranges of the count rates plus or minus two standard deviations (95.5% certainty) of the count rates ($\bar{x} \pm 2\sigma_R$). If these ranges overlap, then the two count rates are the same, statistically speaking. The count rate for the single 20-minute count is 2150±20 cpm to a certainty of 95.5%. This is a range of 2130 to 2170 cpm. The count rate for the 20 one-minute counts is 2142±20 cpm to a certainty of 95.5%, which means we have a range of 2122 to 2162 cpm. Since the two ranges overlap, the two count rates are "not significantly different from each other." The difference in count rates is due to the random error (uncertainty) inherent to radioactive decay. Many more runs or a much longer counting time (lowering σ_R) would be needed to determine if the two count rates are, in fact, different.

Readers should take care not to confuse the meanings of σ and σ_R. Both are expressions of uncertainty and reflect on the precision of the data. Standard deviation (σ) establishes the range within which we can expect another single count to appear. For the data in Table 2.3, it tells us that there's a 68.3% chance that another one-minute count will fall between 2096 counts and 2188 counts. The standard deviation of the count rate (σ_R) establishes the range within which we can expect another count *rate*. For the data in Table 2.3, it tells us that there's a 68.3% chance that the mean count rate of another set of 20 one-minute counts (or a single 20-minute count…) will fall between 2140 cpm and 2160 cpm.

As a practical matter, nuclear scientists often collect data of a nuclide as it decays. For example, while observing the decay of ^{108}Ag ($t_{1/2} = 2.39$ min) we decide to collect data every 20 seconds, convert our counts to counts per minute, and apply corrections. Can we still estimate an uncertainty for each corrected count rate? Yes! We know that the uncertainty of a single count, to confidence level of 95.5% (2σ), is two times the square root of a count. Please note that σ must be calculated using the raw count data, because the uncertainty comes from the measured value. For example, if we observe 2345 counts over a 20-second period, then $2\sigma = 2 \times \sqrt{2345} = 97$ counts. When the count is converted to a rate in units of counts per minute, the count data are multiplied by three. If we want an idea of the precision of the count rate in terms of counts per minute, the value for 2σ must also be multiplied by three.

$$2\sigma_R = \frac{97 \text{ counts}}{20 \text{ s} \times \dfrac{\text{min}}{60 \text{ s}}} = 97 \text{ counts} \times 3 \text{ min}^{-1} = 290 \text{ cpm}$$

Other (such as dead time and background) corrections are applied to the count rate but not to the uncertainty ($2\sigma_R$). The final count rate (cpm) and random error for this example are 7068±290 cpm, or about a 4% uncertainty. Remember when this text stated that count data are usually only good to three significant figures? These numbers suggest that three might be a little generous, at least at this count rate and data collection time.

As we discovered earlier, higher count data (longer count times or a higher activity sample) will have a smaller uncertainty and should be favored in drawing the best-fit line to experimental data. When a computer draws such a line, it typically takes each point as equally valid. You may find that you get better results if you eliminate points with the lowest count rates. They can be eliminated because of their higher uncertainty (low precision). Points that are more than $\pm 2\sigma_R$ from the expected count are sometimes viewed with skepticism by some students but remember that 4.5% of all points are expected to fall outside of this range ($100\% - 95.5\% = 4.5\%$). Points that are beyond 3σ of the expected value (a 99.7% confidence level) *could* be eliminated from most data as they are probably due to mistakes made in collecting the data. However, students are cautioned from simply discarding data because they are outside of some level of precision. Unless you have some other reason to suspect there is a problem with a particular data point, it should be retained.

FIGURE 2.7 A histogram of the 20 count data in Table 2.3.

Radioactive decay data, like those in Table 2.3, should approximately fit a Poisson distribution. If we create a histogram of these data, it could look like Figure 2.7.

Remember that a Poisson distribution should look very similar to a Gaussian (Figure 2.6). Not exactly what we're seeing in Figure 2.7, eh? This is again due to the low amount of data. To start to see a good fit to a Poisson distribution from a histogram, hundreds of data should be collected. Because there is not a lot of data in Table 2.3, the fit is pretty rough.

If we only have a limited set of data, how can we determine that they fit a Poisson distribution? We need to calculate chi-squared (χ^2).

$$\chi^2 = \frac{1}{\overline{x}} \sum (\overline{x} - x)^2 \qquad (2.20)$$

To find χ^2, we'll need to calculate the square of the difference of every individual value in the data set with the mean value ($(\overline{x} - x)^2$), add them up ($\sum (\overline{x} - x)^2$), and then multiply by the reciprocal of the mean value ($1/\overline{x}$). Sounds crazy, but it works. Try calculating this value using the data in Table 2.3. χ^2 for those data should equal 14.9.

Is this good? To interpret a chi-squared value, we'll need to use a table of chi-squared like the one found in Appendix B. To use this table, you need to know that there are 20 data points; therefore, the data have 19 degrees of freedom ($N - 1$). Reading down the first column to 19 there are several values of chi-squared listed on this row of the table to compare with the calculated value. The values in the final and second columns tell us that for 19 degrees of freedom there is one chance in 100 that chi-squared will be smaller than 7.633 and one chance in 100 that it will be larger than 36.19. Our value of 14.9 certainly falls within that range.

A more reasonable set of limits is the 80% chance. There is a 10% chance that chi-squared will be smaller than 11.65 for 19 degrees of freedom. This is obtained by reading down the 0.90 column to 19 degrees of freedom. There is also a 10% chance that chi-squared will be larger than 27.20. These are obtained by reading down the 0.90 and 0.10 columns to 19 degrees of freedom. Therefore, there is an 80% chance that chi-squared will fall between 11.65 and 27.20 for a Poisson distribution. An 80% chance means that eight out of ten times this will happen. Once again, our value of 14.9 fits within the range, giving us confidence that our data fit a Poisson distribution (as well they should!).

There have probably been many times in your life when you did not have a chi-squared table at hand. If you look down the 0.50 column, you will note that chi-squared increases with the number of trials N. The value of the chi-squared is roughly $N - 2$. If you ever need a ballpark estimate of chi-squared for, say, 31 determinations, it would be about 29. The entry in the table for 30 degrees of freedom, that is 31 data at the 50% probability, is 29.34. Darn close.

So far, we've only discussed the inherently random nature of radioactive decay, which leads to a certain amount of uncertainty in data collected from radioactive sources. This is truly random error, but the wonderful thing about it is that we can always quantify it. What if we carefully calculate the random error associated with some decay measurements, but it is clear some additional error is also present? That error would be systematic error and is likely due to some external change during the experiment. Perhaps the source or the detector moved, or some shielding between the source and detector was added or removed. Perhaps there is a flaw in the experimental procedure. It could be any number of factors. A good scientist will take great care in making measurements and consider many possibilities if systematic error is apparent.

Very commonly, we subtract a background count rate (R_b) from our gross count rate (R_g) to obtain a net count rate (R_n). Can we determine the amount of uncertainty in the result, the net count rate? Yes, we can. The standard deviation of the net count rate (σ_{Rn}) is equal to the square root of the sum of the squares of the standard deviation of the gross count rate (σ_{Rg}) and the standard deviation of the background count rate (σ_{Rb}). This is mathematically represented in Equation 2.21.

$$\sigma_{R_n} = \sqrt{\sigma_{R_g}^2 + \sigma_{R_b}^2} = \sqrt{\frac{R_g}{t_g} + \frac{R_b}{t_b}} \qquad (2.21)$$

Alternatively, we can take the square root of the sum of the gross count rate (R_g) divided by the time taken to collect these counts in minutes (t_g) and the background count rate (R_b) divided by the time taken to collect background in minutes (t_b).

Example 2.19: A sample counted for 4000 s gave 2225 counts. A 2000 s background count registered 436 counts. What is the net count rate and its uncertainty, expressed to a 68.3% confidence level?

$$t_g = 4000 \text{ s} \times \frac{\text{min}}{60 \text{ s}} = 66.7 \text{ min} \qquad R_g = \frac{2225 \text{ counts}}{66.7 \text{ min}} = 33 \text{ cpm}$$

$$t_b = 2000 \text{ s} \times \frac{\text{min}}{60 \text{ s}} = 33.3 \text{ min} \qquad R_b = \frac{436 \text{ counts}}{33.3 \text{ min}} = 13 \text{ cpm}$$

$$R_n = 33 \text{ cpm} - 13 \text{ cpm} = 20 \text{ cpm}$$

$$\sigma_{R_n} = \sqrt{\frac{33 \text{ cpm}}{66.7 \text{ min}} + \frac{13 \text{ cpm}}{33.3 \text{ min}}} = 0.9 \text{ cpm}$$

Note that the uncertainty of the net count rate (σ_{Rn}) is always larger than either of the two uncertainties used to calculate it $(\sigma_{Rg}$ and $\sigma_{Rb})$. This should make some intuitive sense as we are subtracting two values (gross count rate and background count rate) that both have error associated with them. The result (net count rate) is a smaller value than the gross counts and will therefore have greater uncertainty. Just out of curiosity, let's calculate the σ_R values for the gross and background count rates in Example 2.19.

$$\sigma_{R_g} = \frac{\sqrt{2225 \text{ counts}}}{66.7 \text{ min}} = 0.7 \text{ cpm} \qquad \sigma_{R_b} = \frac{\sqrt{436 \text{ counts}}}{33.3 \text{ min}} = 0.6 \text{ cpm}$$

Both values are lower than σ_{Rn}, and they contribute nearly equally it. Was this fortuitous, or is there a way we can plan our counting time so that neither the gross count rate nor the background rate excessively contributes to the uncertainty in the net count rate? As usual, the answer is yes, but we'll need to know our expected count rates and how much time is available (t_t). The ratio of the time devoted to collecting background to that used to collect the gross counts is equal to the square root of the ratio of their count rate ratios (Equation 2.22).

$$\frac{t_b}{t_g} = \frac{t_b}{t_t - t_b} = \sqrt{\frac{R_b}{R_g}} \tag{2.22}$$

Of course, the time used to find the gross counts (t_g) is equal to the total time (t_t) minus the time used for counting background (t_b).

Example 2.20: A source has a count rate of about 2000 cpm, and the background rate is 15 cpm. If you have 55 minutes to count both the sample and background, how long should you count each to minimize uncertainty in the net count rate?

$$\frac{t_b}{55\,\text{min} - t_b} = \sqrt{\frac{15\,\text{cpm}}{2000\,\text{cpm}}} \qquad t_b = 4\,\text{min} \qquad t_g = 55\,\text{min} - 4\,\text{min} = 51\,\text{min}$$

The toastier the source, the shorter the time needed to count background relative to time needed to count the source. In other words, calculations of this type are most important for low-count rate sample.

QUESTIONS

2.1 Without using the fact that one unified atomic mass unit is equal to 1.66×10^{-24} g, calculate the mass of exactly 1 g of ^{12}C in unified atomic mass units. Does this value look familiar? Briefly explain.

2.2 Calculate the average atomic mass of naturally occurring Mg.

2.3 Naturally occurring rubidium is a mixture of only two isotopes: ^{85}Rb (84.9118 u) and ^{87}Rb (86.9092 u). If the average atomic mass for Rb is 85.4678 u, calculate the percent abundance of its two isotopes.

2.4 Calculate the number of ^{81}Br atoms that are present in a sample of 1.56 g of pure bromine.

2.5 What is the difference between count rate and decay rate? How are they related, mathematically?

2.6 Derive Eq. 2.3 from Eq. 2.1.

2.7 What fraction of the original activity is left after three half-lives have passed? After five?

2.8 What is the activity of a ^{60}Co source calibrated at 95 kBq 2016 days ago?

2.9 Counts from a radionuclide decrease from 5718 dpm to 515 dpm in 24.0 hours. What is the half-life of this nuclide?

2.10 A sample of ^{99}Mo has an activity of 15,000 dpm at noon today. Calculate its activity 30.0 days from now.

2.11 A sample of ^{204}Tl was calibrated at 0.501 μCi. If its current activity is 1.48×10^5 dpm, when was it calibrated?

2.12 In 1971, Darleane Hoffman reported the detection of 20 million atoms of ^{244}Pu, isolated in an ore sample from a California mine. Assuming two significant figures, what mass of ^{244}Pu would've been necessary 4.6 billion years ago to produce 20 million atoms in 1971?

2.13 If you observe 2554 cpm for a ^{137}Cs source on the first shelf of your detector and the actual activity is 10,756 dpm, what is the percent efficiency for the first shelf of the detector?

2.14 Using a ^{204}Tl source that was calibrated at 361 kBq on May 15, 2013, you observe 347,816 cpm, 236,560 cpm, and 135,284 cpm on the first, second, and third shelves below your detector on May 26, 2017. Calculate the percent efficiency on all three shelves.

2.15 A sample of ^{90}Y records 1500 cpm on the first shelf of the detector. If the percent efficiency of the first shelf is 3.45%, what will the activity of the source be 5.00 days from now?

2.16 Calculate the activity due to the decay of ^{40}K in your body. It will help to know that there are 140 g of potassium in a person weighing 70 kg.

2.17 Calculate your specific activity based on hydrogen, carbon, and potassium. The natural percent abundances of these nuclides are listed in Table 2.4. It might also be useful to know that humans are 18.6% carbon, 9.5% hydrogen, and 0.35% potassium by weight.

TABLE 2.4
Percent Abundances

Nuclide	% Abundance
^{3}H	5.0×10^{-15}
^{14}C	1.2×10^{-10}
^{40}K	0.0117

2.18 Calculate the specific activity (MBq/g) of 0.353 g of NaTcO$_4$. Assume all of the technetium is ^{99}Tc, and it is the only radioactive nuclide present.

2.19 Calculate the specific activity (μCi/g) of naturally occurring platinum.

2.20 An Internet newspaper reported that the Japanese government was allowing fertilizer with a specific activity of less than 200 Bq/kg to be sold on the public market shortly after the Fukushima meltdowns. The implication was that the government was nefariously allowing its people to consume contaminated agricultural products. A common fertilizer used by home gardeners is 20% (by weight) K$_2$O. Assuming two significant figures, calculate the specific activity of the fertilizer in Bq/kg. Why is this a conservative value?

2.21 A piece of wood taken from an Egyptian tomb shows 55.9% of the ^{14}C activity that a living piece of wood did during the approximate time the tomb was constructed. When did the tree containing this piece of wood die?

2.22 Samples of bone and charcoal found near some pottery shards in China are 19,000 years old using carbon dating. What would you expect the specific activity of carbon in these samples to be?

2.23 A meteor found in Antarctica contains 0.892 g of ^{206}Pb for every 1.00 g of ^{238}U. Assuming that it didn't contain any ^{206}Pb when it was formed, determine the age of the meteor.

2.24 A rock is found to contain 0.092 g of ^{40}K and 0.530 g of ^{40}Ar. How old is the rock? What assumption(s) are you making to solve this problem? Do you believe they are valid?

2.25 Five samples from a rock exhibit the nuclide ratios listed in Table 2.5. How long ago was this rock formed? What was the original ratio of ^{87}Sr to ^{86}Sr?

TABLE 2.5
Isochron Data for Problem 2.25

Sample	1	2	3	4	5
^{87}Sr/^{86}Sr	0.7560	0.8213	0.9085	0.9809	1.025
^{87}Rb/^{86}Sr	1.05	3.47	6.53	9.22	10.5

2.26 What assumptions are made when determining the age of a rock using isochron dating?

2.27 What are the partial half-lives for the two major modes of decay for ^{36}Cl?

2.28 A sample of copper containing 9.76×10^{-15} g of ^{64}Cu gives a count rate of 2315 cpm on a detector that is 7.20% efficient toward this energy of beta particle. Assuming the detector can only detect beta decay, calculate the half-life of this nuclide.

2.29 Determine if each of the following are at secular equilibrium, transient equilibrium, or neither.
   a. A rock containing ^{235}U and ^{231}Pa
   b. A sample of ^{47}Ca is isolated and observed for 35 days
   c. A sample of ^{72}Zn is isolated and observed for 2 days
   d. A sample of ^{19}F is isolated and observed for 40 days

2.30 Some cave paintings in northern Spain have been dated by determining the ^{230}Th/^{238}U activity ratio in the calcite that later covered them. The idea is that uranium is naturally separated from the rest of its decay series in concentrated calcium carbonate solutions, which deposited on the paintings shortly after they were made. If the activity ratio is 0.290, how old are the calcite deposits on top of this painting? Assume that the ^{230}Th activity has been corrected to account only for the ^{230}Th formed from ^{238}U decay. Why is this correction important?

2.31 A rock contains 10.1 g of ^{232}Th. What is the activity of ^{212}Po in the same rock?

2.32 A rock contains 1.77 g of ^{232}Th. What is the total activity, from all radioactive nuclides, of this rock?

2.33 What is the mass ratio of ^{230}Th to ^{234}Th in a sample of uranium ore?

2.34 Which of the following are true of secular equilibria?
   a. $t_{1/2(A)} \gg t_{1/2(B)}$
   b. $\lambda_{(A)} \gg \lambda_{(B)}$
   c. The activity of the parent will remain constant over the period of observation
   d. It is reached at $5 \times t_{1/2(B)}$ after isolation of the parent

2.35 A sample of ^{99}Mo with an activity of 7.67×10^{13} dpm is isolated. After 56 hours, what activity of ^{99m}Tc would be observed? Assume that only 86% of all ^{99}Mo decays to ^{99m}Tc, and the remainder decays directly to ^{99}Tc.

2.36 A 7.0 μg sample of ^{32}Si is isolated. What will be the activity of ^{32}P in this sample after four weeks? Which type of nuclear equilibrium does this represent?

2.37 How many grams of radium are in a sample of ore containing 1.00 g of Th? What is the activity of ^{219}Rn in this sample?

2.38 A 7.0 mg sample of ^{48}Cr is isolated. When will its daughter reach maximum activity? Which type of nuclear equilibrium does this represent?

2.39 Is it surprising that all 20 counts listed in Table 2.3 fall within two standard deviations of the mean ($\bar{x} \pm 2\sigma$)? Briefly explain.

2.40 A sample was counted 20 times for one minute each in the same geometry. The results (counts) are shown in Table 2.6. Calculate the mean, standard deviation, RSD, and σ_R for the mean count rate. How many (%) data are within 1σ, and outside of 2σ? Do these data

TABLE 2.6
Count Data for Question 2.40

2018	2098	2217	2075	2087
2112	2207	2117	2045	2062
2056	2039	2139	2153	2222
2122	2048	2143	2225	2110

appear to fit a Gaussian distribution? Do they fit a Poisson distribution? Briefly explain your answers.

2.41 Calculate the RSD for the 15 data given in Table 2.7 for "Run 1." Repeat with the data from "Run 2." Do you observe a difference? Briefly explain.

TABLE 2.7
Count Data for Question 2.41

Run 1 (counts)	85	83	97	82	84	80	94	92
	85	97	104	88	93	83	94	
Run 2 (counts)	1967	1908	1954	1916	1928	1887	1904	1964
	1896	1897	1922	1849	1880	1855	1837	

2.42 A sample is counted on two different G-M counters. Counter 1 gave 5000 counts over 200 s, while counter 2 recorded 5250 counts over the same time interval. Is the difference in the two count rates from these two detectors statistically significant? Two other counters (3 and 4) gave 4321 and 4799 counts, respectively, when counted with the same sample, again for 200 s each. Is the difference in the count rates for these latter two detectors statistically significant?

2.43 A sample counted for 15 minutes gave 9000 total counts. A 30-minute background count registered 1200 counts. What is the net count rate and its uncertainty, expressed to a 95.5% confidence level?

2.44 If only 30 minutes of counting time are available, what is the optimum division of time between counting the sample and the background to minimize the error if the background is approximately 30 cpm and the total rate is 100 cpm?

3 Energy and the Nucleus

3.1 BINDING ENERGY

Back in Chapter 1, we learned that the strong force holds the nucleus together. The strong force is energy, and energy is required to pull a nucleus apart (thank goodness!). The energy required to separate a nucleus into individual nucleons is the **nuclear binding energy**. Nuclear binding energy could also be defined as the amount of energy released by the reverse process, i.e., when a nucleus is assembled from its individual nucleons. Energy doesn't just suddenly appear. It has to come from somewhere. In our theoretical construction of a nucleus from protons and neutrons, it comes from mass. You may have learned that mass is always conserved or that energy is always conserved. However, mass and energy can be converted one into the other, therefore they are *collectively* conserved. If matter is destroyed, energy is created. Likewise, it is possible to create matter from energy. Sounds like *Star Trek*, but it's true! The mathematical equation that relates matter and energy is very simple, and very well known.

$$E = mc^2 \qquad (3.1)$$

Energy is equal to mass times the square of the speed of light (2.998×10^8 m/s). When using Equation 3.1, the appropriate unit for mass is the kilogram (kg), which makes the energy unit the joule ($J = kg \cdot m^2/s^2$). Nuclear scientists like to think on the atomic scale and prefer to use the unified atomic mass unit (u) as a unit of mass. How does that convert to energy?

$$1 \text{ u} = 931.5 \text{ MeV}$$

One unified atomic mass unit (u) of mass is equal to 931.5 million electron volts (MeV) of energy. The electron volt (eV) is also a unit of energy, just like the joule. The joule is rather large when thinking on the atomic scale ($1 \text{ J} = 6.242 \times 10^{18}$ eV), so the electron volt is more convenient. An **electron volt** is defined as the amount of energy required to move one electron across a 1 volt potential—in other words, not much.

So, what about nuclear binding energy? We can calculate it by looking at how much energy is released when a nucleus is assembled from protons and neutrons. Let's start small, like an alpha particle ($^4_2$He). The mass of a proton is 1.007276 u, and the mass of a neutron is 1.008665 u, so the mass of an alpha particle should be:

$$(2 \times 1.007276 \text{ u}) + (2 \times 1.008665 \text{ u}) = 4.031882 \text{ u}$$

At this point, it'd be nice to compare this mass to the mass of an actual alpha particle. If we look up ^{4}He in Appendix A or in a chart of the nuclides, an *atomic* mass is given, which means electrons are

DOI: 10.1201/9781003303787-3

included in the mass. To make comparisons easier, let's add the two electrons to our alpha particle, making it a ^{4}He atom.

$$4.031882 \text{ u} + (2 \times 0.000549 \text{ u}) = 4.032980 \text{ u}$$

The electrons are lightweights, but they do make a difference. This is the expected atomic mass, and it is clearly larger than the actual atomic mass of 4.002603 u, meaning that some mass is converted to energy during the assembly of a ^{4}He atom. This difference in mass is called the **mass defect**.

$$4.032980 \text{ u} - 4.0026032 \text{ u} = 0.030377 \text{ u}$$

It should now be apparent why a crazy number of significant figures were necessary at the beginning of these calculations. Once the masses are subtracted, the number of significant figures can go way down (from seven to five in our case!). A good rule of thumb is to express masses six places past the decimal in problems like this. Converting the mass defect into energy gives the nuclear binding energy for ^{4}He.

$$0.030377 \text{ u} \times \frac{931.5 \text{ MeV}}{\text{u}} = 28.30 \text{ MeV}$$

That wasn't so hard; let's try another one!

Example 3.1 Calculate the mass defect and nuclear binding energy for ^{56}Fe.

We can save a little time by combining the proton and electron masses, i.e., using the atomic mass of ^{1}H (1.007825 u).

$^{56}_{26}\text{Fe}_{30}$

26($^1_1$H)	26(1.007825 u) = 26.203450 u
30(^{1}n)	30(1.008665 u) = 30.259950 u
expected atomic mass	56.463400 u
actual atomic mass	55.934937 u
mass defect	0.528463 u
convert to energy	× 931.5 MeV/u
binding energy	492.3 MeV

Wow! The mass defect and binding energy are enormous for ^{56}Fe when compared to those of ^{4}He. But we really should've expected it. Every nucleon will lose a little mass in forming a nucleus, the more nucleons that are glued together the more total mass will be lost—to make more glue. The more mass lost, the greater the binding energy. This also makes sense from the perspective of taking the nucleus apart. The bigger it is, the more energy should be required to pull it completely apart.

Is there any way to compare the nuclear binding energy values for ^{4}He and ^{56}Fe? Yes! Dividing the nuclear binding energy (or the mass defect) by the number of nucleons, gives the average amount of energy it takes to pluck a single nucleon from that nucleus. In other words, how tightly held is each nucleon? This quantity is known as the **binding energy per nucleon**, which can allow comparisons between nuclei.

$$^4_2\text{He} \quad \frac{28.30 \text{ MeV}}{4 \text{ nucleons}} = 7.074 \frac{\text{MeV}}{\text{nucleon}} \qquad ^{56}_{26}\text{Fe} \quad \frac{492.3 \text{ MeV}}{56 \text{ nucleons}} = 8.790 \frac{\text{MeV}}{\text{nucleon}}$$

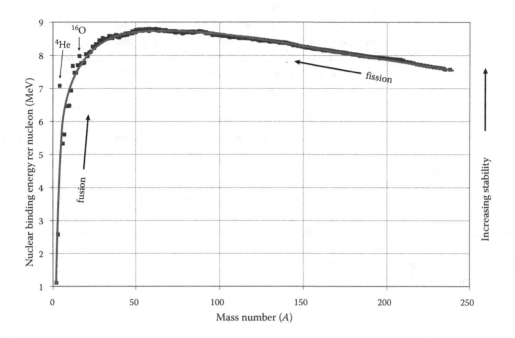

FIGURE 3.1   Relative stabilities of the naturally occurring nuclides.

It's a heck of a lot harder to pry a nucleon off of a ^{56}Fe nucleus than off a ^{4}He nucleus. If these calculations are repeated for all naturally occurring nuclides, a graph like Figure 3.1 could be produced. Those nuclides with relatively high values for their binding energy per nucleon could be considered more stable than those with lower values. In fact, if two light nuclides with low values were to combine (fusion) or one heavy nuclide were to split (fission) into two, a great deal of energy could be released as nuclides with greater stability are formed. These two processes will be discussed in more detail in Chapter 11. For now, it is simply important to understand that changes in binding energy per nucleon are the energetic driving force behind fission and fusion.

Roughly speaking, binding energy per nucleon is strongest for nuclides with mass numbers between 50 and 100, and it is weaker for nuclides above and below that range. Remember that the strong force is about 100 times stronger than the Coulomb repulsion (electrostatic) of protons. When we add another nucleon to a very small nuclide, we generally get a lot of strong force with it, and very little Coulomb repulsion (because there are few protons in the nucleus), the net result being higher binding energy per nucleon. This works until the nuclide starts getting too big and Coulomb repulsion becomes more and more significant relative to the additional strong force, leaving the nucleons held less tightly.

Another way to look at Figure 3.1 is by focusing on the fact that, unlike the Force (in *Star Wars*), the strong force has a limited, rather fixed range. In fact, the average nucleon only feels the strong force from its nearest neighboring nucleons. Imagine you're a proton (because you like to stay positive!) in a relatively small ($A < 50$) nucleus. You feel the attraction to the other nucleons near you, but there's a good chance you're sitting on the surface of the nucleus because there aren't that many nucleons in the nucleus. Being on the surface means you have fewer neighboring nuclides, so you are not held all that tightly. Binding energy per nucleon is relatively low. Now imagine you're a proton in a nucleus with 50 to 100 other nucleons—ahhh, the sweet spot! Percentagewise, there are now more nucleons below the surface, feeling the warm embrace of the strong force on all sides. Overall, this results in stronger attraction and more binding energy per nucleon. Binding energy per nucleon is relatively high and subatomic life is good. Now, imagine you're a proton in a larger nucleus ($A > 100$). You still feel a strong attraction to the nucleons near you, but the nucleus is now packing a lot more positive charge, increasing the pressure to push the nucleus apart, once again lowering binding energy per nucleon.

There are some interesting anomalies in the low mass end of Figure 3.1. Notice that ^{4}He sits well above the curve, indicating that its nucleons are bound especially tightly, and therefore ^{4}He has some unusual stability. This is due to the fact that ^{4}He has two protons and two neutrons—both are "magic numbers" (filled nuclear shells) that impart extra stability. Other nuclides with filled shells, such as ^{16}O, also exhibit unusually high nuclear stability.

Another way to understand the instability of large nuclides toward fission and the instability of small nuclides toward fusion is to look at the changes in mass for the nucleons. Similar to the calculations done above for binding energy per nucleon, we can also calculate the average mass of a nucleon for any given nuclide. Simply divide the mass defect by the mass number (number of nucleons) and subtract from the mass of a neutron or proton. These calculations are performed here for our two example nuclides of ^{4}He and ^{56}Fe.

Average mass of a proton:

$$^4\text{He}: 1.007276 \text{ u} - \frac{0.030377 \text{ u}}{4} = 0.999682 \text{ u}$$

$$^{56}\text{Fe}: 1.007276 \text{ u} - \frac{0.528463 \text{ u}}{56} = 0.997839 \text{ u}$$

Average mass of a neutron:

$$^4\text{He}: 1.008665 \text{ u} - \frac{0.030377 \text{ u}}{4} = 1.001071 \text{ u}$$

$$^{56}\text{Fe}: 1.008665 \text{ u} - \frac{0.528463 \text{ u}}{56} = 0.999228 \text{ u}$$

The average mass of the nucleons is lower in ^{56}Fe than in ^{4}He. Where did that extra mass go? It is converted to energy and is used to hold nucleons in ^{56}Fe a little tighter than in ^{4}He. If we, again, repeat these calculations for all naturally occurring nuclides, a graph similar to Figure 3.1 can be generated (Figure 3.2).

Figure 3.2 is a mirror image of Figure 3.1! This shouldn't be too surprising as Figure 3.1 looks at average energy and Figure 3.2 looks at average mass. If energy is higher, there must be less mass around; and if mass is higher, there's less energy around—the two interconvert nicely in nuclei. Figure 3.2 also explains instability toward fission and fusion, but from a different perspective. When a large nuclide undergoes fission, the products' nucleons will have less mass. Likewise, when very small nuclides combine to form larger ones, each nucleon loses a little mass (a curious weight-loss plan!). Instability toward fission and fusion means that the individual nucleons in certain nuclei can lose mass when they undergo one of these processes. Once the nuclei hit the mass number range of ~50 to ~100, no more mass can be lost by the nucleons, and they are stable toward fission and fusion. That mass lost during fission and fusion means energy (often a lot!) is released.

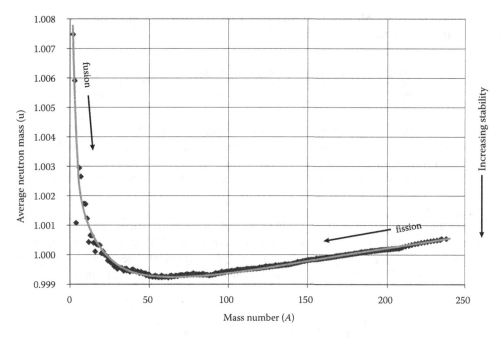

FIGURE 3.2 Average mass of a neutron in the naturally occurring nuclides.

3.2 TOTAL ENERGY OF DECAY

We'll get to the energetics of fusion and fission in Chapter 11, but the conversion of matter to energy also applies to decay processes. How much energy are we talking about? How can we calculate it? It is simply a matter of figuring out how much mass is lost during the course of the decay, then converting to energy. To calculate the mass lost, add up the masses of the products and subtract that from the mass of the parent.

Example 3.2 How much energy is released when a ^{222}Rn nuclide decays?

First, write out the balanced decay equation.

$$^{222}_{86}\text{Rn} \rightarrow \, ^{218}_{84}\text{Po} + \, ^{4}_{2}\text{He}$$

Look up the masses and do the math.

$$\left[222.017578 \text{ u} - \left(218.008966 \text{ u} + 4.002603 \text{ u}\right)\right] \times \frac{931.5 \text{ MeV}}{\text{u}} = 5.597 \text{ MeV}$$

For decay processes this energy is the **total energy of decay**. Notice that it is a little different from the experimentally determined 5.590 MeV listed in Appendix A. Remember that masses are also determined by experiment and all experimentally determined values have some error associated with them. Where does this energy go? Most of it goes into the kinetic energy of the alpha particle. According to Appendix A, the most common alpha particle emitted by ^{222}Rn has an energy of 5.4895 MeV. Sounds about right. The rest is kinetic energy (recoil) of the daughter. We'll look at this in more detail in Chapter 5.

Keep in mind that 5.597 MeV is the amount of energy released for a *single* decay of ^{222}Rn. What if a whole mole of ^{222}Rn nuclides decayed?

$$\frac{5.597 \text{ MeV}}{\text{decay}} \times \frac{1.60 \times 10^{-13} \text{J}}{\text{MeV}} \times \frac{6.022 \times 10^{23} \text{decay}}{\text{mol}} = 5.39 \times 10^{11} \frac{\text{J}}{\text{mol}}$$

For comparison purposes, let's consider a common chemical reaction that cranks out a lot of heat. Burning a mole of methane (CH_4, the main component of natural gas) produces only 9×10^5 J, almost a million-fold less than the decay of ^{222}Rn! Energy is produced by this chemical combustion reaction. Is mass lost? You bet. If you do the math correctly, it should come out to ~10^{-8} g. It is such a small amount that it would be difficult to measure. Therefore, mass is essentially conserved in chemical reactions.

Let's take this one step further. If we could harness all of the energy from the decay of one mole of ^{222}Rn as heat energy (it is easy!), how much ice could it melt? We'll have to dredge up some thermochemistry to answer this question. Remember that ice at 0°C still needs heat applied to melt to liquid water at 0°C. This is called the heat of fusion which is equal to 6.02 kJ/mole for water.

$$5.39 \times 10^{11} \text{J} \times \frac{\text{kJ}}{1000 \text{ J}} \times \frac{\text{mol}}{6.02 \text{ kJ}} \times \frac{18.0 \text{ g}}{\text{mol}} \times \frac{\text{kg}}{1000 \text{ g}} = 1.61 \times 10^6 \text{ kg}$$

Over one and a half million kilograms! That's a lot of ice. Radioactive decay clearly produces a lot of heat. Believe it or not, this helps explain why the Earth's core is so hot. Given the age of the Earth (~4.5 billion years), and the rate at which it is cooling (as a rock in space … we're not getting into global warming here!), it should've cooled off a lot more than it has. The reason why is that there are, and have been, lots of radioactive materials inside the Earth which have been busy decaying. All the energy from these decays is absorbed by the planet as heat, which helps the planet's core remain toasty warm.

Radioactive decay within the planet also explains where helium comes from. Because of its low atomic mass, helium can escape into space when released into the atmosphere. So where do we get all the helium to fill our balloons? When a radioactive nuclide undergoes alpha decay the alpha particle quickly picks up a couple of electrons and becomes a helium atom. If this happens below the surface of the Earth, the helium can become trapped underground. The helium we fill balloons with comes from wells, usually drilled looking for natural gas or oil.

3.3 DECAY DIAGRAMS

Nuclear decay is always **exothermic**, in other words, it always gives off energy. Nuclear scientists sometimes use the term "**exoergic**" to mean the same thing. Nuclides always decay from a less stable nuclide to a more stable one. The atomic number (Z) also changes (except for γ decay), so we can illustrate decay by plotting energy vs. Z such as in Figure 3.3.

FIGURE 3.3 Generic decay diagrams for alpha, beta, positron, electron capture, and gamma decay.

The horizontal lines in Figure 3.3 represent nuclidic states—a nuclide in a particular energy state. The arrow begins at the parent and points toward the daughter. Notice that the daughter is always lower in energy than the parent, indicating that energy is given off during the decay process. The vertical displacement between parent and daughter is the total energy of decay, also known as the energy of transition, and corresponds to the mass lost in the decay. The horizontal position of the daughter relative to the parent varies depending on the type of decay. Two protons are lost during alpha decay, so the daughter is two units of Z lower than the parent. A neutron is converted to a proton during beta decay, so the daughter is one unit higher than the parent. The opposite nucleon conversion happens during positron and electron capture decay, so the daughter ends up one lower in Z. There is no change in atomic number during gamma decay. Drawing these diagrams helps us to see the (sometimes complex) energy changes involved in nuclear decay. These diagrams are also known as energy diagrams or decay schemes.

Let's take a look at some examples, starting with the three flavors of beta decay.

Example 3.3 ^{35}S decays to ^{35}Cl by emitting a beta particle. Draw a decay diagram and write out the balanced equation. Also, calculate the total energy of decay.

The decay equation is:

$$\ce{^{35}_{16}S} \rightarrow \ce{^{35}_{17}Cl} + \ce{_{-1}e}$$

Appendix A tells us the atomic mass of ^{35}S is 34.969032 u and the atomic mass of ^{35}Cl is 34.968853 u. The total energy of decay (E) is therefore:

$$\left(34.969032 \text{ u} - 34.968853 \text{ u}\right) \times \frac{931.5 \text{ MeV}}{\text{u}} = 0.167 \text{ MeV}$$

According to Appendix A, the total energy of this decay is 0.1672 MeV! Our result is pretty darn good. Appendix A also gives the energy of the beta particle as 0.1674 MeV. In this particular decay, it should be the same as the total energy of decay. Is this a problem? No, not really. Subtle differences like this are usually an indication that the two numbers were measured independently, and, again, since all measured values have some error associated with them, small discrepancies can be ignored. The decay diagram can now be drawn and is pictured in Figure 3.4, using the energy values from Appendix A.

Careful examination of Figure 3.3 reveals something strange about positron decay. It is the only one with a bent arrow. As we'll see later, the mass of two electrons must be created in every positron

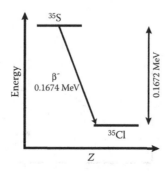

FIGURE 3.4 Decay diagram for ^{35}S.

decay, which requires 1.022 MeV of energy. This energy value is obtained by multiplying the mass of two electrons (5.486×10^{-4} u) by our handy mass/energy conversion constant of 931.5 MeV/u.

$$\left[2 \times \left(5.486 \times 10^{-4}\ \text{u}\right)\right] \times \frac{931.5\ \text{MeV}}{\text{u}} = 1.022\ \text{MeV}$$

Since this is required for every positron decay, we account for it by drawing a line straight down from the parent nuclide. The slanted arrow then represents the energy carried off by the positron. Why is mass created in positron decay, but not in other forms of decay? The answer will have to wait until Chapter 5, stay tuned!

Example 3.4 ^{17}F undergoes positron decay to ^{17}O. Write out the balanced equation and draw a decay diagram. If the total energy of decay is 2.7605 MeV and the atomic mass of the daughter is 16.999132 u, calculate the atomic mass of ^{17}F.

The decay equation is:

$$^{17}_{9}\text{F} \rightarrow {}^{17}_{8}\text{O} + {}_{+1}\text{e}$$

The decay diagram is shown in Figure 3.5. The positron energy value is from Appendix A. Notice that this value differs from the total energy of decay by 1.02 MeV.

The mass of an atom of ^{17}F can be calculated by taking advantage of the fact that the total energy of decay is related to the difference in mass between parent and daughter, just as it was for regular beta decay, except we're doing this problem backward, starting at total energy of decay and working back to the masses. The total mass lost in this decay is:

$$2.7605\ \text{MeV} \times \frac{\text{u}}{931.5\ \text{MeV}} = 0.002963\ \text{u}$$

Which allows us to calculate the mass of the parent:

$$16.999132\ \text{u} + 0.002963\ \text{u} = 17.002095\ \text{u}$$

The actual mass is reported as 17.0020952 u in Appendix A. Another excellent result! Next, let's take a look at electron capture.

FIGURE 3.5 Decay diagram for ^{17}F.

FIGURE 3.6   Decay diagram for ^{41}Ca.

Example 3.5 ^{41}Ca undergoes electron capture decay to ^{41}K. Write out the balanced equation and draw a decay diagram. If the total energy of decay is 0.4213 MeV and the atomic mass of ^{41}Ca is 40.962278 u, calculate the atomic mass of ^{41}K.

The balanced equation is given below—remember that the electron in electron capture decay equations is an orbital electron.

$$^{41}_{20}Ca + {}_{-1}e \rightarrow {}^{41}_{19}K$$

The decay diagram, shown in Figure 3.6, is similar to that for positron decay, except that a straight arrow is drawn and no energy value is placed next to the decay mode. With electron capture, there is no emitted particle (that is easily observed); therefore, no energy value is listed next to "EC."

Calculating the atomic mass of the daughter is very similar to the previous example. Keep in mind that the mass of the daughter is always less than the parent.

$$0.4213 \text{ MeV} \times \frac{u}{931.5 \text{ MeV}} = 0.0004523 \text{ u}$$

$$40.962278 \text{ u} - 0.0004523 \text{ u} = 40.961826 \text{ u}$$

Once again, we have excellent agreement with the value given in Appendix A (40.9618258 u). Let's crank it up a notch. What if we wanted to include more than one decay in our decay diagram? No problem, just draw it in.

Example 3.6 Draw the decay diagram for the decay of ^{32}Si to ^{32}S. If the total energy of decay for the beta decay of ^{32}Si to ^{32}P is 0.2243 MeV, and that for ^{32}P to ^{32}S is 1.7105 MeV, and the mass of ^{32}S is 31.972071 u, calculate the atomic mass of ^{32}Si.

The problem doesn't ask, but the two decays can be represented as:

$$^{32}_{14}Si \rightarrow {}^{32}_{15}P + {}_{-1}e$$
$$^{32}_{15}P \rightarrow {}^{32}_{16}S + {}_{-1}e$$

FIGURE 3.7   Decay diagram for ^{32}Si.

The decay diagram is shown in Figure 3.7. The beta particle energies are from Appendix A and match the total energies of decay reasonably well.

Combining the two decay energies and converting to mass allows us to calculate the atomic mass of ^{32}Si.

$$31.972071 \text{ u} + \left[\left(0.2243 \text{ MeV} + 1.7105 \text{ MeV} \right) \times \frac{\text{u}}{931.5 \text{ MeV}} \right]$$
$$= 31.974148 \text{ u}$$

For our final example, let's look at gamma decay. Even though no nucleons change identity and no particle is spit out of the nucleus during gamma decay, the nucleus still manages to lose mass. Energy, in the form of a gamma photon, is produced, and therefore mass must be destroyed in the process.

Example 3.7 Write out the decay equation and draw a decay diagram for the decay of ^{77m}Br to ^{77}Br. If the mass of ^{77m}Br is 76.921494 u and ^{77}Br is 76.921380 u, calculate the total energy of decay.

Hopefully, this is starting to get a bit easier—the decay equation is:

$$^{77m}_{35}\text{Br} \rightarrow {}^{77}_{35}\text{Br} + \gamma$$

Total energy of decay is calculated the same way it is for all forms of beta decay—from the mass difference between parent and daughter.

$$\left(76.921494 \text{ u} - 76.921380 \text{ u} \right) \times \frac{931.5 \text{ MeV}}{\text{u}} = 0.106 \text{ MeV}$$

Energy values for gamma rays are usually expressed in units of kiloelectron volts (keV). Conversion is pretty straightforward, knowing that there are 1000 keV in 1 MeV.

$$0.106 \text{ MeV} \times \frac{1000 \text{ keV}}{\text{MeV}} = 106 \text{ keV}$$

FIGURE 3.8   Decay diagram for ^{77m}Br.

A little research on ^{77m}Br reveals that the only gamma ray observed in its decay has an energy of approximately 106 keV. We can now draw its decay diagram (Figure 3.8), with confidence.

In most of the examples we've seen so far, the total energy of decay is taken away from the nucleus by particles or photons. As noted earlier, some of the energy of alpha decay is left behind with the nucleus as recoil. This introduces a minor complexity to the energetics of alpha decay. So, we won't be looking at an example of alpha decay here—it will be discussed in Chapter 5.

It should also be noted that the decay examples used in this chapter were carefully selected for their simplicity. Most alpha, beta, positron, and electron capture decay processes initially lead to one or more excited state(s) of the daughter, instead of directly to the ground state as pictured in the examples shown here. This additional level of complexity is also covered in Chapter 5. Finally, it is suggested that the beta and positron particles have the energies shown in Figures 3.4, 3.5, and 3.7. This is not entirely correct. As you will learn in Chapter 5, this is the *maximum* energy these particles can have.

3.4   MASS AS ENERGY

Because of the simple relationship between matter and energy (Equation 3.1) and the facile ways the two interconvert in the nucleus, nuclear nerds often like to express mass in terms of energy. It's not as intuitive, but it makes some of the calculations we've seen in this chapter a bit more efficient. For example, we can use the fact that 931.5 MeV equals one unified atomic mass unit to express the masses of our favorite subatomic particles in energy units. It's handy to have a few more significant figures on the mass/energy conversion factor, so we'll use 931.494 MeV/u in this section. Let's convert the mass of a proton (1.007276 u) to units of MeV.[1]

$$1.007276 \text{ u} \times \frac{931.494 \text{ MeV}}{\text{u}} = 938.272 \text{ MeV}$$

Likewise, the mass of a neutron is 939.565 MeV and the mass of an atom of ^{1}H is 938.783 MeV. Armed with these values we can now calculate nuclear binding energy using units of energy.

Example 3.8   Calculate the nuclear binding energy for ^{56}Fe.

If we do this correctly, the answer should be the same as Example 3.1. The first step is the same, find the expected mass, in units of MeV.

$$26(938.783 \text{ MeV}) + 30(939.565 \text{ MeV}) = 52,597.3 \text{ MeV}$$

---

1   Nuclear scientists often use units of MeV/c^2 instead of just MeV. This comes from solving Equation 3.1 for mass ($m = E/c^2$).

Now we need subtract the actual mass of an iron-56 atom in units of MeV.

$$52,597.3 \text{ MeV} - \left(55.934937 \text{ u} \times \frac{931.494 \text{ MeV}}{\text{u}}\right) = 492.2 \text{ MeV}$$

Yes, excellent agreement with Example 3.1! However, critical analysis of Examples 3.1 and 3.8 shows that it takes the same amount of work, mathematically speaking, to reach the solution. We were promised more expeditious calculations in this section! Hang on, we're not there just yet. Example 3.8 demonstrates that we can approach nuclear binding energy from either a mass or an energy perspective, and end up in the same place.

When two mass values are subtracted in this chapter, they are almost always very close to one another, necessitating use of a large number of significant figures. A convenient shortcut is use of **mass excess** (Δ). Mass excess is how much an atomic mass differs from the nuclide's mass number. For example, ^{14}N has a mass of 14.003074 u, so it has a mass excess of 14.003074 u − 14 u = 0.003074 u. ^{56}Fe has a mass of 55.934937 u, so it has a mass excess of 55.934937 u − 56 u = −0.065063 u. Mass excess values are commonly tabulated in online charts of the nuclides, and are also available for select nuclides in Appendix A. If you want to be a retro nuke nerd, you should get yourself a printed copy of the *Nuclear Wallet Cards*. The tabulated values are in units of energy (keV or MeV). Let's convert the two values determined in this paragraph to MeV.

$$^{14}\text{N} \quad \Delta = 0.003074 \text{ u} \times \frac{931.5 \text{ MeV}}{\text{u}} = 2.863 \text{ MeV}$$

$$^{56}\text{Fe} \quad \Delta = -0.065063 \text{ u} \times \frac{931.5 \text{ MeV}}{\text{u}} = -60.61 \text{ MeV}$$

The main expedience to mass excess values is using them to calculate energy changes for nuclear processes.

Example 3.9 Using mass excess values, determine how much energy is released when a ^{222}Rn nuclide decays.

A repeat of Example 3.2! All we need to do is look up the values in Appendix A and subtract. Here's a reminder of what the decay equation looks like.

$$^{222}_{86}\text{Rn} \rightarrow {}^{218}_{84}\text{Po} + {}^{4}_{2}\text{He}$$

$$16.374 \text{ MeV} - (8.352 \text{ MeV} + 2.425 \text{ MeV}) = 5.597 \text{ MeV}$$

Excellent agreement with Example 3.2, and a bit quicker both to write out and to solve! Let's see if this approach also works with beta decay by repeating Example 3.3.

Example 3.10: Using mass excess values, determine how much energy is released when ^{35}S decays.

$$^{35}_{16}\text{S} \rightarrow {}^{35}_{17}\text{Cl} + {}_{-1}\text{e}$$

$$-28.846 \text{ MeV} - (-29.014 \text{ MeV}) = 0.168 \text{ MeV}$$

Using mass excess to determine energy also works for nuclear reactions, but we'll have to wait until Chapter 10 to see it in action.

QUESTIONS

3.1 Using Eq. 3.1, show that 1 u = 931.5 MeV.

3.2 Calculate the mass defect for ^{72}Ge. Define mass defect and explain what it represents.

3.3 Calculate the binding energy per nucleon for ^{239}Pu. Would you expect this nuclide to undergo fission, fusion, or neither? Briefly explain.

3.4 Calculate the average mass of a proton in ^{6}Li, ^{66}Zn, and ^{266}Bh.

3.5 Calculate the average binding energy per nucleon and the average mass of a neutron in ^{99}Tc. Would you expect this nuclide to undergo fission, fusion, or neither? Briefly explain.

3.6 Calculate the total energy of decay for ^{226}Ra. How many kilograms of ice could a mole of ^{226}Ra decays melt? How many moles of methane would need to be burned to melt the same mass of ice?

3.7 Using information from Question 2.17, calculate how much mass you lose each month due to radioactive decay of ^{3}H, ^{14}C, and ^{40}K.

3.8 Your body is 18.6% (by weight) carbon. Using 1.2×10^{-10} as the percent natural abundance of ^{14}C, calculate the activity (dpm) of ^{14}C in your body. How much energy (J) is being released in your body every minute as a result of ^{14}C decay?

3.9 Radioactive materials have sometimes been used directly as a power source. Thermal energy, produced during the decay process, is converted to electrical energy using a thermocouple. How much energy (J) could be generated from the decay of a mole of ^{238}Pu? How much energy (J) is generated per minute by 2.0 kg of ^{238}Pu?

3.10 Define nuclear binding energy, binding energy per nucleon, and total energy of decay.

3.11 Using information provided in Appendix A, draw the decay diagrams for the following nuclides:

   a. ^{33}P   b. ^{55}Ni   c. ^{49}V   d. ^{172m}Lu   e. ^{190}Pt

3.12 $^{165}_{68}$Er undergoes isobaric decay to $^{165}_{67}$Ho . The atomic mass of the parent is 164.930723 u, and the daughter is 164.930322 u. Using only the information given here, write out the decay equation, draw a decay diagram, and calculate the total energy of decay.

3.13 ^{89m}Y decays to its ground state. The total energy for the decay is 0.9091 MeV, and the atomic mass of the daughter is 88.905848 u. Using only the information given here, write out the decay equation, draw a decay diagram, and calculate the atomic mass of ^{89m}Y.

3.14 ^{90}Sr decays to ^{90}Y with a total energy of decay of 0.546 MeV. This is also the maximum energy of the beta particle emitted. If the mass of the parent is 89.907738 u, calculate the mass of the daughter and draw a decay diagram. Use only the information given here.

3.15 Using information given in Appendix A, draw a decay diagram for ^{13}N. Show how its atomic mass can be calculated from the mass of the daughter and its total energy of decay.

3.16 Draw a decay diagram for the main decay pathways for ^{68}Ge to a stable nuclide.

3.17 Calculate the nuclear binding energy for ^{72}Ge and compare to the mass defect determined in Question 3.2.

3.18 Define mass excess and briefly explain how it is a convenience in calculations.

3.19 Using mass excess values from Appendix A, calculate the total energy of decay for each of the nuclides shown here.

   a. ^{17}F   b. ^{42}Ar   c. ^{194}Hg   d. ^{239}Pu

3.20 ^{112}Sb can decay via positron to ^{112}Sn with a total energy of decay of 7.06 MeV. Using values from Appendix A, what is the mass excess of ^{112}Sb?

4 Applications of Nuclear Science I

Power and Weapons

Now that we've covered some of the basics of nuclear science, we can begin to discuss a few of its applications. This chapter, and Chapter 9, are specifically designed for those that wish to limit their study of the science and would like to focus on how modern society uses its knowledge of ionizing radiation and nuclear processes. Some of the topics covered in this chapter are covered in more detail in Chapters 11 and 12, so, those of you continuing your study of nuclear science may wish to skip this chapter for now and come back to it when you are ready to begin looking at these applications. This "double coverage" of some of the topics below results in some redundancies between this chapter and later ones—which hopefully serves only to reinforce important concepts rather than prove a distraction.

The first three chapters have taught us a bit about ionizing radiation, nuclear decay, and the tremendous energies resulting from nuclear transformations. The penetrating power of ionizing radiation is used in hospitals, clinics, and private medical practices throughout the world to diagnose and treat disease, and to preserve food (Chapter 9). The awesome amount of energy released by nuclear fission is used to provide electricity from hundreds of nuclear power plants around the world and has been developed into horrific weapons of mass destruction. This chapter provides an introduction to understanding nuclear reactors and weapons, in the context of what we've already learned about that weird thing we call a nucleus.

4.1 NUCLEAR POWER

Approximately 10% of the world's electricity is produced from roughly 440 nuclear power reactors. About 150 more are under construction or are in the planning stages (at the time of this writing), with more than 300 additional plants being proposed for future construction (by 2040). Much of the new construction is occurring in China, India, and Russia. There was a great deal of nuclear power plant construction in the 1960s and '70s, but very little in the following three decades. During the first couple of decades of the 21st century, there's been a renewed interest in nuclear plant construction.

The renewed interest in nuclear power stems from increasing demand for electricity tempered by concern over global climate change. Unlike the other major forms of electricity production, nuclear generates almost no carbon dioxide; and, until the Fukushima meltdowns of 2011, was increasingly perceived as safe, reliable, and cost-effective. While perceptions are not always fully reflective of reality, they often have an influence on decision-making, and must be acknowledged. After Fukushima, some perceptions shifted, dampening some of the enthusiasm for the nuclear plant renewals and new construction. During the same time there was dramatic drop in natural gas prices, encouraging construction of natural gas power plants. At the time of this writing (2022), new plant construction was slightly lagging the number of old plants shutting down. Since more than a decade

DOI: 10.1201/9781003303787-4

has passed since the last major nuclear power plant accident, perspectives seem to be shifting again, with increasing numbers of people looking more favorably toward nuclear power.

Approximately 20% of the U.S. and 14% of Canada's electrical production is nuclear, while in many European countries, and in South Korea, it accounts for roughly 30% (or more) of all power generation. France leads the world with a little less than 70% of all electrical generation coming from nuclear, although the French government plans to decrease to 50% nuclear by 2035. In the United States, nuclear power plant construction came to a virtual standstill in the 1980s and remained that way for almost 30 years. The long hiatus was a result of the relatively high cost of building new nuclear power plants, the accidents at Three Mile Island and Chernobyl, and the federal government's inability to come up with a consistent plan to deal with spent fuel. As a result, the existing U.S. plants tend to be rather old. Because they have a limited lifetime, some are decommissioned, while others have sought, or are seeking license renewals to continue beyond their originally intended period of operation.

The impact of the 2011 meltdowns at the Fukushima plants is being felt most strongly in Japan. Prior to the massive earthquake and tsunami in March 2011, Japan had roughly 50 nuclear power plants in operation, producing about 30% of its electrical needs. By mid-2022, only ten plants had returned to operation, accounting for only 6% of the nation's electricity. Fukushima's effects were also felt in Europe, where Germany decided to return to its 1998 plan to phase out nuclear power.[1] In the short term, Germany's (and Japan's) nuclear phase out has largely been replaced with increased fossil fuel generation. Despite occasional shortages of fossil fuels, German public opinion remains strongly opposed to construction of new nuclear plants. Switzerland and Belgium plan a more gradual phase out. They will run their existing reactors but not replace them with new ones. As already mentioned, France plans to reduce nuclear's share of electrical generation from about 70% to 50% by 2035. Sweden reduced its fleet of reactors in 2019, cutting nuclear's share of electricity generation from around 40% to about 30%, but is now (mid-2022) somewhat ambivalent concerning the future of nuclear. South Korea also appears to be shifting away from a post-Fukushima plan to slowly phase out nuclear. Other nations with nuclear power plants appear to be moving forward with plans for construction of new nuclear plants, and license renewals for older plants. In China, India, and Russia, a total of 32 new plants were under construction in mid-2022. Additionally, about ten countries without any nuclear generation (as of 2022) are building or planning construction of nuclear power plants.

The U.S. government began to quietly encourage the licensing and construction of new plants in the late 1990s and early 2000s. This change in attitude resulted from increasing costs of oil and natural gas during this period, which currently account for about 37% of all electricity production in the United States, and increasing concerns that burning fossil fuels is negatively affecting the global environment. Ironically, nuclear power, which has long been opposed for environmental concerns (potential for a meltdown, dangers from waste), is now being promoted as a relatively green power source because it does not produce greenhouse gases. Regardless of how it is viewed, new nuclear power plant construction in the United States became a reality in the early 2010s. Four plants were under construction in early 2017 in South Carolina and Georgia, and were expected to be operating by 2020. Construction of the two South Carolina plants was abruptly halted in August 2017 following the bankruptcy of the main contractor, Westinghouse. Construction of the two plants in Georgia was delayed, but they are expected to start commercial operation in 2023.

The focus of the remainder of this section will be on how nuclear reactors work, as well as the science underlying such contentious issues such as reactor safety and cost, proliferation of nuclear weapons, and nuclear waste.

---

1 Germany had canceled the phase-out in 2009.

4.1.1 NUCLEAR FISSION

Before we can understand how a nuclear power reactor works, we need to learn a little bit about nuclear fission. **Nuclear fission** is the splitting of a nucleus into two (or more—rare!) fragments and usually occurs when a neutron hits the nucleus of certain, large atoms. Nuclear scientists often imagine the nucleus as a more fluid object, rather than a collection of hard spheres (the nucleons). In fact, it is more like a drop of water (more on this in Chapter 6) suspended in air or some Silly Putty® in the hands of a small child.[2] The nucleus can elongate and deform, taking on shapes (like an egg, Frisbee®,[3] football,[4] or even a bow tie!) other than a perfect sphere. In particularly large nuclides, these deformations are a way to relieve some of the pretty intense proton–proton repulsions. When struck by a neutron, enough energy can be added to certain nuclei that they stretch and separate into two, roughly equal halves. The trick is to provide enough energy. With the relatively slow neutrons (i.e., "thermal" neutrons—see Chapter 10) present in almost all nuclear reactors, only three nuclides commonly undergo nuclear fission: ^{233}U, ^{235}U, and ^{239}Pu. Nuclides that undergo fission with thermal neutrons are called **fissile**.

An example of a fission reaction involving ^{235}U is:

$$^{235}_{92}U + ^{1}_{0}n \rightarrow ^{132}_{50}Sn + ^{102}_{42}Mo + 2^{1}_{0}n$$

This is an example of a nuclear reaction because a particle is colliding with a nucleus, making new stuff. Note that it is balanced because the atomic and mass numbers add up to the same values on both sides of the arrow. Notice also that two (new) neutrons are released in this particular reaction. This is fairly common in fission reactions, with two or three neutrons typically being released by each fission reaction that takes place. The two product nuclides (^{132}Sn and ^{102}Mo in the example above) are called **fission products** (also known as fission fragments). Fission is a notoriously messy reaction in that lots of different fission products can be formed. In fact, over 400 different nuclides are observed in the fission of ^{235}U. In other words, if you have a lot of ^{235}U undergoing fission reactions, like in a nuclear reactor, there are going to be lots of different fission products being made.

How much energy is released by a fission reaction? On average, about 200 MeV are released in a fission reaction! Holy cow! That's 50 to 100 times more energy than is released by a typical nuclear decay, which we already know is fairly energetic. This quantity can easily be calculated by finding the difference in mass on either side of the arrow, i.e., determine how much mass is lost in this reaction.

Mass of reactants	^{235}U	235.043930 u
	$1^{1}_{0}n = 1(1.008665\ u) =$	+1.008665 u
		236.052595 u
Mass of products	^{132}Sn	131.91774 u
	^{102}Mo	101.91030 u
	$2^{1}_{0}n = 2(1.008665\ u) =$	+2.01733 u
		235.84537 u

Mass lost = 236.052595 u–235.84537 u = 0.20722 u

The amount of energy released per reaction can then be calculated in the usual manner, by converting matter to energy.

$$\text{Total energy released} = \frac{931.5\,\text{MeV}}{u} \times 0.20722\ u = 193.0\ \text{MeV}$$

---

2 Silly Putty is a registered trademark of Crayola LLC of Easton, Pennsylvania.
3 Frisbee is a registered trademark of Wham-O Inc. of Carson, California.
4 Most of you outside the U.S. could substitute "rugby ball" here.

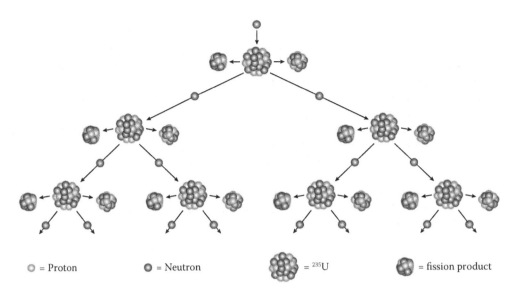

FIGURE 4.1 Nuclear chain reaction.

The fact that, on average, two to three neutrons are produced in each fission reaction is quite useful since these neutrons can cause further fission reactions to take place, provided there are more fissile nuclides around, and they get hit by these neutrons. To encourage this kind of behavior, you'd want to arrange lots of fissile nuclides very close to each other. If every fission reaction produces two to three neutrons, and most of these cause additional fission reactions, a **fission chain reaction** will result. This is illustrated in Figure 4.1. One neutron initiates fission in a single ^{235}U nuclide at the top of the figure, which also produces two neutrons. These two neutrons then initiate fission in two more ^{235}U nuclides, producing two neutrons each. Now there are four neutrons that initiate four more fission reactions, which, again, produce two neutrons each. Now there are eight neutrons that could initiate eight more fission reactions. The number of fission reactions that take place will rapidly increase with each subsequent generation of fission reactions. This could quickly get out of control! If there were just enough fissile nuclides around for a self-sustained fission chain reaction, then we'd have a **critical mass** of that nuclide. If more than the critical mass is present, then it is called a **supercritical mass**. A supercritical mass can be arranged in different ways to get different results. A very compact arrangement would increase the odds that every neutron produced will cause another fission reaction in a relatively short time. This will lead to near exponential growth in the number of reactions, and a rather sudden release of a great deal of energy, i.e., a really big explosion. If the supercritical mass is spread out in such a way that, on average, only one of the neutrons released per fission causes another fission reaction, then the reaction rate can be controlled to release a constant amount of energy over a long time. This is the basis of nuclear power reactors.

4.1.2 NUCLEAR REACTORS

The ^{235}U in a nuclear reactor is contained in small cylindrical pellets, about the size of the end of your little finger, which are placed end to end in a long, skinny metal pipe that is then sealed at both ends (**fuel rod**), and these rods are then distributed in the reactor parallel to each other and separated so there's space between them.

A schematic view of a typical nuclear reactor is shown in Figure 4.2. It works the same way that a fossil fuel power plant does. Fossil fuel plants burn materials that contain carbon—like coal, natural gas, or oil. The energy released from the chemical reaction (combustion) that takes place heats

FIGURE 4.2   Schematic drawing of a nuclear power reactor (PWR).

water to steam, which drives a steam turbine that produces electricity. Nuclear reactors use nuclear reactions (fission) to heat the water to steam, which drives a steam turbine that produces electricity. The fundamental difference between fossil fuel and nuclear plants is simply how the water is heated. The goal of a nuclear power plant is to maintain the rate of fission reactions at a constant level. If there are too many neutrons flying around, causing too many fission reactions, too much heat is generated. If the reactor gets too hot, the fuel rods (and the metal assemblies that hold them) will melt (reactor **meltdown**). On the other hand, if there are too few neutrons, the chain reaction will shut down. To maintain the perfect number of neutrons, some will have to be absorbed. This is done with **control rods**, which can move in and out between the fuel rods.

An odd thing about neutrons is that the probability that they will cause a particular nuclear reaction to take place depends on how fast they are moving, i.e., how much energy they have. When they are produced by a fission reaction, they are moving too fast to efficiently cause fission reactions in ^{235}U. This is another reason to put the uranium fuel into thin rods and separate them. This forces the too-fast neutrons to travel through matter before it runs into another ^{235}U atom. We want to be sure that the matter they have to travel through will allow them to pass, but slow them down. This is called moderating the neutrons, and the matter used is called a **moderator**. The most common moderators are water (sometimes called "light-water" by nuclear geeks), heavy water (water with ^{2}H, also known as deuterium, instead of the usual ^{1}H in regular water), and carbon (typically in the form of graphite).

The moderator slows the neutrons so they are more likely to cause fission, and the control rods soak up extra neutrons to keep the fission reactions occurring at a reasonable rate. Now all we need is some way to transfer the heat from the fuel rods. The **coolant** does this and can be any material that transfers heat well, but the most common (by far) in nuclear reactors is water. Water is nice because it also functions as the moderator, it's cheap, and it has a high heat capacity. Heat capacity is the amount of energy it takes to warm up a certain amount (like one gram) of a substance by one degree. It takes more energy to warm up water than equivalent amounts of most anything else, so water can hold more heat per gram than other materials.

Careful examination of Figure 4.2 will show that the water heated by the nuclear reactions is not the same water (steam) that drives the turbines. The water heated by the nuclear reactions is kept under enough pressure that it remains a liquid. It is pumped through the reactor **core** (where the fuel is), then through a heat exchanger to transfer some of its thermal energy to another closed water loop. This second loop is operated at lower pressure, so the water is heated to steam, which then drives the turbines to produce electricity. This secondary loop of water then needs to be cooled (condensed back to liquid), so it runs through another heat exchanger, transferring its excess thermal energy to a third loop of water (labeled "cooling water" in Figure 4.2). Unlike the first two loops, the third loop is not closed, as it transfers excess thermal energy to the environment—a river, lake, ocean, or the air. The immense, hourglass-shaped cooling towers that are so often associated with nuclear power plants simply transfer this waste heat to the air. The white clouds that are often seen emerging from them is not smoke, but water vapor.

The reactor depicted in Figure 4.2 is called a pressurized water reactor (PWR), because water is used as a coolant and it is kept under pressure in the reactor core. Roughly 70% of the world's operable nuclear reactors are a PWR design. Another popular design (~14%) is the boiling water reactor (BWR—Figure 4.3). The boiling water reactor doesn't excessively pressurize the water that is pumped into the core, so it boils. The steam that is generated drives the turbines directly instead of transferring its thermal energy to a second water loop. The steam still needs to be condensed and the waste heat transferred to the environment. Notice that another difference is that BWRs have a smaller containment structure which is enclosed within the reactor building. The reactor building doesn't necessarily have the distinctive silo shape like most other nuclear reactors and it also acts as secondary containment—another barrier between the reactor core and the environment.

Dealing with waste heat is also an issue for fossil fuel power plants. Coal or natural gas plants also need to be located near large bodies of water to soak up the waste heat, or have cooling towers to dump the waste heat into the air. So, the large cooling towers that so often symbolize nuke plants are neither unique nor necessary to them. If waste heat from any type of power plant is not being dumped into the air, then it is likely going into water nearby. Dumping waste heat into a body of water can have deleterious environmental consequences. To avoid these problems, the cooling water can be pumped into an artificial holding pond before releasing it to natural waters.

As the previous paragraphs imply, not all the energy released by the fission reactions can be converted to electricity, i.e., the process is not 100% efficient. In fact, only ~35% of the energy released by the nuclear reactions is successfully converted to electrical energy. This is comparable to efficiencies obtained in most fossil fuel power plants.

Because the BWR has less plumbing than a PWR (no heat exchanger), it might seem that boiling water reactors would be cheaper to build. While this is true, boiling water reactors are a bit more expensive to operate. As the water flows through the reactor it becomes slightly contaminated with radioactive materials. Because this water leaves the reactor building, extra shielding is required on any pipes that contain it, and special procedures are necessary for people that work around the turbines to minimize their dose. In the end, BWRs and PWRs cost about the same to build and run.

FIGURE 4.3 Schematic for a boiling water reactor (BWR).

4.1.3 Nuclear Fuel

Of the three nuclides that commonly undergo fission with thermal neutrons, only ^{235}U can be dug up out of the ground in large enough quantities to power a reactor. Naturally occurring uranium ores are common—the largest proven reserves are currently in Australia, Kazakhstan, Canada, Russia, and Namibia. The uranium is isolated through physical and chemical processes from some of the other elements present in its ores. It is isolated as U_3O_8, which is also known as "yellow cake," probably because it is a bright yellow powder. Naturally occurring uranium is only 0.72% ^{235}U, almost all of the rest (99.27%) is ^{238}U, which does not undergo fission with thermal neutrons.

Most nuclear reactors need the uranium to be ~4–5% ^{235}U. Increasing the percentage of one isotope relative to others is called **enrichment**, and typically requires a physical (rather than chemical) process. For many years, the process of choice for uranium enrichment was **gaseous diffusion**. The U_3O_8 is chemically transformed to uranium hexafluoride (UF_6). UF_6 is a solid under normal conditions (1 atm, 20°C), but sublimes to a gas if the temperature is raised just a bit or the pressure is lowered. Because the $^{235}UF_6$ molecules are a little lighter than the $^{238}UF_6$ molecules, they move (diffuse/effuse) a little faster. If we pass UF_6 gas through a very long pipe, the gas that emerges at the far end will have a higher percentage of $^{235}UF_6$ relative to $^{238}UF_6$. The pipe has to be pretty darn long, as the difference in mass is quite subtle. The relative diffusion rates can be calculated using Graham's Law (Equation 4.1).

$$\frac{rate_1}{rate_2} = \sqrt{\frac{molar\ mass_2}{molar\ mass_1}} \tag{4.1}$$

$$\frac{rate_{^{235}UF_6}}{rate_{^{238}UF_6}} = \sqrt{\frac{molar\ mass_{^{238}UF_6}}{molar\ mass_{^{235}UF_6}}} = \sqrt{\frac{352.04\ g/mol}{349.03\ g/mol}} = 1.0043$$

$^{235}UF_6$ only diffuses 1.0043 times (or 0.43%) faster than $^{238}UF_6$! Extremely large facilities were constructed to accommodate the tremendous amount of plumbing required. The first such plant built in the United States was the K-25 facility near Oak Ridge, Tennessee. It was four stories tall, half a mile long (0.8 km), and 1000 feet wide (300 m). A huge building!

Today, enrichment using **gas centrifuges** is more common. UF_6 is placed in a cylinder that is spun around at high speed. The heavier $^{238}UF_6$ tends to collect near the cylinder wall, while the lighter $^{235}UF_6$ tends to collect near the central axis of the cylinder (Figure 4.4). Gas centrifuges are preferred for uranium enrichment because they require only about 3% of the power when compared to an equivalent gaseous diffusion plant. Gas centrifuge plants are also much smaller than gaseous diffusion plants and are therefore attractive to those who wish to be somewhat secretive about their enrichment activities.

Once enriched, the uranium is converted back to an oxide (UO_2) and compressed into small pellets for nuclear fuel. If the uranium is to be used in a weapon, it needs to be greater than 80% ^{235}U, which means more enrichment, followed by a chemical transformation to uranium metal. Note that once uranium exceeds 20% ^{235}U is called **highly enriched uranium**, or **HEU**. The 20% threshold may seem oddly arbitrary, but once exceeded, it is relatively easy to get up to weapons-grade uranium (> 80% ^{235}U).

The uranium left over from the enrichment process is ~99.7% ^{238}U and is referred to as **depleted uranium** (or **DU**). Because of its exceptionally high density (1.7 × greater than lead!), low cost, and because it is a relatively low radiological danger, depleted uranium is sometimes used as ballast in boats, counterweights, or, somewhat ironically, radiation shielding.

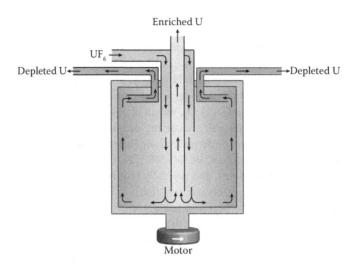

FIGURE 4.4   Flow schematic for a gas centrifuge.

Depleted uranium has also been used extensively in armor piercing shells, especially by the United States in its Gulf Wars. There has been some concern that use of DU weapons may result in a significant health hazard to those exposed to the uranium-containing vapors or shrapnel when they were used to destroy Iraqi tanks. However, because of the relatively long half-life of ^{238}U, a relatively large amount must be concentrated in one human for it to have a significant radiobiological effect. In fact, uranium will cause kidney failure (due to its chemical toxicity) at levels much lower than when it becomes a concern due to its radioactivity. There is also a concern that shrapnel left in the desert may be an environmental hazard. Again, its long half-life and its natural ubiquity in our environment should tend to ameliorate this concern.

Can other fuels (besides ~4–5% ^{235}U in ^{238}U) be used in nuclear reactors? Certainly! Any fissile material will work, with ^{235}U, ^{233}U, and ^{239}Pu being the most popular. ^{239}Pu is both produced and "burned" (undergoes fission) in almost all nuclear power reactors. It is produced in reactors when ^{238}U reacts with a thermal neutron. Instead of undergoing fission, ^{238}U forms ^{239}U.

$$^{238}_{92}\text{U} + {}^{1}\text{n} \rightarrow {}^{239}_{92}\text{U} + \gamma$$

^{239}U then decays to ^{239}Pu through two beta emissions.

$$^{239}_{92}\text{U} \rightarrow {}^{239}_{93}\text{Np} + {}_{-1}\text{e} \qquad t_{1/2} = 23 \text{ min}$$
$$^{239}_{93}\text{Np} \rightarrow {}^{239}_{94}\text{Pu} + {}_{-1}\text{e} \qquad t_{1/2} = 2.4 \text{ d}$$

Likewise, ^{232}Th (naturally occurring!) can absorb a neutron, then decay to form ^{233}U.

$$^{232}_{90}\text{Th} + {}^{1}\text{n} \rightarrow {}^{233}_{90}\text{Th} + \gamma$$
$$^{233}_{90}\text{Th} \rightarrow {}^{233}_{91}\text{Pa} + {}_{-1}\text{e}$$
$$^{233}_{91}\text{Pa} \rightarrow {}^{233}_{92}\text{U} + {}_{-1}\text{e}$$

Nuclides like ^{238}U and ^{232}Th, which can produce a fissile nuclide in a nuclear reactor, are called **fertile.**

Reactors can be designed for ^{239}Pu production. Very simply, a slab of ^{238}U (depleted uranium) is placed in a reactor to soak up neutrons to produce enough ^{239}Pu to make its separation from the remaining ^{238}U worthwhile, without allowing too much of the ^{239}Pu to undergo fission. In theory, ^{239}Pu could also be separated from the reactor fuel pellets. However, this is a rather awkward way to obtain weapons-grade material because it would involve shutting down and refueling the reactor.

Highly enriched uranium or ^{239}Pu are generally used for nuclear weapons (see Section 4.2), however, separating ^{239}Pu from a slab of uranium is much easier than enriching ^{235}U to greater than 80% (which is necessary for weapons). Instead of a tedious physical process, plutonium can easily be separated from uranium using a simple chemical process. Its ease of isolation has made ^{239}Pu the nuclear weapons material of choice and the proliferation of nuclear weapons all too easy.

The dramatic reduction of the U.S. and former Soviet nuclear weapons arsenals following the end of the Cold War has generated a great deal of highly enriched uranium and ^{239}Pu. Much of the nuclear materials from the decommissioned U.S. weapons are being securely stored at the Pantex plant near Amarillo, Texas, and at the Y-12 complex near Oak Ridge, Tennessee. These materials can be used for reactor fuel. For security purposes, the uranium is blended with depleted uranium to bring the ^{235}U down to ~4–5%, and the ^{239}Pu can be mixed with reactor-grade uranium to produce a mixed oxide (MOX) fuel. European and Japanese reactors have been burning MOX fuel for many years.

A high-profile program called "Megatons to Megawatts" was initiated between Russia and the United States in 1993. Under this program, the United States bought highly enriched uranium from Russian warheads and military stockpiles and paid Russia to convert it to reactor fuel for use in U.S. reactors. Approximately 500 tons of Russian highly enriched uranium was converted over the history of the program, which concluded in 2013. At one time, roughly half of the fuel in U.S. nuclear reactors came from the Russian military. This meant that roughly 10% of all electricity in the U.S. at that time originated from material that could've been targeting a U.S. city for destruction. A wonderful irony!

4.1.4   Reactor Safety

While their safety record is excellent when compared to fossil fuel and hydroelectric plants, people generally view nuclear power plants as dangerous.[5] A small portion of this concern is justified, but almost all of it is based on fear and ignorance. Since concerns related to waste and proliferation are discussed below, safety issues related to the mining of uranium and the potential of a catastrophic accident are emphasized here.

Most of the world's uranium is currently mined in Kazakhstan (45% of all U mined in 2021), Namibia (12%), Canada (10%), and Australia (9%). Like other forms of mining, uranium mining is hazardous. In addition to the physical risks shared by all miners, uranium mining also has potential radiological hazards. If not properly ventilated, the air in uranium mines contains relatively high concentrations of radon. Breathing air with too much radon over a long period of time can increase the long-term risk of lung cancer. Uranium miners take preventive measures, and their radiation exposure is carefully minimized and monitored, therefore they face little additional risk when compared to other miners. Unfortunately, this was not true in the early days of uranium mining. In the 1950s, miners were exposed to, and ingested doses that would be unthinkable today. Decades later these uranium miners had cancer rates that were significantly higher than national averages.

Perhaps a bigger risk from uranium mining has to do with the mine tailings. Tailings are the solid material left over at the mine after the uranium has largely been removed. In some cases

---

5   In some countries, there is also an acceptance of the necessity of nuclear power (they have few other resources to generate electricity), as well as greater local appreciation of the economic benefits.

these tailings have been left on the surface, and can leach radioactive nuclides (and heavy metals) into local water supplies. While many of these sites have been, or are now being cleaned up, hundreds of people have been exposed, and for a small fraction, this exposure could result in a premature death.

Chernobyl is the only major nuclear power plant accident known to cause significant human fatalities. Despite the magnitude of this accident, and the fear it has engendered, only about 30 deaths immediately followed the accident. Those that died were emergency response personnel who received very large doses while trying to contain the disaster. A couple hundred, to a few thousand, premature deaths are also expected—primarily due to thyroid conditions in those who were children living near the reactor at the time of the accident. As of 2012, approximately 1800 thyroid problems have been attributed to the Chernobyl accident, and only about a dozen of these have resulted in death. The reader should note there is considerable uncertainty in the predicted numbers (couple hundred to a few thousand), and that the actual numbers are relatively low. Despite the enormous scale of the Chernobyl accident, relatively few people have died from exposure to the radioactive materials that were released.

No deaths are attributable to radiation doses from the accidents at Three Mile Island or Fukushima. Based on the doses, very few, if any, are expected. However, the population displacements caused by radioactive contamination of large tracts of land around Fukushima and Chernobyl have caused significant emotional stress. There's little doubt that these dislocations have caused real health effects, including early deaths. Some scientists have suggested that, given this disparity of negative health effects, future evacuations following a large-scale radioactive contamination be more carefully considered.

So, yes, nuclear power has proven hazardous, and may continue to do so for future millennia if our solution to the waste disposal issue proves inadequate (vide infra). How does nuclear power stack up against other forms of electrical generation? Tens of thousands of people die each year due to coal's use as the most significant form of power generation in the world today. Coal mining remains one of the most dangerous occupations, and the transportation and burning of the large amounts of coal necessary to keep the plants running remains dangerous for the general public. The burning of coal and other fossil fuels adds to the dramatically increasing concentration of CO_2 in the Earth's atmosphere, which is clearly contributing to global climate change. If just some of the recent meteorological extremes were aggravated by global warming, then the safety record of fossil fuels is much worse than stated at the beginning of this paragraph. A NASA study in 2013 estimates that nuclear power should be credited with saving 1.84 million lives between 1971 and 2009, because the presence of the nuke plants prevented more fossil fuel plants from being built, resulting in less air pollution. Coal also naturally contains trace amounts of uranium, thorium, and mercury. If not trapped, they can come out of the stack to be distributed downwind. It is an irony that people living downwind of a coal burning plant can receive a greater radiation dose than those living near a nuclear power plant, although neither is significant. However, the tons of mercury that have been pumped into the environment as a result of burning coal are now rendering many fish species hazardous for human consumption. If fossil fuels were held to the same standards of safety as nuclear power (e.g., careful isolation of all hazardous wastes from the environment), their costs would be driven much higher, making nuclear more economically attractive.

Hundreds of nuclear plants around the world have been operating for decades with excellent safety records. Even so, it might be helpful to more closely examine the three most significant nuclear power plant accidents in recent history—Three Mile Island (TMI) in 1979, Chernobyl in 1986, and Fukushima in 2011. All three accidents were the result of a loss of water coolant, which caused melting of the fuel rods (only partial for TMI). Once the water coolant is lost, the fission chain reaction shuts down. This is because the water also moderates the neutrons, and the fission chain reaction shuts down on its own because of the lower probability of fast neutrons inducing a fission reaction. This happened at TMI and Chernobyl when the reactors lost coolant.[6] However, even after the fission reactions have stopped, there is still enough heat being generated by the decay of the fission products that melting of the fuel rods can occur. Exacerbating this scenario is the exothermic reaction of the zinc cladding on the outside of the fuel rods with steam to form hydrogen gas. It is unfortunate that some of the early nuclear proponents once stated that meltdowns were not possible, as it has reduced the credibility of the industry.

The meltdown at TMI was largely the result of a pressure relief valve that was stuck open and drained coolant from the reactor vessel. Subsequent confusion from conflicting indicators and a bevy of alarms resulted in about half of the fuel melting. The TMI reactor core material did not breach the reactor vessel, however, ~10^{12} Bq of radioactive gases (mostly Xe and Kr) were vented during the crisis, then again over a year later. Because it was all in the gas phase and it was composed of elements that are chemically inert, it dispersed quickly resulting in minimal doses to people living nearby. Radiation levels outside of the plant were carefully monitored, and there was never a significant danger to the public. As a result, mandatory evacuation orders were never issued, which was fortunate, as the resulting panic would've undoubtedly resulted in some injury and loss of life.

In an odd twist of fate, the TMI accident occurred shortly after the release of the movie *The China Syndrome*, which depicted an eerily similar loss of coolant accident in a nuclear power plant in Southern California. The movie was fictional and suggested that the ramifications of a nuclear reactor core meltdown would be dire: "...render an area the size of Pennsylvania permanently uninhabitable,[7] not to mention the cancer that would show up later." Many confused the movie with the actual events at TMI (which, coincidentally occurred *in* Pennsylvania). This serendipitous coalescence of fiction and reality helped to harden public attitudes in the United States, and elsewhere, against nuclear power.

The Chernobyl accident was the worst of the three, primarily because the reactor building was significantly damaged, directly exposing the core to the atmosphere. This type of reactor suffered from a known flaw in its design that allowed for rapid increases in power output under certain conditions. Unfortunately, these conditions occurred at a time when the reactor operators had bypassed safety systems to conduct an experiment. The rapid rise in core temperature resulted in two steam (or a steam and a hydrogen) explosions that blew the roof off the reactor building and started several fires. It was the large explosions, direct exposure of the glowing red-hot reactor materials to the atmosphere, and the fires that led to the extensive distribution of radioactive materials. Approximately 1.4×10^{19} Bq of radioactivity was released. Much of the radioactive material was dropped on Northern, Eastern, and Central Europe, although measurable quantities were spread throughout much of the Northern Hemisphere.

Extinguishing the fires, and covering and cooling the exposed core gave some of those involved high, acute radiation doses. Twenty-eight died as a result of these exposures in the months that followed the accident. Doses to the general public living near the accident were exacerbated by the government's lethargy in ordering an evacuation, and its failure to adequately monitor foodstuffs (particularly milk) that were potentially quite contaminated. This was the cause of many of the roughly 1800 cases of thyroid problems mentioned earlier.

Three reactors in Fukushima, Japan, suffered fuel meltdowns following a massive earthquake and tsunami. The reactors all weathered the magnitude 9.0 earthquake remarkably well. There is no indication that they were damaged by the quake. However, they were not adequately protected against the ~14 meter tsunami that flooded the station 45 minutes later. The tsunami knocked out almost all backup power generators, batteries, and electrical switching stations on the site. Without power, the

---

6   The Fukushima reactors all automatically shut down the fission chain reaction (scrammed) when the earthquake hit. It wasn't until ~45 minutes later that the tsunami hit and problems began.

7   The Chernobyl accident caused the greatest evacuation, and covered an area of 4300 km², which is only 3.6% the size of Pennsylvania. Many of the areas within the Chernobyl exclusion zone have since opened for resettlement.

FIGURE 4.5 Damage to the structure surrounding Fukushima Daiichi reactor #4. (Photo courtesy of Tokyo Electric Power Co.)

pumps could not circulate enough coolant through the reactor core, and they overheated from the decay of fission products in the fuel rods. Over the following three days, these reactors suffered fuel meltdowns as the water in the core boiled off, exposing the fuel rods. When this happened, some of the hot zinc fuel cladding reacted with the steam to form hydrogen gas. Enough hydrogen gas built up in the buildings surrounding the reactor cores to cause three large explosions (Figure 4.5). Some of the volatile fission products (noble gases, ^{131}I, ^{137}Cs, and ^{134}Cs) were also dispersed because containment had been breached from the excessively high temperatures and pressures in the reactors.

Approximately 9.4×10^{17} Bq of radioactive materials were released from the reactors. The vast majority was released to the air, and some contaminated land nearby. Despite the severity and scale of the Fukushima meltdowns, the Japanese government did a good job evacuating nearby areas and monitoring food and water. As a result the doses to the general public are much lower than they were for Chernobyl. It appears that the doses are so low that it is fairly certain that no adverse health effects will be seen. Likewise, the doses to the workers at Fukushima appear[8] to have been much more limited than at Chernobyl. So much so that it is likely that just a handful of workers (~6 out of 20,000) face only a slightly increased risk of future cancer.

The primary lasting effect of Fukushima appears to be psychological and economic. The Japanese public initially showed a strong reluctance toward nuclear power in the years since the accident. Restarting the ~50 reactors in Japan has been a slow and difficult process, primarily due to strong public opposition. As mentioned earlier, only ten Japanese reactors were producing electricity by mid-2022. Normally responsible for generating roughly 30% of Japan's electricity, their capacity is being made up from conservation and increased burning of fossil fuels (more expensive for Japan). The stabilization of the damaged reactors, the cleanup of contamination, as well as beefing up safety features of other reactors also proved costly.

Operation and design of current and future nuclear plants are imminently conscious of mistakes made by operators as well as the design flaws of the TMI and Chernobyl reactors, and the Fukushima site. Perhaps the most significant is incorporation of passive shutdown and cooling systems, which could safely stabilize a reactor without power or human intervention. Some new designs incorporate a "core-catcher" to safely contain a core meltdown, preventing contamination outside of the reactor. While these systems can be integrated into new reactor designs, operators are also working to better

---

8   It has been reported that some of the data for a few workers were attenuated by covering dosimeters with a thin sheet of lead.

anticipate potential emergencies at existing reactors so that future significant releases of radioactive materials are prevented.

4.1.5 NUCLEAR WASTE

When fuel rods are first removed from the reactor core, they are still pretty toasty—very radioactive and still producing a fair amount of thermal energy (hot!), both of which are due primarily to the presence of short-lived fission products. The fuel rods are stored underwater at the reactor site while the relatively short half-life nuclides decay, and the rods cool off. The water acts as shielding for the radiation as well as a coolant. After they've cooled off a bit, the rods can be placed in dry cask storage. These casks are designed to allow air-cooling (convection) of the rods, provide shielding, and are generally indestructible.

What is spent fuel composed of? Remember that the fuel starts out as ~4–5% ^{235}U in ^{238}U. When the ^{235}U burns in a nuclear reactor, the percentage of ^{235}U gradually decreases. When fuel rods can no longer be used (**spent**), they still contain roughly 1% ^{235}U and now have ~1% ^{239}Pu (from neutron reaction with some of the ^{238}U, followed by decay), which could still be burned in a reactor. Only ~3% of the spent fuel is fission products, and the remaining 95% is still ^{238}U (Figure 4.6). Since the fission products generally have shorter half-lives than the uranium and plutonium, and the ^{235}U and ^{239}Pu can still be used as fuel, some have suggested that separations be performed on the spent fuel. This is known as **reprocessing** and is the approach some countries[9] have taken toward spent fuel. It should be noted that reprocessing is much more expensive than mining uranium and enriching it to new fuel, and, as discussed below, presents a proliferation problem. The major motivations for reprocessing are environmental. Reprocessing will significantly reduce the volume of waste needing to be stored, allow more of the waste to become cold (no longer radioactive) sooner, and makes more extensive use of the fuel.

Since the spent fuel can be recycled, its life and death are often referred to as the **nuclear fuel cycle** (Figure 4.7). As mentioned in Section 4.1.3, the uranium is first mined, undergoes physical and chemical separations, is usually enriched, then is fashioned into fuel pellets (fuel fabrication). These activities are often referred to as the "front-end" of the fuel cycle. The uranium is then used to produce electricity in a reactor (Section 4.1.2). When the spent fuel is removed it can be reprocessed for additional atom splitting in a reactor or buried. Activities with the fuel after it's removed from the reactor are often referred to as the "back-end" of the cycle. When reprocessing is used, the fuel cycle is often referred to as "closed" because the circle is complete.

Some countries[10] plan to simply dispose of their spent fuel as radioactive waste. This is called the "once-through" or "open" fuel cycle because the fuel is only in the reactor once, and the circle is not complete. Under these circumstances, it's probably better to refer to the spent fuel as "nuclear waste" since it is destined for disposal. The U.S. federal government promised companies running

■ U-235
■ Pu-239
■ Fission Products
■ U-238

FIGURE 4.6   Composition of spent nuclear fuel.

---

9   Most notably China, France, India, Japan, Russia, and the United Kingdom.
10   Such as Canada, Finland, Spain, Sweden, and the United States.

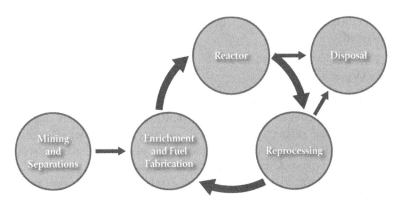

FIGURE 4.7 The nuclear fuel cycle.

nuclear reactors that it would take all of the nuclear waste and bury it in a geologic repository. For many reasons (some discussed later), the U.S. government was unable to meet its own deadlines to open the repository, and all spent fuel currently remains on site at U.S. reactors. Deciding on the type(s) and location(s) of repositories seems to be a difficult task for the U.S. government. Interim storage facilities are now (2022) being explored to allow the government to fulfill its obligation to the reactor companies without committing to a more permanent solution.

While the United States vacillates, most other countries[11] are moving forward with their nuclear waste disposal plans. Regardless of whether reprocessing is planned, all nuclear waste seems destined to be buried deep underground. Why? Many options have been studied, from dumping in the ocean near a subduction zone, to launching it into the sun. From scientific and engineering studies, a geologic repository appears to be the safest way to dispose of nuclear waste.

The best support for a geologic repository originally came from a uranium mine. The French used to get some of their uranium from the African nation of Gabon. Back in the early 1970s, a shipment from the Oklo mine in southeastern Gabon arrived at the French enrichment plant with lower than normal percentages of ^{235}U. French scientists started looking into this and found that other isotopic ratios at the Oklo site were not what they should be. Through a great deal of analysis, they reasoned that about 1.7 billion years ago, a fission chain reaction took place. In other words, Oklo was the site of a natural nuclear reactor. All the conditions were right. At that time, the ^{235}U natural abundance was ~4%, and there was plenty of water around to moderate the neutrons. It looks like the Oklo reactor ran for several hundred thousand years (off and on, depending on the presence of water), burning about 6 tons of ^{235}U and generating about 1 ton of ^{239}Pu (which underwent fission, or decayed to ^{235}U) and 15,000 MWy of power—roughly equivalent to 4 years output of a modern reactor. It also appears that almost all of the radioactive nuclides produced by the Oklo reactor stayed right where they were made, or traveled only a short distance, suggesting that geologic repositories may be an effective long-term storage solution for radioactive wastes. This is a pretty strong suggestion, as the Oklo reactor was located near the surface where there's a lot of water movement. If we can just marginally improve on Nature, we should be safe.

The Swedes opened such a repository in 1988 near Forsmark, Sweden. The repository is located under the Baltic Sea in 1.9 billion year old granite. Finland opened a similar repository in 1995. Another opened in 1999 near Carlsbad, New Mexico, in the United States. It is called the "Waste Isolation Pilot Plant" (WIPP), and is constructed in a 200 million year old salt formation ~2000 feet underground. None of these repositories has accepted spent nuclear fuel, but they do accept less

---

11 Sweden, Finland, and France are the furthest along in this process.

radioactive (low-level) wastes from research and other activities involving radioactive materials. As suggested above, a good repository is geologically stable and dry. If WIPP had any water flowing through it, the salt would've dissolved, and it is known that the salt at WIPP is over 200 million years old.

Another consideration is security. Having all this waste in one, isolated, place is safer than several, more accessible places. However, it should be emphasized that the waste needs to cool off for 30–50 years before it is placed in a geologic repository. Once it's stuck in the ground, it's a little more difficult to ensure it can be kept cool enough. Also, radiation levels drop very quickly in the first couple of decades after removing the fuel from the reactor, making it safer to transport. While the United States has decided to leave all the waste at the reactors where it was used (much to the consternation of the utilities that own the reactors), other countries have opted to consolidate their casks in temporary, aboveground waste storage while waiting for the waste to cool enough to be buried.

A common argument against nuclear power is that its waste remains highly radioactive for a very long time. The amount of time given varies, but is typically something rather unimaginable such as tens or hundreds of thousands of years. This simply isn't true. After ~300 years, 98% of the radioactivity is gone, i.e., it is only 2% as toasty as when it was pulled out of the reactor.[12] After ~1000 years it is no more radioactive than if we had left the uranium in the ground. As can be seen in Figure 4.6, the waste is mostly (~96%) uranium—pretty much the same as the uranium ore that was mined in the first place. Putting this stuff back in the ground (where it came from) should then make some sense. There are still billions of tons of uranium ore in the soil, some of it not buried particularly well, yet we do not fear it, why then should we fear carefully placing the waste into meticulously selected, geologically stable, repositories? What about the plutonium? Well, plutonium is chemically very similar to uranium. If we put it into the ground in a place where the uranium has been sitting still for the last couple billion years, the plutonium should also stick. Extensive research into the geochemistry of plutonium confirms that it will.

Finally politics are also a consideration. In most countries, volunteer communities are sought among those with suitable geology. Communities often volunteer because of the economic boost a repository will bring to a local area. Repositories can also be forced in areas with less political clout—such was the case with the U.S. repository at Yucca Mountain.

Unfortunately for the U.S., WIPP is unsuitable for nuclear waste because the heat from waste will draw water to it, bathing the containers in salt water. Under these conditions, they would quickly corrode, and potentially release their contents to the environment. Yucca Mountain, near Las Vegas in Nevada, was selected in 1987 as the sole site for a geologic repository for spent fuel and other more radioactive (high-level) wastes. Yucca Mountain shares many of WIPP's positive attributes except that it is not a salt formation, rather it is volcanic tuff, formed from eruptions millions of years ago. Additionally, Yucca Mountain is part of the Nevada Test Site, where nearly 1000 nuclear weapons were tested both above and below ground for over 40 years—until the test ban in 1992. This site is large, isolated, secure, and, at the time of its selection, Nevada was not politically prominent. Opening of Yucca Mountain has been delayed, and it is now uncertain whether it will ever receive waste. One of the major causes of the delay has been concern that the geology might not be as stable or dry as originally thought, especially over the long term. The other cause for delay has been Nevada's increasing political power, partly due to rapid growth in nearby Las Vegas. This has often inhibited key legislation that would facilitate opening of the site, and removed most funding for anything related to the site for several years. Statewide opposition is fairly strong, and is at least

---

12 It is still hot enough that you wouldn't want to snuggle up next to a big pile of the stuff, but not anywhere near as "highly radioactive" as some believe.

partially fueled by the fact that the putative repository was forced on Nevada in a rather clumsy political maneuver, rather than agreed on by all parties.

Despite the fact that there is a lot of it (thousands of tons) and that it is somewhat toasty, spent nuclear fuel is well characterized, rather compact, and quite uniform (it's pretty much the same stuff no matter what reactor it comes from). The other major source of high-level nuclear waste in the United States has none of these positive attributes. It is a legacy of the Cold War with the Soviet Union—making thousands of nuclear weapons generated a lot of radioactive trash, as well as contaminated buildings and grounds. By far, the biggest challenge is cleaning up the 177 underground waste tanks (Figure 4.8) at the Hanford site in south central Washington state. Hanford was started as part of the Manhattan Project and provided the plutonium for the first test weapon as well as the bomb dropped on Nagasaki. During the Cold War, the Hanford site was responsible for producing the plutonium and tritium (^{3}H) for the U.S. nuclear arsenal. Eight nuclear reactors were built along the banks of the Columbia River and plutonium separation facilities were placed further inland.

Separation of plutonium is accomplished through a chemical process. At the end of this process, nitric acid solutions containing fission products, as well as traces of uranium, plutonium, other actinides, and lanthanides, needed safe disposal. The tanks were built to accommodate these wastes. The tanks were made of steel, and would not hold acidic solutions (low pH) for long, so enough sodium hydroxide (NaOH, also known as lye) was added to make the waste solution basic (high pH), and therefore, not corrosive to the tanks. Tank volume occasionally became scarce, so the water (not radioactive) was evaporated out of many, allowing additional waste to be added. So much water was evaporated that the tanks now all contain a lot of solids suspended in aqueous slurries (sludge) or hardened into salt formations (saltcake). Radioactive materials are part of or entrapped in both.

Different waste streams were added to different tanks, and sometimes the contents of one tank were pumped into another. Records of these activities are incomplete, leaving each tank with a unique and partially unknown chemical and radionuclide composition. One tank contains a fair amount of organic compounds (containing mostly carbon and hydrogen). The ionizing radiation causes the formation of hydrogen gas (H_2) from these organic compounds, and would build up in the sludge, and then occasionally "burp" out of the tank. Concerns were raised that the hydrogen could cause the tank to explode during a burp, so a method of stirring the tank contents was developed to allow the slow continuous release of the hydrogen gas.

FIGURE 4.8 A 1.3 million gallon waste tank under construction. (Photo courtesy of U.S. Department of Energy.)

Most of the materials in these tanks are not radioactive. The three most common ions present are sodium (Na^+), hydroxide (OH^-), and nitrate (NO_3^-). Various methods of separating the hot (radioactive) from the cold (not radioactive) have been researched and some implemented. The idea is to place the cold stuff into a relatively inexpensive waste form, such as concrete (or grout), on site and **vitrify** the relatively small volume of radioactive material. Vitrification means isolating the radioactive stuff in glass, and is a relatively expensive process. The radioactive material is distributed throughout a large glass cylinder (called **logs**) on site then shipped to a high-level repository such as Yucca Mountain. Glass is the preferred waste form for long-term storage of radioactive nuclides, as even if water flows over it, it is unlikely any of the radioactive material will leach out.

The engineers that designed the tanks thought they would only be used temporarily, not the 70+ years some have been in use. Some of the tanks have cracked and leaked some of their contents into the ground nearby, creating a more difficult cleanup scenario.

Similar waste tanks also existed at Department of Energy (DOE) facilities near Idaho Falls, Idaho, Oak Ridge, Tennessee, Aiken, South Carolina, and West Valley, New York. These sites only contained a few tanks each, most have already been remediated or will soon be cleaned up.

Hospitals, universities, and private companies produce small amounts of very low-level radioactive wastes. These wastes can often be disposed of in local landfills, or landfills specially designed for the disposal of hazardous material, i.e., extra effort is taken to isolate the contents of the landfill from the surrounding environment.

4.1.6 Cost of Nuclear Power

When nuclear power plants were first being planned, Lewis Strauss declared "Our children will enjoy in their homes electrical energy too cheap to meter." At the time, Strauss was chair of the U.S. Atomic Energy Commission, so most people understood his comment to be about power from nuclear *fission*. He *may* have been referring instead to power from nuclear *fusion*, but it is not clear. What is clear is that he was speaking for himself, and not the nuclear power industry. When nuclear power subsequently became one of the more expensive ways to generate electricity, the quote damaged the credibility of the industry. The high capital costs associated with nuclear power are probably the single biggest factor inhibiting the construction of new plants, especially in the United States. As previously stated, increases in costs of oil and natural gas (prior to 2008), coupled with increasing concerns that the extensive burning of fossil fuels is adversely affecting the global climate, rekindled interest in nuclear power worldwide through the turn of the century. However, the drop in natural gas prices following the widespread application of hydraulic fracturing, along with the relatively low cost in constructing natural gas power plants, has made generation of electricity from natural gas a popular option.

The average U.S. wholesale cost of generating electricity from all sources in 2005 was $50 per MWh (megawatt hour—the production of one million watts of power for one hour). The same costs for U.S. nuclear power plants varied from $30/MWh to $140/MWh in 2005. Why such a huge spread? The single most important factor in the cost of nuclear power is the time it takes to build the plant. Nuclear is very capital intensive—a lot of money is required up front. Once it is up and running, nuclear plants are relatively cheap to operate because their fuel costs are low compared to fossil fuels—a tremendous amount of energy is generated from very little fuel. Construction time is an issue because most of the capital costs are covered through loans. If the plant takes a long time to build, the interest on the loans can become as large as the loan itself. These costs are recovered only after the plant is built, and are passed on to the ratepayers. This was especially bad through the periods of high inflation like the 1970s and 1980s.

Because of its high upfront costs, nuclear tends to fare better in more highly regulated markets with significant baseload electricity demand. A more regulated market insures the return on the high investment costs in building a nuke plant over the long term. Deregulated markets encourage more short-term (fossil fuel) thinking. Fossil fuel plants, while cheaper to build, experience higher operating costs, and are subject to greater volatility in fuel prices. Nukes do better with higher baseload because they are most cost effective running at full capacity 24/7. Fossil fuel plants are a little easier to switch on and off to meet peak, or intermittent, demand.

The accidents at TMI (1979) and Chernobyl (1986) also contributed to increasing costs of nuclear power generation in the 1980s. Designs to plants under construction at that time were changed, resulting in significant delays and cost overruns. Many existing plants were shut down so new safety equipment could be installed and/or measures instituted. Cost escalations through the 1980s were the most dramatic in the United States, but were also observed elsewhere. Reactors starting construction in the 1990s and 2000s in Japan, India, and South Korea generally show decreasing construction costs.

The U.S. government estimates that new nuclear plants constructed in the United States will end up producing electricity at about $95/MWh in 2022, but there remains significant uncertainty in the construction costs. This is nearly two times the price of electricity from a natural gas plant (roughly $55/MWh). It is not surprising that electrical generation from natural gas has nearly quadrupled in the U.S. from 1990 to 2016. This trend is also observed worldwide. Nuclear's cost is also higher than other non-fossil fuel sources such as solar and wind (even without government subsidies). Some European countries are working toward replacing their nuclear capacity with solar, wind, and other renewable sources.

4.1.7 PROLIFERATION OF NUCLEAR WEAPONS

Since the beginning of the nuclear age, scientists and politicians have tried to separate weapons from nuclear power. Unfortunately, the two are closely related, and has resulted in nine countries[13] now possessing nuclear weapons. Highly enriched uranium (> 80% ^{235}U) or relatively pure ^{239}Pu are required to build nuclear weapons—fortunately they are difficult to come by. The focus for nonproliferation efforts has been to control access to these materials. The "Megatons to Megawatts" program mentioned in Section 4.1.3 is a good example of a nonproliferation success. Weapons-grade uranium from the former Soviet Union was converted to nuclear fuel and burned in U.S. reactors.

In addition to taking weapons-grade nuclear material out of circulation, nonproliferation efforts have focused on two key components of the nuclear fuel cycle: uranium enrichment and spent fuel reprocessing (Figure 4.7), as both can allow production of weapons material. Uranium enrichment facilities can be used to bring the ^{235}U percentage up to ~4–5% for use in a reactor, but the process can easily be continued until the uranium is weapons-grade. Therefore, anyone with a uranium enrichment facility has the potential to produce highly enriched uranium. Since every nuclear reactor contains ^{238}U, they also produce ^{239}Pu. Any chemical reprocessing of the spent fuel can lead to the isolation of ^{239}Pu—although this is a rather inconvenient and rarely used method to get ^{239}Pu. More likely, anyone with a nuclear reactor can simply place a slab of ^{238}U into the reactor to let it soak up neutrons and produce ^{239}Pu. That way there's no need to sacrifice the fuel rods to get the plutonium. The United States government's opposition to the reprocessing of spent fuel was established primarily on nonproliferation grounds.

One key to limiting proliferation of nuclear weapons is therefore some sort of control of enrichment and reprocessing facilities. There aren't many of these facilities, but their numbers grew as the nuclear industry rapidly expanded worldwide through the 1970s and early 1980s. As nuclear power expands into new countries, new enrichment and reprocessing facilities are considered. Consolidation of these facilities where redundant and placement of them under international control

---

13 The United States, Russia, France, the United Kingdom, China, India, Pakistan, Israel, and North Korea all have nuclear weapons as of this writing.

has been suggested as a way to ensure they are used only for energy purposes. Unfortunately, national sovereignty and security concerns tend to inhibit this type of cooperation.

Another key is the strengthening or replacement of the United Nation's treaty on the nonproliferation of nuclear weapons (NPT). The NPT went into effect in 1970 and required the nations with nuclear weapons to disarm and prohibited other nations from developing nuclear weapons. In 1970, only five nations (the United States, the United Kingdom, France, the Soviet Union, and China) possessed nuclear weapons. Unfortunately, four new nations (India, Pakistan, Israel, and North Korea) have since created nuclear weapons.[14] These four nations are the only nations not currently signatories to the NPT. Equally unfortunate, the five original nuclear powers still have significant stockpiles of nuclear weapons. Some have suggested that the United States begin a program of unilateral disarmament, allowing the U.S. to gain some moral leadership while encouraging other nuclear powers to also disarm. They argue that since the end of the Cold War, the nuclear arsenal no longer has anything to deter, so why should it be preserved?

The debate over nuclear power is a complex and sometimes contentious one. The interested reader is encouraged to also look at some of the resources listed in the bibliography. Many contain more detailed discussions of this topic.

4.2   NUCLEAR WEAPONS

4.2.1   Fission Bombs

In theory, an atomic (fission) bomb is a relatively simple device to build; fortunately, getting the fissile material to put into it is rather difficult. The material should have a high probability of undergoing neutron-induced fission, produce a fair number of neutrons with each fission reaction (on average), and have a reasonably long half-life—after we go to all the trouble of building a bomb, it'd be nice to be able to store it for a while. Finally, the fissile material should be accessible—either something we can dig up or easily make in a nuclear reactor. Three nuclides that fit all the criteria well are ^{233}U, ^{235}U, and ^{239}Pu. ^{235}U occurs naturally, but requires high enrichment (> 80%) for use in weapons. ^{233}U and ^{239}Pu can be made in a nuclear reactor, and then chemically separated (Section 4.1.3).

Despite intense security surrounding the Manhattan Project, the Soviet Union obtained a great deal of information by recruiting a number of spies from those working on the project. It was only four years after the United States first tested an atomic bomb that the Soviets also detonated one. No spying is necessary today, as fairly good descriptions of nuclear weapons are posted on the Internet.

The basic concept behind a **fission bomb** is also simple—it is based on the accelerating fission chain reaction that occurs when a supercritical mass of fissile material exists. The idea is to quickly create a supercritical mass, and then try to keep it contained for as long as possible. Making the supercritical mass quickly means that very little of the fissile material undergoes fission (and releases energy!) before the supercritical mass is formed. Keeping it contained means maximizing the number of fission reactions (energy released) before the supercritical mass blows itself apart, ending the chain reaction. The result should be a really big explosion.

The simplest fission bomb design is the "**gun-type**" shown in Figure 4.9. In this weapon, two subcritical masses of ^{235}U are separated in the bomb. One is "shot" (like a gun shooting a bullet) into the other by a chemical (conventional) explosive. When the two masses come together, they form a supercritical mass. To help start the fission chain reaction, a neutron **initiator** is placed where the center of the supercritical mass ends up. The initiator's purpose is to inject a large number of neutrons into the supercritical mass at the very moment it is formed. This can be accomplished with a small amount of a polonium–beryllium mixture.

---

14 South Africa also created a small number (~6) of nuclear weapons but dismantled them before signing the NPT in 1991.

FIGURE 4.9   Schematic of a gun-type fission weapon.

Polonium is used because a couple of its isotopes (usually ^{210}Po) decay fairly "cleanly" via alpha; i.e., no (or few) other emissions are observed. ^{235}U also decays via alpha, so why bother with polonium? These isotopes of polonium have shorter half-lives than ^{235}U, and therefore have higher specific activities than ^{235}U. In other words, they crank out more alpha particles per unit mass than the ^{235}U. So why are alpha particles important? If some could be converted to neutrons at the moment the supercritical mass is formed, a lot more fission reactions can take place, and a much larger explosion will result. At the moment the supercritical mass is formed, the polonium gets mashed together with some ^{9}Be. The alpha particles given off at that moment by the polonium can then react with the beryllium to form ^{12}C and a neutron.

$$^9_4\text{Be} + ^4_2\text{He} \rightarrow ^{12}_6\text{C} + ^1\text{n}$$

The first atomic bomb used in warfare was a gun-type weapon, called Little Boy, and was dropped by the U.S. on Hiroshima on August 6, 1945. The Manhattan Project scientists were so confident it would work; they didn't even bother to test a gun-type weapon. Facilitating the decision was the lack of highly enriched uranium—they only had enough to build one bomb. The yield of the bomb was equivalent to 15,000 tons of TNT. Nuclear weapon yield is typically measured in terms of equivalent tonnage of the conventional explosive trinitrotoluene (TNT). For reference, the explosion of one ton of TNT produces 4.2×10^9 J of energy. Little Boy was therefore referred to as a 15-kiloton device.

The casing around a gun-type bomb can be made from any sturdy material. Remember, its purpose is to try to keep the supercritical mass together as long as possible (even if it is just a small fraction of a second) to maximize the number of ^{235}U fission reactions, thereby increasing the weapon's explosive power (yield). Depleted uranium (~99.7% ^{238}U) is an excellent choice. Not only is it a strong metal, it will also interact with the fast neutrons produced by the chain reaction in favorable ways. It can reflect some neutrons back into the ^{235}U, and it can undergo fission with others. Both of these types of interactions will increase the yield of the weapon. The proper term for the ^{238}U used in this way is a **tamper**.

The other kind of fission bomb is the implosion-type (Figure 4.10). It is constructed with a hollow, or "low-density"[15] sphere of ^{239}Pu, which is surrounded by chemical explosives. The idea is to detonate the conventional explosives at exactly the same time, causing the hollow plutonium sphere to collapse to a small solid sphere that is now a supercritical mass. Again, a ^{238}U tamper surrounds the plutonium sphere to increase the yield.

The Manhattan Project scientists knew that a ^{239}Pu bomb required a different design than ^{235}U. The issue is one of timing. A gun-type weapon does not form the supercritical mass fast enough for ^{239}Pu. Because it is more likely to undergo fission reactions and because it produces more neutrons

---

15 If a subcritical mass is compressed, its density will increase, and it can become supercritical. Imagine a marshmallow made from plutonium.

Hollow ^{239}Pu sphere ^{238}U Tamper

Chemical explosive   Neutron initiator

FIGURE 4.10 Schematic of an implosion-type fission weapon.

per fission (on average), ^{239}Pu will begin to blow itself apart before the supercritical mass is completely formed in a gun-type device. Potentially, this could result in the bomb destroying just itself rather than an entire city. The collapse of a sphere is a faster way to create a supercritical mass. In order for it to work, however, the timing of the detonations of all the conventional explosives has to be perfect. It's a bit like trying to squeeze a water balloon with your hands—tough to do without some coming out between your fingers.

The Manhattan Project scientists were so worried about the timing issue, they decided to test an implosion-type device. The codename for the test was "Trinity" and was set up in the Jornada del Muerto (Journey of the Dead Man) desert in southern New Mexico. On July 16, 1945, the first nuclear weapon was detonated, ushering in the atomic age. The power of the explosion exceeded all expectations. The awesome power of the blast moved project director J. Robert Oppenheimer to quote the following passage from the sacred Hindu text, the Bhagavad-Gita:

If the radiance of a thousand suns
Were to burst at once into the sky,
That would be like the splendor of the Mighty One...
I am become Death,
The shatterer of Worlds.

A second implosion-type weapon was dropped on Nagasaki on August 9, 1945, bringing World War II to a horrific conclusion. The Nagasaki weapon was given the name Fat Man and had a yield of 22 kilotons.

As described in Section 4.1.3, ^{239}Pu is formed in the fuel rods in nuclear reactors. It can also be made by placing a slab of ^{238}U in the reactor. To optimize ^{239}Pu production, it is best to run the reactor at low power (low neutron flux) with fuel that has more ^{238}U and less ^{235}U (lower enrichment than normal). These conditions optimize the conversion of ^{238}U to ^{239}Pu and minimize the deleterious reactions of ^{239}Pu with neutrons, either undergoing fission or forming ^{240}Pu.

$$^{239}_{94}\text{Pu} + {}^1\text{n} \rightarrow {}^{240}_{94}\text{Pu} + \gamma$$

When the plutonium is chemically separated from the uranium, all isotopes are separated together, therefore there is always some ^{240}Pu mixed in with the ^{239}Pu. This is bad for a weapon because ^{240}Pu is not fissile and has an annoying tendency to decay via spontaneous fission—no neutron required (Section 11.1)! If, while the supercritical mass is being assembled just prior to detonation, some of the ^{240}Pu undergoes spontaneous fission, a few neutrons will be produced, which can cause a short chain reaction. This will cause the weapon to blow itself apart before many fission reactions have occurred, significantly lowering the yield. In the business, this is known as a **fizzle**, and is believed to have happened when North Korea tested its first nuclear weapon in 2006. To avoid this sort of embarrassment, a plutonium device should be < 7% ^{240}Pu.

Fission weapons have been designed with yields ranging from 0.01 to 500 kilotons. The low end of the spectrum represents an extremely inefficient device, i.e., only a fraction of the fissile material undergoes fission, the rest just gets spread over the blast area or is carried away in the fallout. Little Boy was only about 1.3% efficient, meaning that only 1.3% of the ^{235}U underwent fission. Fat Man was about 17% efficient.

Yield is generally related to the amount of fissile material in the weapon. At a certain point, it is no longer possible to pack more fissile material into a bomb without simply spreading the excess around. In other words, only a certain amount of material can undergo fission before the bomb starts blowing itself apart and the chain reaction shuts down. A maximum of roughly 500 kilotons on fission weapon yield is a practical limit. To achieve more destructive power, the very nature of the bomb needs to change.

4.2.2 FUSION BOMBS

Fusion weapons (also known as hydrogen bombs, H-bombs, or thermonuclear devices) are *much* more powerful. Yields as high as 60,000 kilotons (that's 60 megatons!) have been made. They are over 4000 times more powerful than the bomb dropped on Hiroshima. As their various names imply, these weapons take advantage of the relatively high amount of energy released per unit mass for fusion reactions. To get fusion to take place, very high temperatures and pressures are needed.

A fission device is used as an **initiator** for fusion weapons. It produces the necessary heat, as well as gamma rays and neutrons, which are all helpful in making fusion possible. The fission device is also referred to as the **primary** (Figure 4.11) as it is set off first, and its detonation causes the fusion part of the weapon (the **secondary**) to explode.

A **neutron reflector** is placed behind the primary. Its purpose is to reflect neutrons back into the fissile material of the fission device, and toward the secondary. It is made from beryllium because this element has a low probability of absorbing neutrons, but a relatively high probability of scattering them, making it a pretty good neutron "mirror." Note that not all neutrons are reflected by the beryllium; it simply has a higher probability of reflecting neutrons than most other materials.

A ^{238}U tamper is placed between the primary and secondary. The fast neutrons produced by the primary can (a) pass right through the tamper, (b) be reflected by the tamper back into the primary, or (c) cause the ^{238}U in the tamper to fission. All of the above are helpful in making a more powerful explosion. The ^{238}U casing also acts as a tamper.

How does the secondary work? As shown in Figure 4.11, it is simply a cylinder of fissile material (like ^{239}Pu) wrapped with lithium deuteride (LiD) packed in the device with Styrofoam™[16] (and

FIGURE 4.11 Schematic representation of a fusion weapon.

---

16 Styrofoam is a registered trademark of the Dow Chemical Company.

you thought packing peanuts were invented for grandma to ship you fragile stuff!). The tremendous heat and gamma radiation created by the primary turns the Styrofoam™ into a super-hot plasma, squeezing and heating the secondary. The neutrons that make it through the tamper cause the fissile material inside the secondary to fission, adding more heat (~10^8 K!) and pressure to the LiD. Neutrons being released by the fissile material in the primary and secondary react with the lithium to form tritium (^{3}H):

$$\,^6_3\text{Li} + \,^1\text{n} \rightarrow \,^4_2\text{He} + \,^3_1\text{H}$$

The tritium can now fuse with the deuterium.

$$\,^2_1\text{H} + \,^3_1\text{H} \rightarrow \,^4_2\text{He} + \,^1\text{n}$$

The neutrons produced by this fusion reaction are very high energy (14 MeV), and will get some of the ^{238}U casing to fission, as well as cause more fission to occur in the primary, creating an even bigger explosion. The total yield will depend on how much LiD and fissile material are present, and whether the casing is made from ^{238}U. The really high-yield bombs can get rather large and heavy, and would be impractical as a weapon as there would be no easy way to deliver them to a target.

Even though fusion weapons use fission triggers, these weapons are often considered relatively "**clean**,"[17] whereas fission nukes, especially those with low yields, are called "**dirty**." Adding certain materials can also make a fusion weapon dirty. Most commonly, use of ^{238}U as a tamper or casing material will increase levels of ionizing radiation in the blast area and the fallout. With more fission reactions taking place, there are more fission products to spread around. The addition of natural cobalt (which is 100% ^{59}Co) to a fusion weapon results in the formation of ^{60}Co from the reaction with a neutron.

$$\,^{59}_{27}\text{Co} + \,^1\text{n} \rightarrow \,^{60}_{27}\text{Co} + \gamma$$

^{60}Co emits high-energy gamma photons during its decay and has a half-life short enough to give it a high specific activity, but long enough to make it frightening to move back to contaminated areas for quite a while after the attack.

$$\,^{60}_{27}\text{Co} \rightarrow \,^{60}_{28}\text{Ni} + \,_{-1}\text{e} + 2\gamma$$

Such a weapon was called a "doomsday device" because it could effectively evacuate a large area for many years, depending on how much the populace fears ionizing radiation. Many forms of popular fiction picked up on this moniker, although the fictional devices could usually destroy the entire planet, something that is not possible with a single real weapon.

---

17 It should be a little odd to consider nuclear weapons as "clean." They are very deadly and cause unbelievable environmental destruction and human suffering. Additionally, they spread their highly radioactive fission products over a broad area. The term is only meant to be relative to other nuclear weapons.

4.2.3 Other Bombs

The **neutron bomb** (or enhanced radiation device) is a small H-bomb designed for maximum neutron production while decreasing blast and heat and is therefore considered "clean." This is done by minimizing the amount of fissile material in the primary and secondary (cuts down on blast and fission products) and replacing any ^{238}U in the bomb with non-fissionable materials. The purpose behind a neutron bomb was to act as a deterrent to an enemy with lots of armored vehicles (such as the former Soviet Union) from attacking friendly countries (Western Europe). The intense neutron radiation would easily penetrate the armor but be absorbed by the humans inside them. Supposedly this is a great way to disable a large armored invasion near a friendly city. As clean and tidy as they sound, they would still cause a fair amount of local damage and leave a good deal of radioactive contamination. They were criticized because some have suggested that such a weapon might be too tempting to use in an otherwise conventional war, which could then escalate the conflict to a full-fledged nuclear war.

More recently "**bunker buster**" nukes have been proposed. A nuclear weapon would be placed inside a special casing—designed to hit the ground at high speeds and penetrate the surface to the depth of an underground bunker before detonating. Such weapons exist, with conventional explosives. They are designed like a spear (long and narrow), to maximize ground penetration before detonation. However, if a bunker is built deep enough, and with sufficiently thick walls and ceiling (hardened), it could easily survive a conventional bunker buster bomb. To destroy such a bunker, some have proposed replacing the conventional explosives with a nuclear warhead. Unfortunately, nuclear warheads have more girth, making the whole weapon too wide to penetrate very deep underground. This means the weapon will detonate near the surface and will cause radioactive fallout above the surface, possibly spreading to nearby areas (civilian populations, neighboring countries, etc.). It also means that a nuclear bunker buster may not be any more effective than a conventional one at destroying its target. Again, some have suggested that developing such a weapon could make it easier for a conventional war to escalate to massive nuclear destruction.

Discussion of another (non-nuclear) kind of **dirty bomb** has sprung up since the late 1990s. Back then, Chechen rebels unsuccessfully tried three times to combine radioactive materials and chemical explosives in their struggle against Russia. The technical name for this type of weapon is "radiological dispersal device" or RDD. The chemical explosive could be as simple as the powder gleaned from fireworks, or as powerful as the fertilizer/nitromethane mixture used in the 1995 Oklahoma City bombing. Generally speaking, the purpose behind setting off a dirty bomb is not to cause widespread death and destruction, but rather to cause panic and emotional distress, as well as economic dislocation. It is very unlikely that anyone would die from radiation exposure resulting from a dirty bomb, but it would certainly play into the general public's fear of ionizing radiation. It would also contaminate an area that may have to remain evacuated until it is cleaned-up.

4.3 NUCLEAR FORENSICS

In an era when potential development of nuclear weapons may be a pretext to war, it might be a good idea to have some more conclusive evidence of weapons production before invading. **Nuclear forensics** involves the analysis of radioactive materials or environmental samples containing radioactive materials and/or their byproducts. The goals of analysis are to determine what the material might be, where it came from, and how it is being used. The main areas of interest for nuclear forensics are smuggling of nuclear materials, nuclear terrorism (e.g. dirty bombs), and nuclear weapons proliferation. Perhaps the most dramatic illustration of nuclear forensics in popular culture was in the 1991 Tom Clancy novel (and 2002 movie) *The Sum of All Fears*. In this tale, a terrorist group sets off a small nuclear bomb at a football game in the hopes of starting a nuclear war between the United States and Russia. Fortunately, nuclear scientists quickly analyze the fallout, and, from certain nuclide ratios, determine the nuclear material in the bomb originated in the U.S.

From the moment uranium ores are taken from the ground, they have a specific signature. The percentage uranium in the ore and the other minerals/ores present are often characteristic of a specific site. After mining, the ore is processed in ways that again vary from place to place, creating material that is unique in its precise elemental and nuclidic composition. The material can then go into a reactor, which will again create a characteristic composition because different reactor designs cause the fuel to burn a little differently. The spent fuel may also be reprocessed, which again makes its exact composition unique, depending on the reprocessing techniques used. The material may also go into a weapon, but again be subjected to somewhat unique processes depending on which country is preparing the material. Finally, the material may sit for a time at any stage above, and decay, again creating a material with unique composition. One of the main tasks of a nuclear forensic analyst is therefore a detailed analysis of the exact composition of any material under investigation.

For example, weapons-grade uranium is at least 80% ^{235}U and weapons-grade plutonium must be greater than 93% ^{239}Pu. The main "impurities" are ^{238}U and ^{240}Pu, respectively. Both impurities are difficult to remove because chemical processes cannot separate isotopes, therefore more cumbersome physical processes (like gaseous diffusion) must be used. As a result, different countries adopted different standards for the percentage of fissile materials in their weapons, providing the analyst with a signature. The United States, for example, uses 93% ^{235}U in uranium-based weapons and keeps ^{240}Pu to only 5–6% in its plutonium-containing bombs.

Along with analyzing the remnants of the fissile material in a weapon, there are other ways to learn about a nuclear bomb after it has blown up. The ratio of the fissile material to fission products can indicate the yield of the device. A higher yield would indicate a more sophisticated design, suggesting the weapon came from the arsenal of one of the more established nuclear powers. Since a burst of neutrons is created at the moment of the explosion, some may interact with the immediate surroundings. Thus, it may be possible to determine whether the bomb was in a wooden or metal crate when it blew. Application of nuclear forensic analysis can also determine the origin of the radioactive material in a dirty bomb (RDD). This material could come from the nuclear fuel cycle, or (more likely) from a hospital, clinic, or industrial source.

Nuclear forensics is also useful in illuminating details behind the smuggling of nuclear materials. Since one of the hardest parts of building a nuclear weapon is obtaining the fissile material, attempts to smuggle such material have become more common in popular fiction as well as in the real world. Most such cases are frauds—bad guys pretending to have weapons-grade uranium or plutonium, but in reality only having ^{60}Co, ^{137}Cs, depleted uranium, natural uranium, or ^{241}Am. It's still radioactive, but not useful for a nuclear weapon. From 1985 to 2011 there were more than 800 seizures of radioactive materials. Of these, fewer than 3% have involved weapons-grade stuff. Even then, they were in quantities too small to build a single device. One positive result of these seizures has been increasing accountability and security for all radioactive materials. Also reassuring is that there were no seizures from 2011 to 2022.

Let's say that you live in a country that neighbors one that is highly secretive, and you think they might be building a uranium enrichment facility, nuclear reactors, a spent fuel reprocessing center, and/or nuclear weapons. They won't answer your calls, so you resort to spying. If you have access to soil, plants, animals, air, or water samples that are leaving your neighbor's country, you might be able to figure out what's going on through nuclear forensics. Once operational, these facilities will have emissions that can tell you what they're up to. Certain elements have a tendency to stick to soils, or bioaccumulate in flora and/or fauna. For example, iodine (radioactive, if your neighbor is up to no good) will collect in the thyroid gland of mammals.

If a particular sample contains a nuclide with radioactive daughters that have sufficiently long half-lives, then knowledge of nuclear equilibria (Section 2.6) can be used to figure out how "old" it is, i.e., how long ago the parent was isolated. This only works so long as the ratio of parent to daughter is still changing, i.e., it has not yet reached equilibrium. Once at equilibrium, the ratios no longer change, and it's no longer possible to determine how old the sample is.

Example 4.1 A sample of ^{235}U provided for analysis has a ^{231}Pa to ^{235}U atomic ratio of 1.89×10^{-8}. How long ago was the ^{235}U isolated?

This is an example of secular equilibrium since the half-life of the parent (^{235}U) is very long (7.04×10^8 a) relative to our time of observation (likely just a few hours). We're hoping this sample is not yet at equilibrium. To find out, we'll use Equation 2.11 to calculate the ^{231}Pa/^{235}U ratio at equilibrium.

$$\frac{N_{\text{Pa-231}}}{N_{\text{U-235}}} = \frac{t_{1/2(\text{Pa-231})}}{t_{1/2(\text{U-235})}} = \frac{3.28 \times 10^4 \text{ a}}{7.04 \times 10^8 \text{ a}} = 4.66 \times 10^{-5}$$

It looks like we're not even close to equilibrium! We can now proceed, using Equation 2.9 to determine how much time has elapsed. First, let's rearrange it so we can use the ratio given in the problem.

$$N_{\text{Pa-231}} = \frac{\lambda_{\text{U-235}}}{\lambda_{\text{Pa-231}}} N_{\text{U-235}} \left(1 - e^{-\lambda_{\text{Pa-231}} t}\right) \qquad \frac{N_{\text{Pa-231}}}{N_{\text{U-235}}} = \frac{t_{1/2(\text{Pa-231})}}{t_{1/2(\text{U-235})}} \left(1 - e^{-\lambda_{\text{Pa-231}} t}\right)$$

$$1.89 \times 10^{-8} = \frac{3.28 \times 10^4 \text{ a}}{7.04 \times 10^8 \text{ a}} \left(1 - e^{-\left(\frac{\ln 2}{3.28 \times 10^4 \text{ a}}\right) t}\right)$$

$$4.06 \times 10^{-4} = 1 - e^{-\left(\frac{\ln 2}{3.28 \times 10^4 \text{ a}}\right) t}$$

$$-4.06 \times 10^{-4} = -\left(\frac{\ln 2}{3.28 \times 10^4 \text{ a}}\right) t$$

$$t = 19.2 \text{ a}$$

QUESTIONS

4.1 Define the following: nuclear fission, fissile, fission chain reaction, and critical mass.
4.2 What is the fundamental difference between fossil fuel and nuclear power plants?
4.3 Briefly explain what moderators and control rods do in nuclear reactors.
4.4 What is the difference between a PWR and a BWR?
4.5 Explain why enrichment requires a physical, rather than chemical, process.
4.6 Which would diffuse faster: ^{1}H$_2$O or ^{2}H$_2$O? How much faster?
4.7 Briefly describe how a gas centrifuge enriches uranium.
4.8 ^{233}U can be produced in nuclear reactors in a manner similar to the production of ^{239}Pu. What naturally occurring element must be added to a reactor to produce ^{233}U? Write out all nuclear reactions and decays that are part of ^{233}U production.
4.9 Define the following: depleted uranium, spent fuel, HEU, and reprocessing.

4.10 In the event of another nuclear power plant disaster like Chernobyl, state and local politicians have distributed KI tablets to people living near such facilities. The idea is to take the pill to saturate the thyroid with iodide, thereby preventing thyroid uptake of the radioactive iodide released by the accident. Give two reasons why the distribution of these pills could be considered a political placebo, rather than populace protection. Give one reason why distribution of the pills may help protect public health.

4.11 Airborne radioactive materials released from the Fukushima nuclear plants in the spring of 2011 reached the west coast of the United States about two weeks later. This caused a minor panic as people rushed out to buy potassium iodide pills (see Question 4.10 above) even though radiation levels were only a small fraction above normal background. If you had been living in California at that time, would you have taken KI? Briefly explain.

4.12 Assuming all the radioactive material released by the Chernobyl accident was ^{137}Cs, what mass (kg) was released?

4.13 Calculate the percent abundance of ^{235}U on Earth 1.7 billion years ago. Assume the uranium is only composed of ^{235}U and ^{238}U.

4.14 If you were either for or against nuclear power before reading this chapter, give four to five bullet points outlining the opposite viewpoint. If you had no opinion, flip a coin—heads, write in favor, tails opposed.

4.15 From a public safety perspective, does it make sense to replace nuclear with natural gas power plants?

4.16 What does a tamper do in a nuclear weapon? A neutron initiator?

4.17 Do you think that unilateral nuclear disarmament by the United States would be a positive step for the nonproliferation of nuclear weapons? Explain.

4.18 What mass of ^{239}Pu was present in the bomb dropped on Nagasaki? What happened to the plutonium that did not fission?

4.19 What is an RDD? What are its likely effects?

4.20 What would the ratio of ^{231}Pa to ^{235}U be in a sample of ^{235}U that was last purified 27.4 years ago?

4.21 Homeland security agents intercepted an 8.96 kg chunk of metal from a nefarious foreign organization. The sample is alleged to be ^{235}U. Your analysis shows an alpha count rate about 10 times lower than would normally be expected for bare uranium metal. The density is also a little low for uranium. Finally, the energies of the most prominent alpha particles observed are 4.197 and 4.147 MeV. What can you report to the homeland security agents? What tests would you recommend be performed next?

5 Radioactive Decay
The Gory Details

We were briefly introduced to the most common decay modes in Section 1.3. This chapter explores them in more depth and discusses some of the more unusual ways unstable nuclides can decay.

5.1 ALPHA DECAY

This form of decay is most commonly observed for nuclides heavier than ^{208}Pb. The following equation can generically represent alpha decay.

$$\ce{^A_Z X} \rightarrow \ce{^{A-4}_{Z-2} Y^{2-}} + \ce{^4_2 He^{2+}}$$

Basically, the nucleus is blowing chunks; in this case a ^{4}He nucleus, otherwise known as an alpha particle. The parent is just too big, and the best way it can move toward a stable nuclide is to get rid of four of its nucleons. In terms of stuff spit out by an unstable nucleus (like beta particles, positrons, and photons), alpha particles are quite large. Notice that, in the equation above, charge is also accounted for. Assuming all of the electrons stay on the daughter, it should have a 2– charge because two protons just left the nucleus. The 2+ charge on the alpha particle provides charge balance. It is not always necessary to indicate the electrical charges in decay equations, so long as they are understood.

Alpha decay can also be represented using a somewhat simpler notation.

$$\ce{^A_Z X} \xrightarrow{\alpha} \ce{^{A-4}_{Z-2} Y}$$

Because of the relatively large size and high charge of the alpha particle, it has a high probability of interacting with matter. In other words, it is very easy to shield alpha radiation. Alpha particles are also *monoenergetic*, because the parent and daughter have very specific (quantized nuclidic) energy states. This means that if the energy states of the parent and daughter are always the same, the alpha particle will always have the same energy. The transition from parent to daughter always involves a specific amount of energy, as we've already seen in drawing simple decay diagrams.

The fly in this ointment is the fact that the daughter isn't always produced in the same energy state. If the daughter is initially formed in an excited state, then the alpha particle emitted in that decay will have less energy. The excited states are also quantized, so even if several different alpha energies are possible for the decay of a particular nuclide, only those energies will be observed. How can this be represented in a decay diagram? It'll take a while to answer that. First, let's review a bit of what we learned in Chapter 3 by looking at the decay of ^{246}Cm.

DOI: 10.1201/9781003303787-5

Example 5.1 When ^{246}Cm decays, alpha particles with energies of 5.386 MeV and 5.342 MeV are observed. Gamma photons with an energy of 44.5 keV are also observed. Write out the decay equation and calculate the total energy of decay.

The decay equation (without charges) is written below.

$$^{246}_{96}\text{Cm} \rightarrow \, ^{242}_{94}\text{Pu} + \, ^{4}_{2}\text{He}$$

Calculate the total energy of decay, just like in Chapter 3 (mass values are from Appendix A).

$$\left[246.067224 \text{ u} - \left(242.058743 \text{ u} + 4.002603 \text{ u}\right)\right] \times \frac{931.5 \text{ MeV}}{\text{u}} = 5.475 \text{ MeV}$$

Notice that the total energy of decay is greater than the energy values of both alpha particles observed. This should be expected, as part of the total energy of decay goes into the kinetic energy of the daughter. How much? The answer has to do with momentum. Momentum is mass times velocity and is conserved when a projectile is shot out of a larger object. When a bullet is shot from a gun, there is a certain amount of recoil (or kick) in the gun. The gun wants to move in the opposite direction of the bullet. The amount of recoil will depend on the masses of the bullet and gun. The momentum for each will be the same, but the velocity will be higher for the bullet because it has a lower mass. The same is true for alpha decay (Figure 5.1) where the alpha particle is the bullet and the daughter nucleus is the gun.

The alpha particle moves a lot faster because it has a lower mass. Since kinetic energy is ½ times the mass times the velocity squared, the alpha particle will also have a greater kinetic energy than the daughter. Classical physics tells us that the kinetic energy of the alpha particle (or any smaller object shot out of a larger object) can be calculated using Equation 5.1:

$$K_\alpha = K_T \left(\frac{M}{M+m}\right) \tag{5.1}$$

where M is the mass of the daughter, m is the mass of the alpha particle, and K_T is the energy of transition. The energy of transition is the energy difference between the parent nuclide, usually in the ground state, to the resulting daughter state (could be excited or ground). If the resulting daughter state is the ground state, then K_T is equal to the total energy of decay and K_α represents the maximum energy value for an alpha particle for this decay.

Even though this formula comes from classical physics and is meant to be applied to things like firing balls from a cannon, it works well enough in the subatomic universe. However, you should keep in mind that it is a bit of an approximation, as the subatomic universe follows different rules than our (macro) world.

^{246}Cm ^{242}Pu ^{4}He

FIGURE 5.1 Recoil in the alpha decay of ^{246}Cm.

The kinetic energy of the daughter (or any larger object) can also be determined using the following closely related formula:

$$K_F = K_T \left(\frac{m}{M+m} \right) \tag{5.2}$$

Example 5.2 Using data from Example 5.1, calculate the maximum energy of the alpha particle and the resulting recoil energy of the daughter in the decay of ^{246}Cm.

The maximum energy alpha particle can only result from decay straight to the ground state of the daughter (^{242}Pu). Therefore, K_T is equal to the total energy of decay, which we've already calculated as 5.475 MeV.

$$K_\alpha = 5.475\,\text{MeV} \left(\frac{242}{242+4} \right) = 5.386 \text{ MeV}$$

$$K_F = 5.475 \text{ MeV} \left(\frac{4}{242+4} \right) = 0.0890 \text{ MeV} \approx 89 \text{ keV}$$

Using mass numbers for the masses is sufficient for the level of accuracy needed. Remember, these formulas are not exact for subatomic cannons. If you have any doubts about using mass numbers, go ahead and replace them with exact masses and see if there is a significant effect on the results.

Notice that the two energy values determined in Example 5.2 add up to the total energy of decay. Energy is conserved as it is partitioned between the daughter and the alpha particle.

$$K_T = K_\alpha + K_F = 5.386 \text{ MeV} + 0.089 \text{ MeV} = 5.475 \text{ MeV}$$

What does "maximum energy of the alpha particle" mean? It means that the total energy of decay is exactly partitioned between the alpha particle and the daughter. In other words, a ^{246}Cm decay that emits a 5.386 MeV alpha particle corresponds to a transition from the ground state of ^{246}Cm to the ground state of ^{242}Pu. If ^{246}Cm emits an alpha particle with anything less than the maximum energy, then the transition must end at an excited state of ^{242}Pu, not the ground state. We're finally ready to draw the decay diagram for the decay of ^{246}Cm. It is shown in Figure 5.2.

FIGURE 5.2   Decay diagram for ^{246}Cm.

We've already confirmed that the 5.386 MeV alpha emissions correspond to decay from ground state ^{246}Cm to ground state ^{242}Pu. The lower energy alpha particle emitted must therefore lead to an excited state of ^{242}Pu. In fact, it should be roughly 44 keV higher in energy than the ground state since that is the difference in energy between the two alpha emissions.

$$5.386 \text{ MeV} - 5.342 \text{ MeV} = 0.044 \text{ MeV} = 44 \text{ keV}$$

Fortuitously (it would seem), this is very close to the energy of the gamma photon that is also observed in this decay (44.5 keV). This strongly suggests that it is the observed gamma photon following the emission of a 5.343 MeV alpha particle from ^{246}Cm.

To properly calculate the difference in energy between the ground and the excited states of ^{242}Pu in Figure 5.2, we must first calculate the energy of transition from ground state ^{246}Cm to excited state ^{242}Pu. First rearrange Equation 5.1 so that it is solved for K_T, and then plug in the energy of the alpha particle for this transition.

$$K_\alpha = K_T \left(\frac{M}{M+m} \right) \qquad K_T = K_\alpha \left(\frac{M+m}{M} \right)$$

$$K_T = 5.343 \text{ MeV} \times \left(\frac{242+4}{242} \right) = 5.431 \text{ MeV}$$

The difference in energy for the nuclidic states of ^{242}Pu is the same as the difference in the two K_T values.

$$5.475 \text{ MeV} - 5.431 \text{ MeV} = 0.044 \text{ MeV} = 44 \text{ keV}$$

Again, we have good agreement with the experimentally observed value of 44.5 keV. Keep in mind that all energy values determined above came from experiments (were empirically determined), and all have some error associated with them. Small differences in calculated and observed values can usually be ignored in drawing decay diagrams.

Finally, notice that the arrows from ^{246}Cm do not quite reach the two nuclidic states of the ^{242}Pu daughter. This is to show that the energy of each transition is greater than the energy of the individual alpha particles (the remainder is daughter recoil). Now you know why the alpha arrow didn't touch the daughter nuclidic state in Figure 3.3. Note that other forms of decay in Figure 3.3 all have arrows touching their daughter nuclidic states.

Example 5.3 Show that mass and energy are collectively conserved in the alpha decay of ^{246}Cm.

Before decay takes place, all we have is the mass of the parent. If mass and energy are collectively conserved, then we need to end up with most of that mass plus enough energy to make up the mass that is lost.

mass of parent = mass of daughter + mass of ^{4}He + energy of ^{4}He + recoil energy

$$246.067224 \text{ u} = 242.058743 \text{ u} + 4.002603 \text{ u} + \left(5.386 \text{ MeV} + 0.089 \text{ MeV} \right) \frac{\text{u}}{931.5 \text{ MeV}}$$

$$= 246.067224 \text{ u}$$

Excellent agreement! Note that this only works with the energy for the maximum energy alpha particle. If we wanted to use the 5.342 MeV alpha particle, we'd also need to include the 44.5 keV gamma ray on the right-hand side of the equation. We'd also have to calculate recoil for this transition—there's no guarantee it'll be the same as with the higher energy alpha emission.

Let's take a look at how the N/Z ratio changes in the decay we've been studying.

$$^{246}_{96}\text{Cm}_{150} \xrightarrow{\alpha} {}^{242}_{94}\text{Pu}_{148}$$

$$N/Z = \quad 1.56 \qquad\qquad 1.57$$

It *increases* while forming a lower Z nuclide! As you remember from Figure 1.4, the curvature in the belt of stability means that it is generally better to form a smaller nuclide with a lower N/Z ratio. After a series (or even just one or two) of alpha decays, the ratio will get to the point where beta decay is necessary, because the N/Z ratio is now too high. Examination of Figure 1.7 illustrates this point nicely. The increase in N/Z in alpha decay requires each of the three naturally occurring decay series to be a mix of alpha and beta decay. Notice also that alpha decay does not affect the even or odd nature of the number of protons or neutrons. Both parent and daughter in our example are even–even (*ee*). If the parent had been *eo*, the daughter would also be *eo*.

One final point about alpha decay: Just because a nuclide is larger than ^{208}Pb, doesn't mean it'll decay by alpha. Large nuclides with poor N/Z ratios *can* also decay by beta, electron capture, or positron. In some cases, branched decay (alpha plus one or more flavors of beta) is observed. Likewise, a nuclide doesn't have to be huge to decay via alpha emission. ^{144}Nd (and a few others below ^{208}Pb) decays via alpha, and ^{8}Be splits into two alpha particles when it decays.

5.2 BETA DECAY

There are some problems with the rather simple way beta decay has been represented in this book. Don't worry; our representation has just been incomplete, not inaccurate. As represented so far, beta decay violates three conservation laws:

1. The Law of Conservation of Particles. The beta particle is created by the nucleus, but from what? Flipping a quark? A particle cannot be created all by itself; an antiparticle must also be created at the same time.

2. The Law of Conservation of Angular Momentum. Angular momentum can be thought of simply as spin. Every subatomic particle and every nucleus has a spin value. It can be a whole number, or a whole number divided by 2. This law states that the difference in spin between reactants and products in a nuclear equation must be a whole number. Let's take a look at the spin values of the known participants in the decay of ^{14}C.

$$^{14}\text{C} \rightarrow {}^{14}\text{N} + {}_{-1}\text{e}$$

spin values: 0   1   ½

The sum of the spins on the product side is 1½. Since the spin of the parent is 0, the change in spin will also be 1½, not a whole number, and therefore is in violation of the law. Written in this way, every beta decay will appear to violate the Law of Conservation of Angular Momentum.

3. The Law of Conservation of Energy. Energy can be converted into different forms of energy or even into matter, but it can't just go away or be created from nothing. The problem with

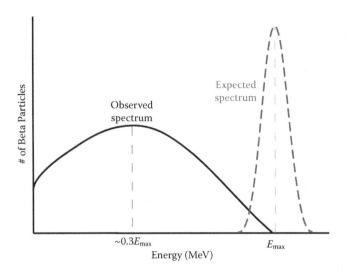

FIGURE 5.3 Expected (dashed curve) and observed (solid curve) energy spectra for beta particles.

beta decay is that the beta particles are not monoenergetic like in alpha decay. A broad energy distribution is observed for all beta particles, instead of a relatively narrow peak, as illustrated in Figure 5.3. It is not possible for the same nuclear transition to take place, but different energies be produced.

The answer to all of our problems would be an antiparticle that is emitted at the same time as the beta particle. This antiparticle should have a spin value of ½, and it should share decay energy with the beta particle in a relatively random fashion. Finally, this antiparticle should have a very low probability of interacting with matter, as it has proven very difficult to detect. An antineutrino ($\overline{\nu}$) has all these attributes. It has no charge, very low mass, and an extremely low probability of interacting with matter. We can now write a complete equation for the decay of ^{14}C.

$$^{14}_{6}C \rightarrow \, ^{14}_{7}N + \, _{-1}e + \overline{\nu}$$

We can also write a generic beta decay equation (including charges) as:

$$^{A}_{Z}X \rightarrow \, ^{A}_{Z+1}Y^{+} + \, _{-1}e^{-} + \overline{\nu}$$

The positive charge on the daughter (Y) results from it having the same number of electrons as the parent (X), yet one more proton. The negative charge on the beta particle balances this out. In general, beta decay equations will not be written in this text with explicit charges or antineutrinos. The reader should understand they are implied. The main reason for leaving the antineutrino off is that it is so difficult to detect, and does not affect normal counting of radioactive material.

Beta decay equations can also be written using a similar shorthand as we saw with alpha decay.

$$^{A}_{Z}X \xrightarrow{\ \beta^{-}\ } \, ^{A}_{Z+1}Y$$

The ultimate example of beta decay is the decay of free neutrons. On their own, neutrons have a half-life of about 10 minutes. It is a good thing they are stable when incorporated into nuclei!

$$_0^1 n \rightarrow \, _1^1 p + \, _{-1}e^- + \bar{\nu}$$

All of the stuff in this equation has a spin value of ½. Is the Law of Conservation of Angular Momentum violated in this decay?

The maximum beta energy (E_{max}) is given in Appendix A and on charts of nuclides, and is often quoted as the energy of the beta particle. Theoretically, this could only happen if the energy of the antineutrino emitted in the same decay is zero. Examination of Figure 5.3 shows that beta particles never have this energy, since it always shares some with the antineutrino. According to Figure 5.3, the most probable energy for a beta particle is ~$0.3E_{max}$.

We've already seen how to draw an alpha decay diagram that incorporates a gamma emission. The same can be done with beta decay. In fact, it is very unusual to observe alpha or beta decay without gamma emission.

Example 5.4 ^{47}Sc decays with the emission of two beta particles with E_{max} values of 0.439 MeV and 0.600 MeV. A gamma ray with an energy of 159.4 keV is also observed. Draw a decay diagram that accounts for all this information.

The complete decay equation is:

$$_{21}^{47}Sc \rightarrow \, _{22}^{47}Ti^+ + \, _{-1}e^- + \bar{\nu}$$

The total energy of decay is calculated in the usual manner, after looking up the atomic masses of parent and daughter (Appendix A).

$$\left(46.952408\,u - 46.951763\,u\right) \times \frac{931.5\,\text{MeV}}{u} = 0.601 \text{ MeV}$$

Since we are using the atomic masses of the parent and daughter nuclides, we don't need to add in the mass of the beta particle—it is already included in the atomic mass of ^{47}Ti (remember it is formed as a monocation). The mass of the antineutrino is so small, it is negligible, so it does not need to be included in the calculation.

The difference in energy between the two beta particles is close to the energy of the observed gamma ray.

$$0.600 \text{ MeV} - 0.439 \text{ MeV} = 0.161 \text{ MeV} = 161 \text{ keV}$$

We're ready to draw! The 0.600 MeV beta particle is very close in energy to the total energy of decay, and therefore corresponds to a transition between the ground states of the parent and daughter. The sum of the lower energy beta particle and the energy of the gamma ray is also pretty close to the total energy of decay, so odds are good that the 0.439 MeV beta emission leads to an excited state of ^{47}Ti which is 0.1594 MeV higher in energy than its ground state. The complete diagram is shown in Figure 5.4.

The most common decay pathways for radioactive nuclides do not always involve a direct transition between ground states, and can involve more than one gamma emission. The most prominent beta particle observed in the decay of ^{43}K has a maximum energy of 0.83 MeV, but

FIGURE 5.4 Decay diagram for ⁴⁷Sc.

FIGURE 5.5 Partial decay diagram for ⁴³K.

its total energy of decay is 1.82 MeV. As part of this pathway, gamma photons with energies of 372.8 keV and 617.5 keV are also observed. These three emitted energies add up to the total energy of decay.

$$0.83 \text{ MeV} + 0.3728 \text{ MeV} + 0.6175 \text{ MeV} = 1.82 \text{ MeV}$$

The primary decay pathway for ⁴³K *could* therefore look like Figure 5.5. Based on the information provided, we can't decide on the order of the two gamma emissions. The only way to know for sure is to consult a higher authority. A convenient source is the interactive chart of the nuclides posted by the International Atomic Energy Agency at www-nds.iaea.org. They provide a decay diagram, which shows that the decay of ⁴³K involves a few other minor pathways as well as affirming Figure 5.5 as correct for the major pathway.

A significant part of the discussion of alpha decay earlier in the chapter was the issue of recoil. Thus far into our look at beta decay, it hasn't been mentioned. Are the calculations above done properly? Let's find out.

Example 5.5 Calculate the recoil energy resulting from beta decay for a free neutron.

This is a good example because neutrons are relative lightweights and the total energy of decay is equal to $E_{\max}$ (0.782 MeV). If recoil is ever observed in beta decay, we should see it here. We can use the same equations as for alpha decay, except m is now the mass of an electron.

$$K_{\text{F}} = K_{\text{T}} \left(\frac{m}{M+m} \right) = 0.782 \text{ MeV} \times \frac{0.000549 \text{ u}}{1.01 \text{ u}} = 0.000425 \text{ MeV}$$

The recoil energy in beta decay is very small compared to the total energy of decay, and can easily be neglected in beta decay energy calculations. The reason beta decay is so different from alpha decay is because the mass of an electron is so much smaller than the mass of an alpha particle. A smaller bullet produces less recoil.

5.3 POSITRON DECAY

Just like beta (minus) decay, positron decay has been simplified in Chapters 1–3. Without the addition of a neutrino to the product side of the decay equation, positron decay would also violate the same conservation laws as beta decay. The complete, generic representation of positron decay is:

$$_Z^A X \rightarrow {}_{Z-1}^A Y^- + {}_{+1}e^+ + \nu$$

A neutrino is formed here, rather than the antineutrino formed in beta decay. This is because the positron itself is an antiparticle; therefore, a particle also needs to be created to properly conserve particles. The positron is the corresponding antiparticle to the electron. Every particle (like protons, neutrons, neutrinos) has a corresponding antiparticle. These antiparticles are collectively known as antimatter. That's the fuel for starships in the stunningly successful *Star Trek* sagas. It is also used to provide the devastatingly explosive force in their photon torpedoes. As is often the case, there is some basis in fact here. When antimatter comes into contact with (regular) matter, they annihilate each other, forming energy. When an electron and a positron annihilate each other, two *annihilation photons* are formed, each with an energy of 0.511 MeV. Let's do the math.

Example 5.6 Calculate the energy released from a positron/electron annihilation.

The equation is:

$$_{+1}e + {}_{-1}e \rightarrow 2\gamma$$

The electron and the positron both have a mass of 5.486×10^{-4} u, and the photons have no mass. Therefore $2(5.486 \times 10^{-4}$ u$)$ is destroyed, and the energy produced is:

$$2 \times \left(5.486 \times 10^{-4}\ u\right) \times \frac{931.5\ \text{MeV}}{u} = 1.022\ \text{MeV}$$

This energy is divided equally between the two annihilation photons, meaning each will be 0.511 MeV. This is a lot of energy released by a small amount of matter. It wouldn't take much antimatter to make a pretty powerful explosive. Interestingly, the two annihilation photons are emitted approximately 180° apart. This forms the basis of positron emission tomography (PET), which is discussed in more detail in Chapter 9. Simultaneous detection of the two photons from annihilation allow for stunning 3-D images of people's innards for medical diagnostic procedures.

Let's get back to the generic positron decay at the beginning of this section. Paying close attention to charge, the daughter (Y) is formed as a monoanion (–1 charge), assuming it has the same number of electrons as the parent. This negative charge is balanced by the positive charge of the positron, so charge is balanced in this decay.

What about mass balance? Since the daughter is produced as an anion *and* a positron is emitted, the total mass on the product side is equal to the atomic mass of the (neutral) daughter plus the mass of two electrons. Compare this with beta decay. There, the difference in mass is simply the difference in atomic masses of the parent and daughter nuclides.

This means we have some additional energy accounting to do in positron decay. The total energy of decay is equal to the (kinetic) energy of the positron plus the energy to "create" two electrons (1.022 MeV, same as you get when they are destroyed!). In beta decay, the total energy of decay can equal the maximum possible beta energy for the transition, something that can never be true in positron decay. In order to calculate maximum possible positron energy for a particular decay, we have to account for the extra mass.

$$\left[\text{at. mass X} - \left(\text{at. mass Y} + \text{mass }_{-1}e + \text{mass }_{+1}e\right)\right] \times \frac{931.5 \text{ MeV}}{u} \qquad (5.3)$$

Example 5.7 ^{18}Ne decays with the emission of two positron particles with E_{max} values of 3.42 MeV and 2.38 MeV. A gamma ray with an energy of 1.04 MeV is also observed. Draw a decay diagram that accounts for all of this information. Also calculate the maximum possible positron energy for this decay.

$$^{18}_{10}\text{Ne} \rightarrow ^{18}_{9}\text{F} + _{+1}e$$

The total energy of decay can still be calculated in the usual manner—from the differences in masses between parent and daughter.

$$\left(18.005697\,u - 18.000938\,u\right) \times \frac{931.5 \text{ MeV}}{u} = 4.433 \text{ MeV}$$

This value differs somewhat from what is recorded in Appendix A. Should we be worried? No way, we did the math correctly—the difference is likely due to experimental errors. The maximum energy for a positron is:

$$\left[18.005697\,u - \left(18.000938\,u + 5.486 \times 10^{-4}\,u + 5.486 \times 10^{-4}\,u\right)\right] \times \frac{931.5 \text{ MeV}}{u}$$
$$= 3.411 \text{ MeV}$$

Notice it matches one of the E_{max} values given in the problem quite closely! This means the 3.42 MeV positron gives ^{18}F in its ground state. We can also guess that the 1.04 MeV gamma photon is emitted following the 2.38 MeV because the difference in the E_{max} values for the two positrons is 1.04 MeV.

$$3.42 \text{ MeV} - 2.38 \text{ MeV} = 1.04 \text{ MeV}$$

As a final check, the maximum energy positron plus the energy to "create" two electrons should equal the total energy of decay.

$$3.42 \text{ MeV} + 1.022 \text{ MeV} = 4.44 \text{ MeV}$$

Close enough. The decay diagram is given in Figure 5.6. The short vertical line from ^{18}Ne represents the 1.022 MeV required by all positron decays. If the total energy of decay is less than 1.022 MeV, positron decay is not possible.

What about recoil? The results should be the same as with regular beta, so the answer is no, it is insignificant. Also like regular beta decay, the shorthand shown below can represent positron decay.

$$^{A}_{Z}\text{X} \xrightarrow{\beta^+} ^{A}_{Z-1}\text{Y}$$

FIGURE 5.6   Decay diagram for ^{18}Ne.

5.4   ELECTRON CAPTURE

Electron capture probably seems like the strangest form of radioactive decay ever, as it is the only one that is not strictly a nuclear change since it requires an orbital electron. Its generic representation is quite similar to positron decay:

$$^A_Z X + \, ^0_{-1}e \rightarrow \, ^A_{Z-1}Y + \nu$$

Remember that the electron on the left side of the arrow is an orbital electron. The neutrino on the product side balances the loss of a particle on the reactant side, and it gets essentially all of the energy of transition. Since the parent uses one of its own electrons to convert a proton to a neutron, the daughter is formed with one less proton and one less electron. Therefore the daughter is neutral, not a cation or an anion. Because the electron is not originally part of the nucleus, some prefer to write electron capture equations with the following shorthand:

$$^A_Z X \xrightarrow{\ \ EC\ \ } \, ^A_{Z-1}Y + \nu$$

Example 5.8 Draw a decay diagram for ^{72}Se. It decays via electron capture with a total energy of decay of 0.34 MeV. A gamma ray with an energy of 46 keV is observed in all decays.

Since the gamma photon is observed in all decays, the electron capture must lead to an excited state of the daughter, which then decays to the ground state by spitting out the gamma (Figure 5.7).

It can sometimes be difficult to decide how to draw the electron capture arrow. In Example 5.8 it is unambiguous, but if information from a printed chart were used, you wouldn't know that the gamma photon is observed in each decay.

FIGURE 5.7   Decay diagram for ^{72}Se.

Positron and electron capture decays both do the same thing. They both produce a daughter with one less proton and one more neutron than the parent. Can we tell if a particular nuclide will prefer to decay by electron capture or positron emission? Sort of. Electron capture is more likely with high Z nuclides, while positron decay is more likely for light nuclides, but that's about the best we can do. For most proton-rich nuclides, either or both forms of decay are observed. The only time we can be definitive is when the total energy of decay is less than 1.022 MeV. In these cases, positron decay cannot occur, leaving electron capture as the only decay option.

To better understand electron capture, we'll need to know a bit about orbital electrons. Electrons occupy a volume of space (orbital) around the nucleus that is (more or less) defined by the solution to the mathematically complex Schrödinger equation. The solution generates a set of four numbers that are unique for each electron in an atom. These numbers are called quantum numbers, and individually represent main energy level, orbital shape, orbital orientation, and electron spin. We'll only be concerned with the quantum number that describes the main energy level. This quantum number is known as the principal quantum number, and is represented by n. It has positive integral values (1, 2, 3, …). Since we're not concerned with orbital shape or orientation, we can represent the different energy levels (shells) as concentric circles around the nucleus. This type of atomic model is called the Bohr model, named after the Danish physicist Niels Bohr who first proposed it. An example is pictured in Figure 5.8. The energy levels $n = 1$, 2, and 3, are often represented by the letters K, L, and M, respectively. The letter designations are older, but are more commonly used in medical applications, so we will continue their use in this text. The L shell is further subdivided into three subshells (L_1, L_2, L_3), and M into five. There is only one subshell for the K shell, so it is simply represented as K.

Which electrons crash into the nucleus? Electron capture is most likely to take place with a K shell (K capture) electron. This makes sense, as the K shell is the lowest energy shell, and they spend the most time closest to the nucleus. Electron capture from the L (L capture) or M (M capture) shells is also possible, but not probable. For example, K capture happens in ~90%, L capture in ~9%, and M capture in ~1% of all electron capture decays of ^{41}Ca. L and M capture increase in probability with Z, which can be rationalized by the fact that all electron shells move closer to the nucleus with increasing Z. The same rationale can be applied to the relative probabilities of positron and electron capture decays. A larger Z means the K shell electrons are closer to the nucleus, so electron capture tends to be more popular with heavier nuclides. In reality, it is a bit more complex. For instance, a large value for the total energy of decay tends to favor positron decay, and high decay energies are more often observed for low Z nuclides.

Superficially, electron capture could be considered a stealth mode of decay. If electron capture occurs without a gamma photon emitted, how can we tell that decay has taken place? According to the generic decay equation at the beginning of this section, only a neutrino is emitted, and neutrinos are extremely difficult to detect. Fortunately, the atom's electrons can tell us. After K capture has occurred, the K shell will be short one electron. This electron vacancy means that the daughter is created with its electrons in an excited state. To get back to the ground state, an electron from a

FIGURE 5.8 The Bohr atomic model.

higher level will drop down to fill the vacancy. When this happens, energy must be emitted. Because the electron energy levels are quantized, the energy emitted for a particular atom is always exactly the same, so long as the electronic transition is between the same two levels. Because the K shell is so much lower in energy than the others, the energy given off when a K vacancy is filled is often an X-ray photon.

Like gamma rays, X-rays are high-energy photons (see Figure 1.1). Gamma and X-ray photons can have the same energy, and to a detector they are indistinguishable. The difference between the two lies solely in how they are generated. Gamma rays are generated in the nucleus, as a result of a transition from an excited nuclear state to one lower in energy. X-rays are generated by electrons in one of two ways. (1) As we've seen here, an outer shell electron fills an inner shell vacancy (there are other ways to create these vacancies). This type of X-ray is termed a **characteristic X-ray**, because they are produced with specific energies. (2) High-speed (free) electrons can interact with matter by slowing down as it travels near a nucleus. This type of X-ray is called **continuous X-ray**, because the amount of energy released will depend on how close the electron gets to the nucleus, producing a broad spectrum of photon energies. It is also known as **bremsstrahlung**, which is German for "slowing-down radiation."

We can finally draw a picture of what happens after electron capture (Figure 5.9). A K shell vacancy is created by the electron capture decay. An L shell electron then fills this vacancy. Keep in mind that any higher energy shell could fill the vacancy, however, an L shell electron is the most probable. As a result of the electronic transition, a characteristic X-ray is emitted that has an energy corresponding to the difference in electron binding energy between the L and K shells in the daughter.

Characteristic X-rays are designated according to the subshells that mark the beginning and end of the electronic transition. The most commonly observed transition is labeled $K_{\alpha1}$, which corresponds to an electronic transition from the L_3 subshell to the K shell. $K_{\alpha2}$ is the $L_2 \rightarrow K$ electronic transition, and $K_{\alpha3}$ is the $L_1 \rightarrow K$ transition. The symbol K_β generically represents a transition from the M shell to the K shell. There is generally little difference in the energies of the various M subshells, so there is also very little difference in the individual K_β values. The energy values for characteristic X-rays correspond to differences in energy between the (sub)shells. These energy values are called **electron binding energies** because they represent the amount of energy required to remove an electron from an atom. These energy levels can be represented in an energy level diagram such as Figure 5.10. This diagram is for the element holmium. Similar diagrams for other elements will vary—typically the numbers will increase with atomic number. This makes sense; it should be more difficult to remove a K shell electron from an element with 92 protons than one with only 12 protons. An approximate energy value is given for the five M subshells in Figure 5.10. They actually range in value from 1.35 keV to 2.13 keV.

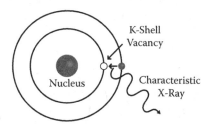

FIGURE 5.9   The aftermath of electron capture—Part I.

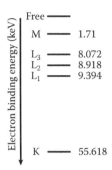

FIGURE 5.10   Energy level diagram for Ho.

Example 5.9 Using Figure 5.10, calculate the energy values of the following characteristic X-rays for holmium: $K_{\alpha 1}$, $K_{\alpha 2}$, and K_{β}.

$K_{\alpha 1}$	$L_3 \rightarrow K$	55.618 keV–8.072 keV = 47.546 keV
$K_{\alpha 2}$	$L_2 \rightarrow K$	55.618 keV–8.918 keV = 46.700 keV
K_{β}	$M \rightarrow K$	55.618 keV–1.71 keV = 53.91 keV

Instead of emitting a characteristic X-ray, an atom can get rid of the energy generated by the electronic transition by spitting out one of its electrons, as illustrated in Figure 5.11. This process is known as the **Auger** (oh-ZHAY) **effect**, and the electron that gets booted is known as an **Auger electron**.

Auger electrons are designated by: (1) the shell or subshell where the vacancy was originally created, (2) the subshell of the electron that drops to fill the vacancy, and (3) the subshell of origin for the Auger electron. For example, Figure 5.11 shows an electron dropping from the L shell to the K. Let's assume that it originated in the L_2 subshell, and that the Auger electron is from the L_3 subshell. The emitted electron would then be called a KL_2L_3 Auger electron. The kinetic energy of the KL_2L_3 Auger electron is equal to the energy gained by the atom when the electron drops from the L_2 subshell to the K shell, minus the energy required to remove an L_3 electron. For holmium (Figure 5.10) it is:

$$KL_2L_3 = 55.618 \text{ keV} -8.918 \text{ keV} -8.072 \text{ keV} = 38.628 \text{ keV}$$

The characteristic X-ray and Auger electron calculations done here are reversed from how they would normally be performed. Typically the energies of the characteristic X-rays and Auger electrons are determined experimentally, and the electron binding energies are calculated from these energies.

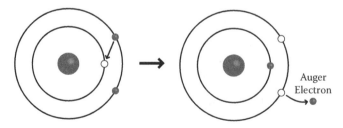

FIGURE 5.11   The aftermath of electron capture—Part II—the Auger effect.

FIGURE 5.12   Fluorescence yield as a function of atomic number. Data from Hubbell, J. H., *J. Phys. Chem. Ref. Data*, 23, 339, 1994.

Multiple Auger effects are possible in a single decay, potentially leaving the daughter as a highly charged cation. Since chemical bonds often involve the sharing of electrons between two atoms, electron capture can be especially disruptive when it occurs within a molecule.

Is it possible to tell whether characteristic X-rays or Auger electrons will be emitted as a result of the electron capture decay? Not really. Both are always observed when electron capture is a decay mode for a particular nuclide. The probability that X-rays will be produced following electron capture is called **fluorescence yield**. It is symbolized by the Greek letter omega (ω). With a K subscript (ω_K), it refers to fluorescence yield when the original vacancy is in the K shell. Figure 5.12 shows a graph of fluorescence yield (ω) vs. atomic number (Z). The probability that characteristic X-rays will be emitted increases with atomic number. In other words, the probability that Auger electrons will be emitted decreases with atomic number. This makes sense; more protons in the nucleus means that electrons in the same shell will be bound more tightly. In this case, the atom is more likely to shed its excess energy by emitting an X-ray rather than try to boot one of the tightly bound electrons.

5.5   MULTIPLE DECAY MODES

We've already seen that proton-rich nuclides can decay by both positron emission and electron capture. Some nuclides, especially those with moderate atomic numbers, in the middle of the belt of stability, and odd numbers of both protons and neutrons, can decay by all three beta modes. Can a decay diagram be drawn for these schizophrenic nuclides? You bet.

Example 5.10: ^{80}Br decays via all three beta modes. Using the information provided here, draw its decay diagram. The total energy of decay for β^- is 2.00 MeV, and it is 1.87 MeV for β^+ and electron capture. The most common beta particles emitted have E_{max} values of 2.00 MeV and 1.38 MeV, while the most common positron has an E_{max} of 0.85 MeV. A 617 keV gamma photon is also prominently observed during β^- decay.

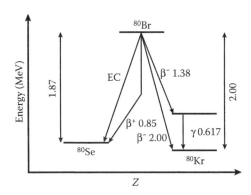

FIGURE 5.13 Decay diagram for ^{80}Br.

The decay diagram is shown in Figure 5.13. All energy values are given in MeV. Note that all values add up to appropriate values, e.g., the 1.38 MeV beta particle combines with the 0.617 MeV gamma photon to roughly equal the total energy of beta decay. Since no gamma ray is associated with electron capture, its arrow is drawn all the way to the ground state of the ^{80}Se daughter. If you're curious, the branch ratios observed for ^{80}Br are 91.7% for β^- decay, 6.1% electron capture decay, and 2.2% positron decay.

Multiple decay modes are also often observed for heavy ($Z > 82$) nuclides with poor N/Z ratios. These nuclides typically decay via alpha and some form of beta. Note that heavy, neutron-rich nuclides are more likely to decay via β^- than heavy proton-rich nuclides decaying via electron capture or β^+ emission. This is likely due to the fact that alpha decay also increases the N/Z ratio (see Section 5.1).

5.6 THE VALLEY OF BETA STABILITY

All three flavors of beta decay involve no change in the mass number—they occur along isobars. Let's take a closer look at a set of isobars, in particular the eight nuclides closest to the belt of stability when $A = 99$ (Table 5.1).

The neutron number, atomic number, and N/Z ratio are tabulated for each isobar. The optimal value for the N/Z ratio is likely close to 1.25 for $A = 99$, since ^{99}Ru is the only stable isobar. The decay mode(s) are given in the next column in Table 5.1. Notice that the isobars above ^{99}Ru have lower N/Z ratios (are proton-rich), and all undergo positron and electron capture decay. The more commonly observed decay mode of the two is listed first for each nuclide. The apparently even mix of the two decay modes is to be expected for the relatively moderate Z values for these nuclides. The isobars listed below ^{99}Ru all have higher N/Z ratios and decay via β^- emission.

Finally, the total energy of decay is listed for each isobar in Table 5.1. Notice that the value for ^{99}Tc is rather small, especially when compared to ^{99}Rh. This suggests that the optimal N/Z ratio for $A = 99$ is probably between 1.25 and 1.30 (the values for ^{99}Ru and ^{99}Tc, respectively), although it is likely closer to 1.25. Notice also that the energy values increase the further the isobar is from ^{99}Ru. We'll soon see that this is not true for all mass number values, but it is often true when A is odd. We can illustrate all of the isobaric decays listed in Table 5.1 in a simplified decay diagram, such as Figure 5.14.

The decays represented in Figure 5.14 are simplified such that only the total energy of decay is represented. In other words, if excited daughter states are formed in some of these decays, we aren't drawing them; we're only looking at the ground states for each isobar. Since ^{99}Ru is the only stable isobar for $A = 99$, any other (radioactive) isobars formed will eventually decay to ^{99}Ru. Because of the increasing magnitude of the decay energies as the isobars move away from ^{99}Ru, the energy

TABLE 5.1
Isobars with $A = 99$

Isobar	N	Z	N/Z	Decay Mode(s)	Decay Energy (MeV)
$_{47}$Ag	52	47	1.11	β^+, EC	5.4
$_{46}$Pd	53	46	1.15	β^+, EC	3.39
$_{45}$Rh	54	45	1.20	EC, β^+	2.04
$_{44}$Ru	55	44	1.25	Stable	Stable
$_{43}$Tc	56	43	1.30	β^-	0.294
$_{42}$Mo	57	42	1.36	β^-	1.357
$_{41}$Nb	58	41	1.41	β^-	3.64
$_{40}$Zr	59	40	1.48	β^-	4.56

changes for a whole set of isobars can be approximated with a parabola. This mathematical model is known as the **semi-empirical mass equation** and was (remarkably!) developed by C. F. von Weizsäcker in 1935. He used the word "mass" because of the simple direct relationship between mass and energy ($E = mc^2$). We could've just as easily graphed mass on the y-axis of Figure 5.14, and it would look just the same. We'll get into this in more detail in Section 6.2.

Energy (or mass) is really the third dimension to the chart of the nuclides (the other two are N and Z—Figure 1.9). If we look at the chart as a whole in this way, we'll see that the stable nuclides are all located at the bottom of an energy valley. The belt of stability is really a **valley of stability**, and Figure 5.14 is a cross-sectional slice of one part of this valley. Figure 5.15 displays this valley for all known nuclides with mass numbers of 46 to 71.

The walls of this valley become increasingly steep as atomic number (Z) decreases. This means that the energy changes become more dramatic for lower Z isobars. This is often cited as a reason why proton-rich, low-Z nuclides tend to decay via positron emission rather than by electron capture. Apparently, positron decay tends to be favored when the energy differences are great. This can be

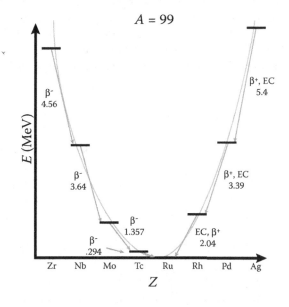

FIGURE 5.14 A simplified decay diagram for some of the $A = 99$ isobars.

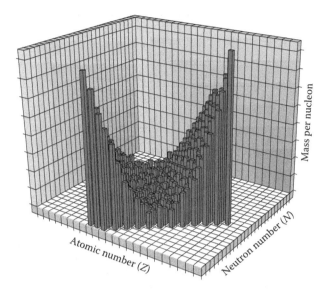

FIGURE 5.15 A 3-D chart of the nuclides from $A = 46$ to $A = 71$.

observed (anecdotally) in Table 5.1. ^{99}Rh has the lowest total energy of decay of all the proton-rich nuclides (high N/Z), and it is the only one to decay via electron capture more often than by positron emission. Don't count on this always being true.

As we scan down through the isobars in Table 5.1, they alternate between even–odd (*eo*) and odd–even nuclides (*oe*). With either N or Z odd (mass number is odd), it is not surprising that there is only one stable nuclide for $A = 99$. It turns out this is always true when the mass number is odd. For all sets of known isobars with an odd mass number, there is only one stable nuclide per set of isobars. Check it out for yourself; it is uncanny! Okay, maybe it's not so strange; perhaps one stable nuclide for each value of A is intuitively expected. It shouldn't be. When A is even, there can be one, two, or three stable nuclides per set of isobars. Check out Table 5.2 for $A = 104$.

When the mass number is even, then the individual isobars must be odd–odd (*oo*) or even–even (*ee*). With both N and Z odd, it is very unlikely a nuclide will be stable. In fact, it'll definitely be unstable for isobars with $A = 104$ (because the exceptions are all low A). On the other hand, an *ee* nuclide would be expected to exhibit greater stability relative to nearby nuclides with odd numbers

TABLE 5.2
Isobars with $A = 104$

Isobar	N	Z	N/Z	Decay Mode(s)	Decay Energy (MeV)
$_{49}$In	55	49	1.12	β^+, EC	7.9
$_{48}$Cd	56	48	1.17	EC, β^+	1.14
$_{47}$Ag	57	47	1.21	EC, β^+	4.28
$_{46}$Pd	58	46	1.26	Stable	Stable
$_{45}$Rh	59	45	1.31	β^-, EC	2.44, 1.14
$_{44}$Ru	60	44	1.36	Stable	Stable
$_{43}$Tc	61	43	1.42	β^-	5.60
$_{42}$Mo	62	42	1.48	β^-	2.16
$_{41}$Nb	63	41	1.54	β^-	8.1

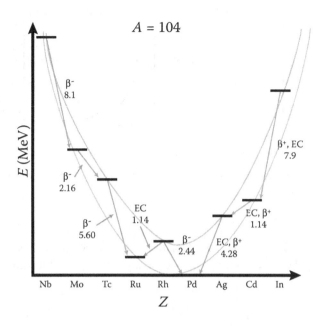

FIGURE 5.16 A simplified decay diagram for some of the $A = 104$ isobars.

of N and/or Z. Scan down through Table 5.2 and the N and Z values. Notice that they alternate between *oo* and *ee*. It's as if there are really two sets of isobars in Table 5.2, those with extra (relative!) stability (*ee*) and those with extra instability (*oo*).

The dualistic nature of even isobars is even more apparent when a simplified decay diagram is drawn (Figure 5.16). There are clearly two parabolas that converge at higher energy values. Even mass numbered isobars are also approximately modeled by von Weizsäcker's equation; we just need two parabolas instead of one. Notice also that the two stable nuclides in this set of isobars do not have exactly the same energy. Even among stable nuclides, stability is apparently relative. So, what is the optimal N/Z value for $A = 104$? It is between the values for the two stable nuclides: 1.26 (^{104}Pd) and 1.36 (^{104}Ru). It is likely closer to 1.26 as ^{104}Pd is lower in energy than ^{104}Ru. As expected, the optimal N/Z ratio for $A = 104$ is slightly higher than for $A = 99$.

As mentioned earlier, sets of isobars with even mass numbers can include one, two, or three stable nuclides. Generic, simplified decay diagrams illustrate all the observed pathways in Figure 5.17.

There are two ways a set of even isobars can have just one stable nuclide: one with a stable *oo* nuclide (Figure 5.17a) and the other with a stable *ee* nuclide (Figure 5.17b). A stable *oo* nuclide should set off an alarm in your head. Remember there are (only) four stable *oo* nuclides ($^{2}_{1}$H$_{1}$, $^{6}_{3}$Li$_{3}$, $^{10}_{5}$B$_{5}$, and $^{14}_{7}$N$_{7}$). In every case, they are low Z and are the only known stable nuclide for their set of isobars. The anomalous nature of their stability can be rationalized as due to the narrowness of the valley of stability for low Z nuclides. These nuclides (the *oo* exceptions) clearly have the best N/Z ratios for their set of isobars, and the adjacent *ee* nuclides are simply too high (too far up the steep walls of the valley) in energy to be stable.

$A = 104$ is an example of the most popular isobaric decay pathways when A is even. Eighty-three sets of isobars are known with two stable *ee* nuclides (Figure 5.17c). A close second are 78 sets with one stable *ee* nuclide (Figure 5.17b), and a distant last are three sets with three stable *ee* nuclides (Figure 5.17d). Therefore, odds are pretty good that when A is even, there'll be either one or two stable *ee* nuclides. Close examination of the two and three stable isobar cases illustrated in Figures 5.17c and d reveals that the stable *ee* nuclides are always separated by one unstable *oo*

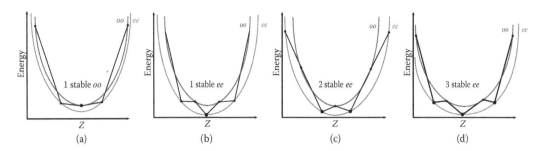

FIGURE 5.17 Simplified decay diagrams representing all possible pathways when mass number (*A*) is even: (a) 1 stable *oo* (b) 1 stable *ee* (c) 2 stable *ee* (d) 3 stable *ee*.

nuclide. Looking over the last couple of pages suggests that two stable isobars will never be adjacent to each other. This is known as the **Mattauch isobar rule**. It is often stated as "stable, neighboring isobars do not exist." Take a look at a two-dimensional printed chart of the nuclides and see if you can find any exceptions.

5.7   ISOMERIC TRANSITIONS

Many nuclear science textbooks define isomers as only those excited nuclear states with measurable lifetimes. This is a rather arbitrary, and somewhat inconsistent definition. "Measurable" has changed over the years, and therefore what was once considered only an excited state is now considered an isomer. It likely stems from the pre-Internet era when the chart of the nuclides was only available in printed form. It is not possible to show every excited state for every nuclide in a printed format. The 17th (2009) printed edition of KAPL's chart generally provides information only on excited states with half-lives greater than one second. To avoid confusion, this chart refers to these as "metastable" states, or simply as "metastates." Recall that this is indicated in the nuclide symbol by adding an "m" at the end of the mass number, for example ^{99m}Tc is a metastate of ^{99}Tc.

We'll use a more inclusive definition for isomer. As stated in Chapter 1, isomers are two nuclides with the same number of protons and neutrons, but are in different energy states. An **isomeric transition** is therefore one isomer changing into another through a nuclear change.

Gamma decay is one type of isomeric transition. We'll have to hold off for a bit before we see the other flavor. In the meantime, let's learn a bit about gamma emission. Since the energy levels of isomeric states are quantized, the energy of the gamma photon emitted is exactly equal to the difference in energy of the two nuclidic states. Like alpha particles, gamma photons are monoenergetic. Unfortunately, many isomeric states are often possible, which can lead to quite a few different gamma photons being emitted as part of a single decay. Also recall that gamma rays have no charge or mass, and are strongly penetrating due to their low probability of interacting with matter.

^{125m}Xe decays according to the diagram in Figure 5.18. First it transitions to another excited state of ^{125}Xe, which is 140 keV in energy below ^{125m}Xe and 112 keV above ground state ^{125}Xe. The second transition is distinct from the first, resulting in the emission of a 112 keV photon. Curiously, a gamma ray with an energy of 252 keV (metastate directly to ground state) is not observed. Why not? It is not "allowed" because of the rather dramatic nuclear changes that would have to accompany such a transition. It has to do with the fractions and the + and – signs written next to each isomeric state in Figure 5.18.

The numbers give the nuclear spin of each state. Each nucleon has a spin of ½, and when they are combined together in a nucleus, the nucleus will also have a spin. When *A* is odd, the ground state nuclear spin is always *x*/2, where *x* = 1, 2, 3, … When *A* is even, the ground state nuclear spin is 0 or a positive integral value. Nucleons have a tendency to pair up with others of their own kind in the

FIGURE 5.18   Isomeric transitions of ^{125m}Xe to ^{125}Xe.

nucleus. When this happens, their spins cancel each other out. As a result, all *ee* nuclides have a 0 spin in the ground state. Nuclides with an odd number of protons or neutrons have spin states that are determined by the energy level occupied by the unpaired nucleon(s). We'll go into this in more detail in Chapter 6. For now, we'll need to just accept the values provided to us.

As is apparent in Figure 5.18, different isomers have different spin states. Differences in spin are the main difference between isomers. In order to transition from one isomer to another, the nucleus needs to change its spin. Generally speaking, the greater the change in spin state, the less likely the transition will take place. In order for ^{125m}Xe to decay directly to the ground state, the spin state must change from 9/2 to 1/2. Unlikely. Instead it first changes to 3/2, then from 3/2 to 1/2.

The + and − symbols in Figure 5.18 refer to **parity**. Nuclear parity is like your right and left hands, they look the same, but they're really mirror images of each other. Likewise a nucleus with + (even) parity looks like a mirror image of an isomer with − (odd) parity. Parity also indicates the direction of the spin; nuclei with even parity spin one way, while those with odd parity spin the other way. Parity plays a role in determining whether a particular isomeric transition can take place, but this is beyond the scope of this text. We will limit ourselves to observing which transitions do not take place, and rationalizing them simply on the basis of the magnitude of change in spin.

When the nucleus changes from one isomeric state to another, energy is released. We've already seen that it can be emitted in the form of a high-energy (gamma) photon. In addition to gamma emission, the atom can also kick out one of its electrons to release the energy. This is called **internal conversion**, and this is why this section is titled isomeric transitions rather than gamma decay. When a nucleus transforms from one isomer to another, it has two options to rid itself of the energy it just created.

The emitted electron is called a **conversion electron**. Conversion electrons usually come from the K shell (K int cov), as these electrons have the highest probability of being close to the

TABLE 5.3
^{125}Te* Isomeric Transitions

Emission	Probability (%)	Energy (keV)
γ	6.66	35.5
K int cov	80.00	3.7
L int cov	11.42	31.0
M int cov	1.90	34.5

nucleus. The kinetic energy of the conversion electron is simply the transition energy minus the electron binding energy. The lowest energy excited state for ^{125}Te is 35.5 keV above the ground state. It undergoes an isomeric transition to the ground state according to the probabilities given in Table 5.3.

The transition energy (total energy of decay) is equal to the energy of the gamma photon (35.5 keV) because it ends in the ground state. The energies of the conversion electrons are all lower than the transition energy by an amount equal to the electron binding energy for that shell. Using the numbers in Table 5.3, we can calculate those energies.

$$\text{K shell conversion electron} = 35.5\ \text{keV} - 3.7\ \text{keV} = 31.8\ \text{keV}$$

$$\text{L shell conversion electron} = 35.5\ \text{keV} - 31.0\ \text{keV} = 4.5\ \text{keV}$$

$$\text{M shell conversion electron} = 35.5\ \text{keV} - 34.5\ \text{keV} = 1.0\ \text{keV}$$

The probabilities of each type of emission following the ^{125}Te* $\rightarrow$ ^{125}Te transition are also presented in Table 5.3. Notice that internal conversion is much more likely than gamma emission. This is partially expected since tellurium has a moderately high atomic number ($Z = 52$). Also note that emission of a K shell electron is most probable, but occasionally electrons from the L and M shell get booted.

The transition used here is rather low in energy. If the transition energy is lower than the binding energy of the electrons, then internal conversion for that shell is not possible. Isomeric transitions that are of greater energy can emit fairly high-energy electrons. Once emitted, they look an awful lot like a beta particle. The only difference between conversion electrons and beta particles is their origin: conversion electrons come from the atom's supply of orbital electrons and result from internal conversion, while beta particles are created in the nucleus by a nuclide undergoing beta decay.

The emission probabilities vary by element *and* by the particular transition. For example, there's another excited state for ^{125}Te that is 145 keV above the ground state. It has a half-life of 57.4 days, and is therefore a metastable state and can be designated as ^{125m}Te. ^{125m}Te has two options for decay: straight to the ground state, or to the other excited state (the one we've been talking about here and have symbolized as ^{125}Te*). Both of those transitions will have different probabilities for emission than those listed in Table 5.3.

The relative probability of internal conversion to gamma emission is given by the **conversion coefficient** (α_T). It is simply the ratio of ejected electrons to gamma photons.

$$\alpha_T = \frac{I_e}{I_\gamma} \tag{5.4}$$

Instead of counting electrons and photons, we can use the probabilities in Table 5.3 to calculate the conversion coefficient for the ^{125}Te* $\rightarrow$ ^{125}Te transition.

$$\alpha_T = \frac{I_e}{I_\gamma} = \frac{80.00 + 11.42 + 1.90}{6.66} = 14.0$$

This tells us that conversion electrons are 14 times more likely than gamma photons to be emitted in this decay. Additionally, we could calculate the conversion coefficients for the individual electron shells.

$$\alpha_{K} = \frac{I_e}{I_\gamma} = \frac{80.00}{6.66} = 12.0 \qquad \alpha_{L} = \frac{I_e}{I_\gamma} = \frac{11.42}{6.66} = 1.71 \qquad \alpha_{M} = \frac{I_e}{I_\gamma} = \frac{1.90}{6.66} = 0.285$$

where α_K is the conversion coefficient for internal conversion of K shell electrons, α_L is for the L shell, and α_M is for the M shell.

5.8   OTHER DECAY MODES

We've covered the major forms of decay: alpha, all three types of beta, and the two types of isomeric transitions. What follows are short descriptions of more unusual decay modes.

5.8.1   SPONTANEOUS FISSION

Spontaneous fission is when a nuclide splits into two large chunks (more or less the same size) and some neutrons without any prompting. This is a significant mode of decay for some of the very heavy nuclides ($A > 230$) and will be discussed in more detail in Chapter 11.

5.8.2   CLUSTER DECAY

Cluster decay happens when very heavy nuclides blow chunks bigger than an alpha particle, but not quite big enough to qualify as spontaneous fission. ^{12}C, ^{14}C, ^{20}O, *inter alia* are (very rarely) emitted from large nuclides in a desperate attempt to quickly form a more stable nuclide.

Superficially, one might think the spontaneous fission and cluster decay should be much more common than they are. Imagine for a moment that you are a really big nuclide, like ^{234}U. Stability is a long way away (Pb) and you'd like to get there quickly. Rather than go through the dozen (or so) alpha and beta decays that make up the normal decay chain, ^{234}U can also undergo cluster decay, spitting out ^{28}Mg to form ^{206}Hg.

$$_{92}^{234}\text{U} \rightarrow \, _{80}^{206}\text{Hg} + \, _{12}^{28}\text{Mg}$$

^{234}U can also undergo spontaneous fission forming a variety of products, e.g., ^{93}Rb and ^{138}Cs along with three neutrons.

$$_{92}^{234}\text{U} \rightarrow \, _{37}^{93}\text{Rb} + \, _{55}^{138}\text{Cs} + 3\,_{0}^{1}\text{n}$$

Yet the branch ratios for ^{234}U defy intuition. It decays via alpha very close to 100% of the time. Spontaneous fission happens only $1.6 \times 10^{-9}\%$ of the time, and cluster decay occurs in roughly $10^{-11}\%$ of all decays. Strange. All three of these decay modes involve basically the same thing, the splitting of the nucleus up into two parts, yet alpha decay is by far the most commonly observed. We'll discuss this in more detail in Chapters 6 and 11, but for now we can realize that there must be an energy barrier to all these processes (since they all release energy—do the math!), and this barrier is much more significant for spontaneous fission and cluster decay. A huge advantage that alpha decay has over these two is that it is a relatively small chunk of the nucleus; therefore, it has the smallest barrier to overcome. It also forms the doubly magic ^{4}He, which has an unusually high binding energy per nucleon. These two factors strongly favor alpha decay over spitting out larger chunks.

As you might imagine, daughter recoil is a big deal in cluster decay. We can tap the same equations (Equations 5.1 and 5.2) to calculate recoil as we used when we looked at alpha decay.

Example 5.11: ^{224}Ra undergoes cluster decay by emitting ^{14}C in $6 \times 10^{-8}\%$ of its decays. Write out a balanced decay equation and calculate the daughter's recoil energy.

Following Examples 5.1 and 5.2:

$$^{224}_{88}\text{Ra} \rightarrow {}^{210}_{82}\text{Pb} + {}^{14}_{6}\text{C}$$

$$\left[224.020212 \text{ u} - (209.984173 \text{ u} + 14.003242 \text{ u})\right] \times \frac{931.5 \text{ MeV}}{\text{u}} = 30.55 \text{ MeV}$$

$$K_\text{F} = 30.55 \text{ MeV} \left(\frac{14}{210+14}\right) = 1.91 \text{ MeV}$$

Notice that recoil is a lot more significant with cluster decay than alpha. Just like recoil for a cannon would be larger than for a pistol.

5.8.3 Proton/Neutron Emission

Can particles smaller than an alpha particle be emitted? Certainly, but these types of decay are generally restricted to very proton-rich nuclides (proton emission) and very neutron-rich nuclides (neutron emission). Examples of each are given below.

$$^{147}_{69}\text{Tm} \rightarrow {}^{146}_{68}\text{Er} + {}^{1}_{1}\text{H}$$

$$^{4}_{1}\text{H} \rightarrow {}^{3}_{1}\text{H} + {}^{1}\text{n}$$

Thulium-147 ($N/Z = 1.13$) can decay by spitting out one of its protons (branch ratio = ~10%), forming erbium-146 ($N/Z = 1.15$, still way too low for this mass number, but better); and hydrogen-4 ($N/Z = 3.00$) ejects a neutron to make tritium (also known as hydrogen-3, $N/Z = 2.00$). As might be expected, these types of decay tend to occur at the fringes of the chart of the nuclides, i.e., only when the N/Z ratio is seriously out of whack. While both modes of decay are fairly unusual, proton emission is much more common than neutron emission. This can be understood in terms of Coulomb repulsion. It's easier to kick a proton out of a nucleus because all the protons are pushing each other away. There's not much of a repulsive force acting on neutrons, so they are only rarely ejected as an independent decay mode. In fact neutron emission is generally observed only for a few highly neutron-rich low-Z nuclides.

Multiple protons or neutrons can also be emitted from especially extremist nuclides—those with particularly poor N/Z ratios. These decay modes are exceptionally rare, but an example would be the release of two protons by ^{12}O ($N/Z = 0.50!$) to form ^{10}C ($N/Z = 0.67$).

$$^{12}_{8}\text{O} \rightarrow {}^{10}_{6}\text{C} + 2{}^{1}_{1}\text{H}$$

Once again, recoil will be significant for these decay modes. Equations 5.1 and 5.2 can still be used.

5.8.4 Delayed Particle Emissions

Like proton and neutron emissions, **delayed particle emissions** are also sometimes observed for nuclides at the extremes of the valley of stability, i.e., up on a high ledge overlooking the valley (what a great view they must have!). In fact, delayed emissions of protons and neutrons are much more common than direct emissions of either. The difference is that a delayed particle emission means a proton or neutron is emitted *after* the parent nuclide has already undergone some flavor of beta decay. This tends to happen when the daughter is formed in a particularly high-energy excited state and is an alternative to an isomeric transition to a lower excited state or to the ground state. In this case an isobar is not formed as the product. For example, almost 7% of all ^{137}I decays emit a beta particle then a neutron (93% of the time it just emits a beta particle). The minor branch (beta plus neutron decay) can be represented as:

$$^{137}_{53}I \rightarrow {}^{136}_{54}Xe + {}^{0}_{-1}e + {}^{1}_{0}n + \overline{\nu}$$

However, a better representation would include the fleeting formation of the excited daughter, followed by neutron emission from the excited daughter.

$$^{137}_{53}I \rightarrow {}^{137}_{54}Xe^* + {}^{0}_{-1}e + \overline{\nu}$$

$$^{137}_{54}Xe^* \rightarrow {}^{136}_{54}Xe + {}^{1}_{0}n$$

By undergoing beta plus neutron decay, ^{137}I lowers its high N/Z ratio (1.58) two ways, first by converting a neutron to a proton (beta) and then by spitting out another neutron. The end result is formation of a stable, *ee* daughter (^{136}Xe, $N/Z = 1.52$).

Example 5.12: ^{19}N undergoes beta plus neutron in approximately 55% of its decays. Assuming a maximum beta energy of 6.12 MeV and a neutron energy of 2.32 MeV, sketch out a decay diagram for this decay mode only.

The balanced decay equation is:

$$^{19}_{7}N \rightarrow {}^{18}_{8}O + {}_{-1}e + {}^{1}n$$

The total energy of decay is:

$$\left(19.01702 \text{ u} - \left(17.999161 \text{ u} + 1.008665 \text{ u}\right)\right) \times \frac{931.5 \text{ MeV}}{u} = 8.56 \text{ MeV}$$

Before we draw, we should realize that daughter recoil will be small, but significant. We'd better use Equations 5.1 and 5.2 to calculate the recoil for the neutron emission.

$$K_T = 2.32 \text{ MeV} \times \frac{18+1}{18} = 2.45 \text{ MeV} \qquad K_F = 2.45 \text{ MeV} \times \frac{1}{18+1} = 0.129 \text{ MeV}$$

The decay diagram is shown in Figure 5.19. Notice the gap left in the neutron emission to indicate daughter recoil.

Like direct emission of these particles, they can also be emitted in bunches. Two, three, or four neutrons can follow beta decay; and, as you may have guessed, tend to be emitted in multiples when the N/Z ratio and decay energy are especially high. Likewise, a couple of protons can be

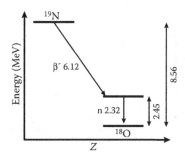

FIGURE 5.19 Beta plus neutron decay of ^{19}N.

emitted following positron decay, when the energy of transition is great and the N/Z ratio is very low. Delayed alpha particles or tritium nuclei can also be observed for some nuclides. As we'll see in Chapter 11, emission of delayed neutrons is an important safety feature in nuclear reactors.

5.8.5 DOUBLE BETA DECAY

Double beta is a rather unusual form of decay, and can occur with a neutron-rich, *ee* nuclide. An example is the decay of ^{82}Se to ^{82}Kr by the simultaneous emission of two beta particles. Note that two antineutrinos are also emitted to satisfy the conservation laws discussed earlier in this chapter.

$$^{82}_{34}\text{Se}_{48} \rightarrow {}^{82}_{36}\text{Kr}_{46} + 2\,{}^{0}_{-1}\text{e} + 2\overline{\nu}$$

Why does double beta decay happen? ^{82}Se is neutron-rich ($N/Z = 1.41$), so it needs to decay via beta. If that happened, it would form $^{82}_{35}\text{Br}_{47}$ (note it is *oo*). Using the masses of these nuclides, let's calculate the energetics of this putative decay.

$$^{82}_{34}\text{Se} \rightarrow {}^{82}_{35}\text{Br} + {}_{-1}\text{e}\,?$$

$$\left(81.916699\ \text{u} - 81.916805\ \text{u}\right) \times \frac{931.5\ \text{MeV}}{\text{u}} = -0.0987\ \text{MeV}$$

This doesn't make sense! The total energy of decay is always positive. Careful examination of the math reveals that it is done correctly. The strange thing is that the mass of the daughter (^{82}Br) is *greater than* the mass of the parent (^{82}Se). Mass should be lost during a decay process, producing energy—not the other way around. In other words, ^{82}Se *cannot* undergo beta decay to ^{82}Br. ^{82}Se is still neutron-rich, and must undergo decay, so it emits two beta particles at once and decays directly to ^{82}Kr.

The energetics for the decays of ^{82}Se and ^{82}Br are illustrated in a simplified decay diagram in Figure 5.20. If more $A = 82$ isobars were included, this figure would look like Figure 5.17b with ^{82}Kr as the lone, stable isobar. Notice that ^{82}Br lies higher in energy (has more mass) than ^{82}Se, making that decay pathway impossible.

Nuclides observed to undergo double beta decay typically have very long half-lives. For example, the half-life for ^{82}Se is 9×10^{19} years! These long half-lives are attributed to the fact that creating two particles and two antiparticles all at once in a nucleus is a difficult thing, and therefore of low probability.

Double beta decay has been observed for ten neutron-rich nuclides: ^{48}Ca, ^{76}Ge, ^{82}Se, ^{96}Zr, ^{100}Mo, ^{116}Cd, ^{128}Te, ^{130}Te, ^{150}Nd, and ^{238}U. Equivalent processes for proton-rich nuclides have been observed for four nuclides. Theoretically, double positron, electron capture/positron, and double electron

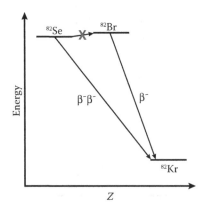

FIGURE 5.20 Simplified decay diagram for ^{82}Se and ^{82}Br.

capture decays are all possible for certain proton-rich *ee* nuclides. However, only double electron capture is observed. ^{78}Kr, ^{124}Xe, ^{130}Ba, and ^{132}Ba all decay via double electron capture with very long ($\sim 10^{21}$ a) half-lives. The decay equation for ^{130}Ba is given here as an example.

$$^{130}_{56}\text{Ba} + 2\,_{-1}e \rightarrow\,^{130}_{54}\text{Xe} + 2\upsilon$$

QUESTIONS

5.1   Define the following: recoil energy, antineutrino, characteristic X-ray, conversion coefficient, and fluorescence yield.

5.2   If the total energy of decay for ^{245}Cm is 5.623 MeV, what is the recoil energy of the daughter of the decay?

5.3   Calculate the atomic mass of ^{240}U from the primary mode of decay for ^{244}Pu. Write out the decay equation and use 4.589 MeV as the kinetic energy of the most energetic alpha particle possible for this decay. What is the recoil energy of ^{240}U?

5.4   Using information from Appendix A, draw a decay diagram for ^{268}Mt. Calculate the recoil energy of the daughter. For whom are the parent and daughter elements named?

5.5   List and briefly explain the three conservation laws that are satisfied in beta decay. What must also be emitted for these to hold true?

5.6   Calculate the atomic mass of ^{198}Au using only the information given here. It decays via beta and the atomic mass of its daughter is 197.966769 u and the total energy of decay is 1.372 MeV.

5.7   The most common beta particles observed in the decay of ^{59}Fe have E_{max} values of 0.466 and 0.274 MeV. The two most prominent gamma rays have energies of 1099 and 1292 keV. If the total energy of decay is 1.565 MeV, and the atomic mass of the daughter is 58.933195 u, what is the atomic mass of ^{59}Fe? Draw a decay diagram consistent with the information given here.

5.8   Explain why ^{37}Ar might be expected to undergo positron decay but doesn't.

5.9 ^{54}Mn (53.940363 u) decays to ^{54}Cr (53.938880 u) via electron capture. In almost every decay a gamma photon with an energy of 835 keV is emitted. Write out the decay equation and draw a decay diagram using only the information given here. Give two possible reasons why a gamma photon is *not* seen in a very small percentage (< 0.1%) of these decays.

5.10   The binding energies for electrons in element X are given in Table 5.4. Calculate the energy of the $M_2 \rightarrow K$ X-ray, and the $M_2M_3M_5$ Auger electron. What is the energy of the KL_2L_3 Auger electron? What kind of Auger electron would have an energy of 12.315 keV?

TABLE 5.4
Electron Binding Energies for Element X

Subshell	Binding Energy (keV)
M_5	3.332
M_4	3.491
M_3	4.046
M_2	4.831
M_1	5.182
L_3	16.300
L_2	19.693
L_1	20.472
K	109.651

5.11 ^{64}Cu undergoes all three forms of beta decay. Draw a decay diagram consistent with the information given here. The total energy of decay for β⁻ is 0.579 MeV, and for β⁺ and EC it is 1.675 MeV. The only beta particle observed has an E_{max} of 0.578 MeV, and the only positron has an E_{max} of 0.651 MeV. A 1346 keV γ photon is observed, but only in a small fraction of the electron capture decays.

5.12 A nuclide decays giving off two different positrons, and a single gamma ray of 677 keV. If the total energy of decay is 6.138 MeV, what are the likely energies of the two positrons emitted?

5.13 Draw the possible decay diagram for the nuclide in the previous question.

5.14 The most probable energy for a positron is ~$0.4E_{max}$. Briefly explain why this is larger than for beta particles (negatrons).

5.15 Calculate the energy resulting from the reaction of 0.01 moles of positrons with an equal number of electrons.

5.16 The photon torpedoes used in *Star Trek* use antimatter as an explosive. Calculate the mass of antimatter necessary to equal the explosive yield of the bomb dropped on Hiroshima. Comment on whether these weapons could someday become a reality. Why is the word "photon" particularly appropriate to these weapons?

5.17 Complete the first four (empty) columns in Table 5.5 without any additional information, and then check your work using Appendix A. Use the appendix to fill in the final column. Finally, draw a simplified decay diagram for this set of isobars, showing only the major mode of decay between nuclides.

TABLE 5.5
Data for Some $A = 25$ Isobars

Isobar	N	Z	N/Z	Mode of Decay	Atomic Mass (u)
$^{25}_{9}$F					
$^{25}_{10}$Ne					
$^{25}_{11}$Na					
$^{25}_{12}$Mg					
$^{25}_{13}$Al					
$^{25}_{14}$Si					

5.18 Using the table from the previous question, what are some conclusions that can be made about a set of isobars?

5.19 Draw a simplified decay diagram for $A = 168$ from tantalum to terbium showing only the major mode of decay between nuclides.

5.20 Why is Mattauch's rule always true?

5.21 Explain the m in ^{60m}Co. What is meant by the fact that its I_e/I_γ ratio is 41?

5.22 Using only the information given in Appendix A, produce your best guess at a complete decay diagram for ^{97m}Tc.

5.23 A total of 94.40% of all ^{137}Cs decays result in the formation of ^{137m}Ba, while the remaining 5.60% produce ground state ^{137}Ba directly. If only 85.1% of all ^{137}Cs decays produce a gamma photon (the remainder produce a conversion electron), what is the conversion coefficient for ^{137m}Ba decay?

5.24 What is the energy of a conversion electron produced from the K shell during the decay of ^{137}Cs?

5.25 What are the differences between conversion coefficient and fluorescence yield?

5.26 Using mass differences, calculate the amount of energy released by all three decay modes given for ^{234}U in Section 5.8.2. Use the fission equation given in the chapter. Why is ^{234}U naturally occurring?

5.27 Calculate the kinetic energies of the daughter and the emitted nucleus in the cluster decay of ^{234}U.

5.28 Using information in Appendix A, draw a decay diagram for ^{5}He and estimate the kinetic energies of the products.

5.29 ^{14}B can decay via beta with a delayed neutron. Using information in Appendix A, write out a balanced decay equation and calculate the total energy of decay for this pathway.

5.30 Using Appendix A, draw a decay diagram for ^{130}Te.

5.31 Explain why double beta decay is only observed for nuclides with an even mass number.

5.32 Sketch out a simplified decay diagram for the $A = 130$ isobars from ^{130}Sn to ^{130}Ce. Note anything unusual.

6 Nuclear Structure

WARNING! This chapter is not intended for the casual reader. Proceed only if you are truly curious about a deeper understanding of nuclear structure and the different forms of decay, or your instructor is telling you that you had better read and understand this stuff.

While Chapter 5 addressed a large number of important questions concerning radioactive decay, there are likely more. What is the semi-empirical mass equation, and what does it tell us about nuclear stability? How do the main decay modes occur? For example, how is it possible for two protons and neutrons (an alpha particle) to suddenly eject itself from a nucleus? What does it mean to be proton-rich—is it just too many positive charges crammed into a small space, or is there more to it? How is it even possible to be neutron-rich—wouldn't more neutrons just add more strong force and make the nucleus more stable? What does the nucleus look like, and how can its structure help us understand decay and reactions? This chapter seeks to answer these questions through an understanding of how the nucleus is put together, but be warned, some understanding of quantum mechanics is necessary.

6.1 NUCLEAR SHAPE AND SIZE

Nuclei are often represented as a hard sphere with a well-defined border, much like a billiard ball (e.g. Figure 1.3). Individual protons and neutrons are also typically represented as hard spheres that are packed into nuclei like a bunch of marbles all glued together (e.g. Figure 4.1). However, we know from quantum mechanics that the subatomic universe does not consist of objects with fixed borders, things are a lot fuzzier. The nucleus can also adopt shapes other than a sphere. Since it can spin rather rapidly, it can be an oblate spheroid. This shape is like a ball of clay that's been smushed on the top and bottom, or the beginnings of a smash burger (Figure 6.1). The Earth is an oblate spheroid, which makes some sense since it is also spinning; it has a larger circumference around the equator than the poles. A nucleus can also be elongated on its axis, forming a prolate spheroid which looks like a rugby ball or an American football.

Just like orbitals for electrons, the nuclear shapes in Figure 6.1 lack well-defined boundaries. They get fuzzy around their edges, which is a reflection of the quantum mechanical uncertainty in locating a subatomic particle. Also like electrons, protons and neutrons occupy quantized energy levels, and are occasionally in a level that might be a bit higher than the other nucleons and can be thought of as more loosely bound to the rest of the nucleus. Especially neutron- or proton-rich nuclides have this problem and this makes it easier for them to decay via particle emission (Section 5.8.3).

DOI: 10.1201/9781003303787-6

Sphere   Oblate Spheroid   Prolate Spheroid

FIGURE 6.1   Common nuclear shapes.

For the purposes of this book, the nucleus can be considered a sphere with a radius r that is approximately equal to the radius constant ($r_0 = 1.4$ fm, 1 fm = 10^{-15} m) multiplied by the cube root of the mass number (A) as shown in Equation 6.1.

$$r = r_0 \sqrt[3]{A} \tag{6.1}$$

Example 6.1 Estimate the volume of an ^{56}Fe nucleus. Repeat for ^{112}Cd.

Start by calculating the nuclear radius using Equation 6.1, then the volume assuming the nuclei are spherical. For ^{56}Fe:

$$r = 1.4\,\text{fm} \times \sqrt[3]{56} = 5.4\,\text{fm} \qquad V = \frac{4}{3}\pi r^3 = \frac{4}{3}\pi \times (5.4\,\text{fm})^3 = 644\,\text{fm}^3$$

For ^{112}Cd:

$$r = 1.4\,\text{fm} \times \sqrt[3]{112} = 6.7\,\text{fm} \qquad V = \frac{4}{3}\pi r^3 = \frac{4}{3}\pi \times (6.7\,\text{fm})^3 = 1287\,\text{fm}^3$$

Notice that ^{112}Cd has twice as many nucleons as ^{56}Fe. While the radius does not double in value, the volume does. Note also that the units of femtometers (fm) are sometimes referred to as fermis (1 fm = 1 fermi).

If we plot nuclear density vs. distance from the center, it would look like Figure 6.2. The amount of matter per unit volume is rather constant as we move out from the center of the nucleus. In other words, it is consistently solid, like a ball bearing, and does not have voids like Swiss cheese or pumice. As we reach the surface, the density does not drop off sharply, as expected for a hard sphere, rather it tapers off. This is consistent with the "fuzzy" surfaces shown in Figure 6.1.

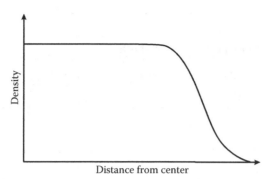

FIGURE 6.2   Mass distribution in the nucleus.

6.2   THE SEMI-EMPIRICAL MASS EQUATION

Section 5.6 stated that the semi-empirical mass equation was a mathematical model for the parabolic arrangements of sets of isobars when graphing energy (or mass) against atomic number (Z), i.e., drawing a simplified decay diagram like Figures 5.14 and 5.16. It is also a way to calculate nuclear binding energy (E_B). Wait a minute, we learned how to calculate nuclear binding energy in Section 3.1. It was simply a matter of looking at the difference between the actual mass of a nuclide and the mass of the sum of its parts, then converting to energy. The semi-empirical mass equation (Equation 6.2) calculates the same thing, using a set of empirically derived constants (a_x) and the values for mass number (A) and atomic number (Z). It quantifies everything that contributes to or detracts from nuclear binding energy. It is "semi-empirical" because it is a combination of fitting experimental data with some theoretical components.

$$E_B = a_v A - a_s A^{2/3} - a_c \left(\frac{Z^2}{A^{1/3}} \right) - a_a \frac{(N - Z)^2}{A} + \delta \tag{6.2}$$

Believe it or not, this equation is in the form of a quadratic. Therefore, for constant values of the mass number (A), a graph of energy vs. atomic number (Z) will give a parabola, just like the ones we saw in Section 5.6 (Figures 5.14 and 5.16).

The first term, $a_v A$, is somewhat inappropriately called the "volume term." Rather than establishing the size of the nucleus, it is an estimate of the strong force—the attractive force between all nucleons in a nucleus. Remember from Section 1.2 that the strong force only acts over very small distances. In fact, a nucleon buried deep inside of a nucleus will only feel the strong force from its nearest neighboring protons and neutrons. Nucleons on one side of the nucleus do not feel an attraction to nucleons on the other side (unless it is a pretty small nucleus, like ⁴He). Notice that the volume term depends only on mass number (A), therefore, the bigger the nucleus, the larger this term will be. In many respects this term acts as a rough estimate of the nuclear binding energy, the larger the nucleus, the greater the binding energy (as we saw back in Section 3.1). It is only a rough estimate because it assumes that all nucleons are completely surrounded by other nucleons. The nucleons on the surface of the nucleus do not feel the warm embrace of as many of their subatomic kin, therefore the volume term overestimates the nuclear binding energy.

The second term, $a_s A^{2/3}$, corrects the first by subtracting an estimate of how much strong force the surface nucleons are missing out on. This quantity is appropriately named the "surface term." Note that this term will be more mathematically significant for smaller nuclei than larger because of the two-thirds power. This is appropriate because smaller nuclei have a larger percentage of their nucleons on the surface. The term comes from a combination of the formula for the surface area of a

TABLE 6.1
Semi-Empirical Mass Equation Constants

a_v (MeV)	a_s (MeV)	a_c (MeV)	a_a (MeV)	a_δ (MeV)
15.56	17.23	0.7	23.285	11

sphere ($4\pi r^2$) and the direct proportionality of atomic radius with the cube root of the mass number (Equation 6.1).

The third term in Equation 6.2, $a_c(Z^2/A^{1/3})$, is known as the "Coulomb term." It is also negative because it represents the repulsive force of the protons in the nucleus. Notice that it will increase rather dramatically as atomic number (Z) increases and is only somewhat tempered by an overall increase in size (dividing by the cube root of the mass number). Notice that adding more neutrons to a nucleus will increase A without affecting Z, thereby decreasing the deleterious effect of the Coulomb term.

We can't just keep adding neutrons, otherwise the fourth term, $a_a(N–Z)^2/A$, which is the "asymmetry term," will start to become relatively large. The term "asymmetry" comes from the fact that larger differences between the numbers of neutrons (N) and protons (Z) will result in greater instability, especially for lighter nuclides. The "asymmetry" simply refers to the imbalance between these two values. This might seem a little too simplistic as stability for heavier nuclides means that their N/Z ratios need to be larger than one. That's why this term is divided by mass number (A)—to dampen its effect for heavier nuclides. The asymmetry term also corrects for the change in breadth of the isobaric parabolas as mass number increases. This minor detail was briefly mentioned back in Section 5.6. Isobars with small values of A exhibit rather narrow parabolas, while those with larger values have broader parabolas.

The final term (δ) is the pairing term, and is affected by the oddness (or evenness) of N and Z. Unlike electrons, protons and neutrons really like to pair up with another of their kind, i.e., even numbers of either protons or neutrons add stability and odd numbers increase instability (Table 1.1). When its mass number is odd, the nuclide is either odd–even or even–odd, and the pairing term (δ) is zero. That's our baseline. When mass number is even, then the nuclide is either even–even and $\delta = +a_\delta A^{-1/2}$, or odd–odd and $\delta = -a_\delta A^{-1/2}$. Notice when the pairing term is positive (nuclide is *ee*), it contributes to nuclear binding energy, and when negative (nuclide is *oo*), it takes a little something away from it. This also results in just one parabola when A is odd, but two parabolas when A is even (Section 5.6).

The values for the constants (a_x) in Equation 6.2 are determined by careful measurements of nuclear masses and adjusting the constants so the nuclear binding energy fits with the measured mass defect. A set of values is given in Table 6.1.[1]

Example 6.2 Using the semi-empirical mass equation, calculate the nuclear binding energy of ^{56}Fe.

This is the same as Example 3.1, only the approach is different. The result should be the same. ^{56}Fe has 26 protons and 30 neutrons. Since it is an *ee* nuclide, δ is $+11 \times A^{-1/2}$.

$$E_B = 15.56\,\text{MeV} \times 56 - 17.23\,\text{MeV} \times 56^{2/3} - 0.7\,\text{MeV} \times \left(\frac{26^2}{\sqrt[3]{56}}\right) - 23.285\,\text{MeV} \times \frac{(30-26)^2}{56} + \frac{11\,\text{MeV}}{\sqrt{56}}$$

$$= 871\,\text{MeV} - 252\,\text{MeV} - 124\,\text{MeV} - 7\,\text{MeV} + 1\,\text{MeV}$$

$$= 490\,\text{MeV}$$

---

1   These values vary somewhat from one textbook to another.

This compares quite well with the value obtained in Example 3.1 (492.3 MeV). Notice that the final two terms (asymmetry and pairing) have a relatively small effect on the final value.

The semi-empirical mass equation calculates nuclear binding *energy* despite the fact that its name suggests that *mass* is the goal. As we've seen, mass and energy are fairly interconvertible, especially at the subatomic level. Remember from Section 3.1, that energy values can be converted to mass using the conversion factor of u/931.5 MeV.

We can also explore the effects of the various terms in Equation 6.2 graphically (Figure 6.3). If nuclear binding energy is calculated using solely the volume term, then divided by mass number to obtain binding energy per nucleon, a constant value (a_v = 15.56 MeV) results. This is represented by the solid black line in Figure 6.3. At first glance this looks like a very poor fit with the experimental values (solid gray line), but remember that the volume term overestimates the amount of strong force in a nucleus. What if we used ~8 MeV for the constant a_v? We'd still have a straight line, but it'd be pretty close to many of the experimental values. A pretty remarkable fit for such a mathematically simple model. If we calculate binding energy per nucleon using both the volume and surface terms, we get the line with the long dashes in Figure 6.3, which is a lot closer to the experimental curve. Notice how adding the surface term affects the smaller nuclei more dramatically than large ones. This makes sense because small nuclei have a larger proportion of nucleons on the surface. Adding the Coulomb term into the mix (line with short dashes) helps get us even closer to the experimental values. Notice that it affects the larger nuclides more dramatically because they contain more protons. Finally, when we use almost all of the semi-empirical mass equation by adding in the asymmetry term (dotted line), the match with what is experimentally observed is pretty darn good. Including the pairing term isn't necessary at this level (the scale of Figure 6.3). As we saw in Example 6.2, this is a minor perturbation, and would only affect *oo* and *ee* nuclides. It would result in almost imperceptible "oscillations" in Figure 6.3 as drawn here.

Figure 6.3 also nicely sums up what we've learned from the semi-empirical mass equation about the main forces at work within the nucleus. First there's the strong force (the volume term), working to hold all the nucleons together in the nucleus. Nucleons on the surface do not feel as much strong force, so when a nucleus is small and has a large percentage of its nucleons on the surface, they cannot be held as tightly. The most disruptive force in the nucleus is the Coulomb repulsion—all of the protons trying to push each other away. As nuclei get larger, this becomes more significant.

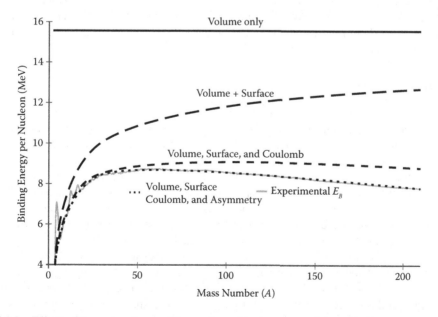

FIGURE 6.3 Effects of the various terms of the semi-empirical mass equation on binding energy per nucleon.

Optimal balance between minimizing the percentage of nucleons on the surface and Coulomb repulsion is achieved when $Z \sim 26$ (iron). Could this be what the Skywalkers represent in the *Star Wars* movies?

Close examination of Figure 6.3 reveals a few nuclides where the match is not so good. ^{4}He, ^{12}C, and ^{16}O have experimental binding energies significantly higher than predicted using the first four terms in the semi-empirical mass equation. These nuclides are all even–even, so including the pairing term in the graph would help, but the main issue for ^{4}He and ^{16}O is that they are doubly magic. Since Equation 6.2 doesn't account for filled shells, we should expect these anomalies. The observed differences for these nuclides are also more pronounced because they are small. There's not a lot of strong force (the volume term is small), so the effect of having a filled shell is relatively significant.

We can now reiterate what we learned about the general shape of Figure 3.1 back in Section 3.1. The low mass end is dominated by the surface term because such a large percentage of the nucleons are on the surface, and being on the surface means less strong force per nucleon. Beyond ^{56}Fe (or so), the Coulomb term becomes more significant as the increasing numbers of protons seek to push the nucleus apart. Balance between the surface and Coulomb terms is best achieved for the nuclides with mass numbers (A) between ~50 and ~100, the sweet spot! These nucleons are bound most tightly and have the highest values of binding energy per nucleon. They also have the lowest masses of their individual nucleons (Figure 3.2).

The semi-empirical mass equation is part of a larger nuclear theory known as the "liquid drop" model. In the absence of gravity a drop of water will form a sphere from all of the intermolecular interactions between the molecules. In the liquid drop model, we imagine the strong force as analogous to these interactions. Water molecules on the surface will feel less attraction, because they are not completely surrounded by other molecules (the surface term). The floating droplet tries to minimize surface area by forming itself into a sphere (as do many nuclei). We'll see later that the liquid drop model can also help us understand how nuclear fission takes place, but it has its limitations. For example, to understand the anomalies in Figure 6.3, and better model nuclear structure in general, it is necessary to use a quantum mechanical model, which we'll call the shell model.

Why do we need more than one model to understand nuclear structure? Electronic structure and spectroscopy are modeled adequately with just quantum mechanics. Why is the nucleus different? The situation is much more complex with the nucleus than it is with electrons. There are two forces, electrostatic (Coulomb repulsion) and the strong force, operating in the nucleus and there are two different nucleons, protons and neutrons, jammed into some pretty close quarters. Because of this extra complexity, and our relatively poor understanding of the nucleus, we use a couple of different models to describe nuclear phenomena.

6.3 THE SHELL MODEL

Quantum mechanics helps us understand the spaces (orbitals) that an atom's electrons occupy as well as the energies of light that are emitted or absorbed when electrons move from one energy level to another. Similarly, quantum mechanics can help us understand the spins and parities observed for nuclei, as well as relative stabilities, possible radioactive decays, and nuclear reactions.

Nucleons follow somewhat different rules than electrons. Remember that the first few electron sublevels that get filled are 1s, 2s, 2p, 3s, 3p, 4s, and 3d, in order of increasing energy. There are a couple of fun rules to help remember the ordering—rules that come from the quantum mechanical model of the atom. Because nucleons are different than electrons, and because they also have the strong force acting between them, the ordering of nucleon states is different. Using a parabolic nuclear potential well (Figure 6.4a), the states increase in energy according to $2(n-1) + l$, where n and l are the principal and angular momentum quantum numbers, respectively. The 1s nucleon state is the lowest with $2(1-1) + 0 = 0$. The 1p is next lowest with a value of one. Notice it is the *one*-p sublevel. Another difference between electrons and nucleons is that n can have *any* positive integer

FIGURE 6.4 Possible nuclear potential wells: (a) a parabola and (b) a rounded square.

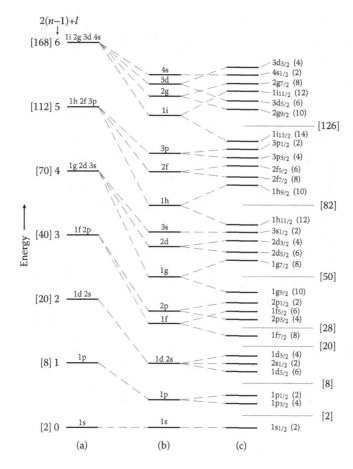

FIGURE 6.5 Nucleon energy levels based on (a) a parabolic nuclear potential well, (b) a rounded square well, and (c) accounting for spin-orbit coupling. Adapted from Mayer, M. M. and Jensen, J. H. D., 1955.

with any value of l. Yep, quantum mechanics is weird. $2(n-1) + l = 2$ for both the 1d and 2s (check it yourself!) states, so they are degenerate in Figure 6.5a. Notice that the number of degenerate sublevels increases at even values of $2(n-1) + l$. The $2(n-1) + l$ values are given just to the left of the energy levels in Figure 6.5a.

To the left of the $2(n-1) + l$ values in Figure 6.5a is another number in square brackets. This is the running total of nucleons after each level is filled. Remember that an s state is filled by two nucleons, p by 6, d by 10, and so on as shown in Table 6.2. The running total is significant because we should see significant gaps in energy after shell closures—i.e., at the magic numbers of 2, 8, 20, 28, 50, 82, and 126. This works great for the first three (2, 8, and 20), but fails after that. This shouldn't be

TABLE 6.2
Number of Nucleons Required to Fill Each Value of l

Value of l	0	1	2	3	4	5	6
Letter Designation	s	p	d	f	g	h	i
Number of Nucleons to Fill	2	6	10	14	18	22	26

too surprising as the parabolic nuclear potential well is pretty simple, and somewhat unrealistic. We know the strong force really only works between neighboring nucleons, so why would the potential continue to deepen (strong force increase) after a nucleon is below the surface? It should not. This is perhaps why this model works for the first three shell closures but not for larger nuclides. Lighter nuclides don't have a lot of nucleons below the surface, so their potentials can be reasonably approximated by a parabola.

The second nuclear potential well (Figure 6.4b) has steeper walls and a flattened bottom, suggesting that the strong force felt by a particular nucleon is more or less constant, except near the surface. Surely, this will be a better model. Notice that moving from the parabolic to the rounded square potential lowers the energy of the sublevels. The steeper walls of the rounded square allow more nucleons to sit at or near the bottom of the well. These nucleons feel the warm embrace of others all around them. Also notice that the sublevel degeneracies are broken by using the rounded square potential (Figure 6.5b). In every case, the sublevel with the highest l ends up lowest in energy. Higher l means that the nucleons are spending more time further from the center of the nucleus (remember the difference in shape between p and s electron orbitals?). In the parabolic well these nucleons are crawling up the sides more, so moving to the rounded square potential drops them more significantly. Sadly, we still don't have large energy gaps at shell closures beyond 20 nucleons. In fact, all of them occur while in the midst of filling a sublevel. For example, the 28th nucleon placed in the nucleus is the eighth to populate the 1f sublevel (out of a total of 14). Another fail. We clearly need to account for something else—something that further breaks up the degeneracies of the sublevels (Figure 6.5c).

The next level of complexity we'll consider adding to our model is spin-orbit coupling, which is basically a combination of the angular momentum (l) with the spin (s)[2] quantum numbers. It turns out that quantum numbers aren't totally independent of each other, especially when they've both got some angular momentum associated with them. Mathematically, we'll mix their values to make a new quantum number, j, which will represent the total angular momentum (Equation 6.3).

$$j = l \pm s \tag{6.3}$$

The 1p sublevel gives us j values of $1 + 1/2 = 3/2$ and $1 - 1/2 = 1/2$. The 1p nucleon levels are therefore split into the $1p_{3/2}$ and $1p_{1/2}$ sublevels. Notice that the higher value of j ends up at lower energy—it turns out that addition of l and s always gives the lower energy value. Notice also that the j values become the subscripts to distinguish the now non-degenerate sublevels. Applying spin-orbit coupling to the 1d sublevels splits them ($2 \pm 1/2$) into the $1d_{5/2}$ and $1d_{3/2}$ sublevels. Notice here that the splitting (vertical separation on Figure 6.5c) is a little wider. As l increases, so does the magnitude of the splitting. This turns out to be pretty handy, as the 1g splits so much that a significant energy gap now opens up between the 50th and 51st nucleons, exactly where we needed it! Likewise, large

---

2   Often represented as m_s for electrons.

splittings of the 1h and 1i sublevels occur, providing large gaps for the 82 and 126 nucleon shell closures. We've finally got a model that accounts for filled levels of nucleons.

The number of nucleons that can occupy each sublevel is given in parentheses to the right of each sublevel in Figure 6.5c. Notice that these values can be calculated by adding one half to the total angular momentum value, then multiplying by two $((j+1/2) \times 2)$. Further to the right in this diagram are numbers in brackets which represent a total number of nucleons reached at that point. As you may have already noticed they are the magic numbers, i.e., the numbers of nucleons in a filled shell.

6.4   APPLICATIONS OF THE SHELL MODEL

It's only a model, and therefore has limitations. Some of the energy levels are very close to one another and their relative ordering can shift as the levels are filled (just like electrons). Also, not all nuclei are perfectly spherical (which is an assumption hidden in our model) as we learned back at the beginning of this chapter. When nuclear shape loses symmetry, the relative ordering of the energy levels in Figure 6.5c changes, especially at higher energies (higher Z or N). In other words, if we observe something that doesn't strictly follow Figure 6.5c, we shouldn't be too concerned. It's still a pretty good model, so let's see what we can do with it.

6.4.1   SPIN AND PARITY

As we learned back in Section 6.2, nucleons love to pair up with another of their kind. Therefore, if we have an even number of protons or neutrons, we'll know they are all paired up (in the ground state). When they pair, their individual spins cancel each other, just like electrons. If the nuclide has even numbers of both protons and neutrons, all nucleon spins are exactly cancelled out, the nuclear spin is zero (Section 5.7) and the nucleus is spherical. If a nuclide has odd numbers of protons or neutrons, the nuclear spin is determined by the unpaired nucleon(s). It turns out the nucleon level occupied by the unpaired nucleon gives us the nuclear spin and parity. Nice, eh?

When there's only one unpaired nucleon, i.e., an oe or an eo nuclide, the nuclear spin and parity can be determined (guessed) from the nucleon state the lone proton or neutron occupies. It is equal to the value of j for the sublevel occupied by the lone nucleon. For example, ^{15}N has seven protons and eight neutrons. All the neutrons and six of the seven protons are paired up and cancel out each other's spin. The seventh proton sits in the $1p_{1/2}$, therefore, we'd expect ^{15}N to be spin 1/2, and it is! Likewise, ^{11}B has a lone proton in the $1p_{3/2}$ and has a nuclear spin of 3/2. Parity is determined from the value of l for that same nucleon state. If l is even (Table 6.2), parity is even (positive). When l is odd, parity is odd (negative). In both examples here, $l = 1$, so parity is odd (negative).

Example 6.3   Using Figure 6.5c, guess the spin and parity for ^{25}Mg.

^{25}Mg has 12 protons and 13 neutrons. The 13th is the penultimate neutron placed in the $1d_{5/2}$ nucleon sublevel, so we'd expect a 5/2 nuclear spin (since $j = 5/2$), and even parity (since $l = 2$). Both are correct, check it out with your favorite chart of the nuclides, or the nuclear wallet cards (www.nndc. bnl.gov/wallet).

Things can get a little squirrelly with higher numbers of nucleons. The energy levels are much closer together,[3] making it easier to reorder a bit as they fill. This is especially true when a high spin state is predicted. For example, the isotones $^{199}_{78}$Pt$_{121}$, $^{203}_{82}$Pb$_{121}$, $^{205}_{84}$Po$_{121}$, $^{207}_{86}$Rn$_{121}$, and $^{209}_{88}$Ra$_{121}$ all have even Z and 121 neutrons. According to Figure 6.5c, the lone neutron should occupy the $1i_{13/2}$, making these nuclei spin 13/2 with even parity. Instead, they are spin 5/2, and have odd parity.

---

3   This is not well-represented in Figure 6.5c due to space restrictions.

Scanning a chart of the nuclides shows that 5/2– is pretty common in this region and that high spin values for ground state nuclei are generally quite rare. What seems to happen in this part of the chart is that the $1i_{13/2}$ sinks a bit and the $2f_{5/2}$ comes up to become the highest energy sublevel before the 126 neutron shell closure. This is not necessarily due to distortions of the nuclear shape, rather an effort to avoid high spin values. $^{201}_{80}Hg_{121}$ differs from the other $N = 121$ isotones because it is spin 3/2 with odd parity. Again, it seems likely that the energy levels near the 126 neutron shell closure have shifted a bit, making the $3p_{3/2}$ now the highest sublevel below the gap.

What if we have more than one unpaired nucleon? Can we still guess the spin and parity of *oo* nuclides? Of course we can, it's just a bit more complicated, and even more fraught with doubt. Let's start with the simplest *oo* nuclide, deuterium ($^2_1H_1$). When we think of protons and neutrons filling up energy levels, we need to remember that they fill up separate sets of levels. Imagine two copies of Figure 6.5c, one for neutrons and one for protons. In this case the lone proton sits in its $1s_{1/2}$ sublevel and the lone neutron sits in its $1s_{1/2}$ sublevel. Separately, each nucleon is spin 1/2 with even parity. To find the nuclear spin, we need to combine the two spin values by adding and subtracting them.

$$J = 1/2 + 1/2 = 1$$

$$J = 1/2 - 1/2 = 0$$

The nuclear spin is represented by J to help distinguish from the quantum number j. The nuclear spin for deuterium is any integer from zero to one, i.e., it is either 0 or 1, we can't be certain which. Looking up ^2H on a chart of the nuclides tells us it is spin 1. Parity is determined a bit more definitively for *oo* nuclides. Simply multiply the two parity values by each other. In this case positive times positive yields positive (even). Spin and parity values are often combined and symbolized as J^π. For deuterium the spin/parity is 1+. Note that *oo* nuclides will always have integral spins, while *oe* and *eo* nuclides will be half-integral in the ground state.

Example 6.4 Guess the spin/parity for ^{28}Al.

^{28}Al has 13 protons and 15 neutrons. The 13th proton is in the $1d_{5/2}$ nucleon sublevel, and the 15th neutron is in the $2s_{1/2}$ sublevel. Add and subtract the spin values.

$$J = 5/2 + 1/2 = 3$$

$$J = 5/2 - 1/2 = 2$$

The nuclear spin is therefore 2 or 3. The parity is positive times positive, which yields positive. As nearly as we can figure, the spin and parity is either 2+ or 3+. A quick lookup tells us J^π is 3+ for ^{28}Al.

Let's be a bit more deliberate in our approach to Example 6.4 by checking the spin values with neighboring isotopes and isotones before we add and subtract them. Both $^{27}_{13}Al_{14}$ and $^{29}_{13}Al_{16}$ are spin 5/2, which agrees nicely with our assumption that the lone proton for all three isotopes (^{27}Al, ^{28}Al, and ^{29}Al) sits in the $1d_{5/2}$ sublevel. Since ^{27}Al and ^{29}Al have even numbers of neutrons, the lone proton dictates the nuclear spin. It turns out that the isotones $^{27}_{12}Mg_{15}$ and $^{29}_{14}Si_{15}$ are both 1/2+, again affirming our assumption that the 15th neutron is $2s_{1/2}$.

As another example, let's apply this more deliberate approach to $^{144}_{61}Pm_{83}$. Its two neighboring isotopes are $^{143}_{61}Pm_{82}$ and $^{145}_{61}Pm_{84}$. Both have spin/parity values of 5/2+, suggesting the unpaired proton occupies the $2d_{5/2}$ orbital, consistent with Figure 6.5c. Its two neighboring isotones, $^{143}_{60}Nd_{83}$ and $^{145}_{62}Sm_{83}$, are 7/2–, suggesting the lone neutron is in the $2f_{7/2}$ instead of the $1h_{9/2}$, which is predicted by Figure 6.5c. The range for the nuclear spin for ^{144}Pm can be calculated with some confidence.

$$J = 7/2 + 5/2 = 6$$

$$J = 7/2 - 5/2 = 1$$

Nuclear spin can then be 1, 2, 3, 4, 5, or 6. That's quite a range. It turns out that the nuclear spin is 5 for ^{144}Pm. As a rule of thumb, the nuclear spin values tend to be either the minimum or maximum values of the calculated range. If they are not the highest or lowest value, then they are typically only one removed from the extremes. As usual, parity is clearly odd (negative) since we are multiplying a positive parity by a negative parity, $J^{\pi} = 5-$.

We can now understand the assertion made in Section 5.7 that when A is odd, the ground state total angular momentum is $x/2$ (where $x = 1, 2, 3, \ldots$) and when A is even, J is 0 or a positive integer. When mass number is odd, the nuclide is oe or eo, i.e., it has only one unpaired nucleon. Nuclear spin is then determined by the sublevel occupied by the unpaired nucleon, and is therefore fractional. When there are odd numbers of both protons and neutrons (oo), there are two fractional spins which combine to give a whole number.

We can also use the shell model to understand excited nuclear states. Excited states can be created by either promoting a nucleon from a lower level to the highest occupied ground state level or by promoting a nucleon from the highest occupied ground state level to an unoccupied level. Let's look at examples of each.

Example 6.5 $^{137}_{56}$Ba$_{81}$ (ground state) has a spin and parity of 3/2+ and $^{137m}_{56}$Ba$_{81}$ is 11/2−. Which nucleon states have the unpaired neutron for these two nuclides?

Both isomers have an even number of protons and 81 neutrons, which should make the ground state 11/2− ($1h_{11/2}$) according to Figure 6.5c. Huh? Must be another example of excessive spin avoidance. The highest energy sublevel below the 82 neutron shell closure must be $2d_{3/2}$ to give $J^{\pi} = 3/2+$. A quick scan of this region of the chart of the nuclides shows 3/2+ is pretty common, while 11/2− is absent for ground states. Moving the $2d_{3/2}$ sublevel up to the top, Figure 6.6 illustrates the neutron configurations for these two isomers between the 50 and 82 neutron shell closures. The only difference between the ground and metastable states is a neutron vacancy in the $2d_{3/2}$ sublevel in the ground state drops to the $1h_{11/2}$ in the metastable state, i.e., one of the ground state $1h_{11/2}$ neutrons is promoted to the $2d_{3/2}$ in the metastable state.

The $2d_{3/2}$ wasn't the only sublevel to shift between Figures 6.5c and 6.6. Notice that the $1h_{11/2}$ dropped down to third place (will that keep them from seeing postseason play?). This was done because there's another excited state between the metastable and ground states of ^{137}Ba. If the vacancy drops just one sublevel, from the ground state to the $3s_{1/2}$, a nuclide that's 1/2+ forms.

FIGURE 6.6 Neutron configurations for ground state (left) and metastable state (right) of ^{137}Ba.

According to your favorite online chart of the nuclides, this intermediate excited state is about 284 keV above the ground state. The metastable state is about 662 keV above the ground state.

Example 6.6 Ground state $^{209}_{83}\text{Bi}_{126}$ has a spin and parity of 9/2– and its lowest energy excited state is 7/2–. Which nucleon states have the unpaired proton for these two isomers?

Both isomers have an even number of neutrons and 83 protons, which makes the ground state 9/2– ($1h_{9/2}$) according to Figure 6.5c. Yay, this one fits the shell model! Figure 6.7 shows the unpaired proton configurations for these two isomers. The unpaired proton occupies the $1h_{9/2}$ sublevel in the ground state and jumps to the $2f_{7/2}$ in the excited state. Delightfully straightforward.

Remember that the shell model has limitations. It tends to work best when the nucleus is spherical, which is the case near shell closures or when $J = 0$ or 1/2. Nuclei more distant from shell closures tend to be prolate (more common) or oblate and are less likely to nicely follow the shell model. Even when the nucleon configuration is close to a shell closure, the energy levels can shift around, especially at higher energies when spin can be reduced. Figure 6.5c is really just a rough guide that helps us better understand experimental data.

FIGURE 6.7 Proton configurations for ground state (left) and excited state (right) for ^{209}Bi.

6.4.2 DECAY

Armed with the liquid drop and shell models of the nucleus, let's see if we can better understand radioactive decay, starting with another look at von Weizsäcker's semi-empirical mass equation (Equation 6.2). Remember from Section 5.6 that Equation 6.2 provides a parabola which approximately fits a group of isobars such as those shown in Figures 5.14 and 5.16. In other words, it predicts beta decay—all three flavors. Let's see how quantitative it is.

Example 6.7 Use Equation 6.2 to predict the total energy of decay for ^{56}Co.

^{56}Co decays to ^{56}Fe via electron capture or positron decay releasing 4.6 MeV. We already calculated the nuclear binding energy for ^{56}Fe in Example 6.2 as 490 MeV. If we similarly calculate the nuclear binding energy for ^{56}Co, the difference should be the total energy of decay.

$$E_B = 15.56\,\text{MeV} \times 56 - 17.23\,\text{MeV} \times 56^{2/3} - 0.7\,\text{MeV} \times \left(\frac{27^2}{\sqrt[3]{56}}\right) - 23.285\,\text{MeV} \times \frac{(29-27)^2}{56} - \frac{11\,\text{MeV}}{\sqrt{56}}$$

$$= 871\,\text{MeV} - 252\,\text{MeV} - 133\,\text{MeV} - 2\,\text{MeV} - 1\,\text{MeV}$$

$$= 483\,\text{MeV}$$

$$490\text{ MeV} - 483\text{ MeV} = 7\text{ MeV}$$

It's a bit off, but remember the semi-empirical mass equation is a simple mathematical model. This is clearly not the best way to calculate energies of decay, but it gives us a rough idea of beta decay within a set of isobars.

The semi-empirical mass equation (Equation 6.2) can also give us an idea of where the parabolas will bottom out. In other words, tell us the optimal N/Z ratio for any value of A (Equation 6.4).

$$Z_A = \frac{A}{2}\left(\frac{81}{80+0.6A^{2/3}}\right) \qquad (6.4)$$

The quantity Z_A is the optimal atomic number for any set of isobars with mass number A.

Example 6.8 Estimate the optimal N/Z ratio for a set of isobars with 104 nucleons.

We did this more deliberately in Section 5.6 and guessed it would be approximately 1.32 or 1.33. Let's see what Equation 6.4 tells us.

$$Z_A = \frac{104}{2}\left(\frac{81}{80+0.6(104)^{2/3}}\right) = 45.2$$

Remember this is the optimal number of protons, which matches closely with rhodium. So far so good. The number of neutrons is then

$$104 - 45.2 = 58.8$$

and the optimal N/Z ratio is

$$58.8/45.2 = 1.30$$

Not bad for such a seemingly simple equation! Try this for yourself on your favorite isobars.

Let's shift to the shell model and use it to take a relatively simple look at the three flavors of beta decay. Figure 6.8 shows somewhat simplified energy level diagrams for the nucleons in three isobars, $^{12}_{5}B_7$, $^{12}_{6}C_6$, and $^{12}_{7}N_5$. These diagrams are "somewhat simplified" in that the spacings between and precise placements of the sublevels are not accurate. However, a couple of aspects of Figure 6.8 are deliberate. Notice first that the proton ("p") sublevels are placed higher in energy than their neutron ("n") counterparts. This is because protons have charge and repel one another. Secondly, the proton sublevels are closer to one another than corresponding neutron sublevels. As we move to higher energy, the difference between corresponding proton and neutron sublevels becomes smaller. Remember that as nucleons occupy these sublevels, their energies can shift.

What does Figure 6.8 tell us about beta decay? Note what happens to the nucleon configurations when ^{12}B undergoes beta minus decay to ^{12}C—a neutron changes into a proton and the resulting configuration is lower in energy (more stable!). This answers one of the questions posed at the beginning of this chapter—"How is it even possible to be neutron-rich?" If a nuclide has too many neutrons,

FIGURE 6.8 Nucleon energy levels for three isobars.

converting one to a proton results in a lower-energy configuration. A similar thing happens when ^{12}N undergoes positron decay to ^{12}C. Converting a relatively high-energy proton to a lower-energy neutron results in greater stability. Beta decay is all about maintaining proper balance between Z and N (ZeN?). Also notice that both parents illustrated in Figure 6.8 have two unpaired nucleons, and the daughter has all of its nucleons paired up. Remember that pairing of nucleons results in greater stability, so we'd expect a little lowering of the sublevel energies in the daughter (^{12}C). While this is not shown (Figure 6.8 is "somewhat simplified"), we can imagine a slight lowering in energy of the sublevel where pairing occurs—like a cat settling into your lap for a nap.

When the imbalance between protons and neutrons goes to the extreme, particle emission is observed (Section 5.8.3). A great example is ^{15}F, which decays via proton emission to ^{14}O. ^{15}F has nine protons and only six neutrons, making it pretty unbalanced. Making matters worse, the unpaired proton sits in the $1d_{5/2}$ sublevel, just above the gap associated with filling a shell. It is as if that proton is dangling on a precipice high above the neutrons—so high, it's easier to break loose of the nucleus rather than convert into one. Physically, this proton sits a bit *outside* the rest of the nucleus and is only very weakly bound, facilitating particle emission as a decay mode. Because one of its protons is loose, ^{15}F is referred to as a **halo nucleus**—its fuzziness extends a bit further than its more stable isobars. Attempting to add an additional proton to ^{15}F would likely be impossible as it would remain unbound.

Likewise, nuclides with an overabundance of neutrons relative to their protons will reach a point where adding an additional neutron will not be successful. These limits are called **drip lines** and they form boundaries well above and below the belt of stability on the chart of the nuclides. The proton-rich drip line is pretty well defined by the most proton-rich nuclides on the chart of the nuclides. In other words, we've pretty much observed all of the possible proton-rich nuclides. However, the drip line on the neutron-rich side of the chart is generally a good distance from the known edge. This suggests that there is a significant number of neutron-rich nuclides yet to be observed, and is an active area of contemporary research.

What do the liquid drop and shell models tell us about delayed particle emission (Section 5.8.4)? Remember that these decay modes are typically observed when a beta minus or positron decay generates a highly excited daughter. If the excited daughter has a proton or neutron in a high-energy sublevel, it will be loosely bound, just like the halo nuclei that undergo particle emission. Spitting out this nucleon then becomes competitive with or preferential to isomeric transition.

Decay pathway preferences can also be qualitatively understood with the help of the shell model. ^{60}Co has a spin and parity of 5+ (is that what you would've predicted?) and decays via beta minus to ^{60}Ni which is 0+ (it is *ee*). A decay diagram is shown in Figure 6.9. Its main observed pathway is to emit a beta particle with a maximum energy of 0.317 MeV forming an excited state of ^{60}Ni that is 4+. That excited state decays via isomeric transition to another one (2+) emitting a 1.17 MeV gamma photon. The 2+ excited state then decays via IT to the ground state emitting a 1.33 MeV gamma photon.

FIGURE 6.9 Decay diagram for ^{60}Co.

Remember from Section 5.7 that changes in nuclear spin primarily determine the likelihood of an isomeric transition. The larger the change, the less likelihood of the transition. We would therefore conclude that a 2.50 MeV gamma photon is not observed in the decay of ^{60}Co because the change in total angular momentum from 4 to 0 (the higher excited state directly to the ground state) is too large. Reality is slightly more complex. A 2.50 MeV is observed,[4] but in only 2×10^{-6}% of all decays of ^{60}Co. We can understand the difficulty a nucleus has changing its total angular momentum in terms of inertia. Angular momentum, like linear momentum, does not like change. The more change in J, the less likely the transition.

Herein lies the origin story of metastable states. When an isomeric transition is forced to go through a relatively large change in spin, longer half-lives result and the isomer becomes metastable. Sounds like a good episode of *The Flash*. As an example, ^{99m}Tc is 1/2– and the only lower energy isomers are 7/2+ and 9/2+ in order of decreasing energy. The latter is the ground state. Both isomeric transitions are impeded by the significant changes in spin, so ^{99m}Tc has a half-life of about 6 hours. What about the changes in parity? It turns out that a change (or no change) in parity has a somewhat variable, but definitely subtle effect on the probability of a transition. For these reasons, it is ignored in this text. Take another look at Examples 6.5 and 6.6. Can you explain why ^{137}Ba has a metastable state at 662 keV above the ground state but not one at 284 keV above the ground state? Can you also explain why the lowest level excited state for ^{209}Bi is not metastable?

Back in Section 5.7 we learned that conversion electrons are more likely to come from the inner electron shells (lower energy levels like K or L) than outer. Quantum mechanics tells us that the wave functions for the nucleons and the electrons need to overlap for the excess nuclear energy to transfer. This overlap is greatest for the K shell electrons, so they have the highest probability (by far!) of getting booted during an internal conversion decay. Also, remember that electrons in p, d, and f levels all have a node (zero probability) at the nucleus, making them especially hard to kick out.

Similarly, the existence of electron capture decay is a validation of quantum mechanics. If the Bohr model (Figure 5.8) were strictly correct, then there's no way an electron in a fixed orbit could ever interact with the nucleus. Because quantum mechanics allows for a non-zero probability that electrons will occupy space within the nucleus, electron capture can happen. This also helps us understand why electron capture is more common among higher Z nuclides. The electrons are held more tightly by the increased number of protons, therefore their wave functions are more likely to overlap with the nucleus, and electron capture is facilitated.

Change in angular momentum also affects the probability of other decay modes. Notice that two beta pathways are illustrated in Figure 6.9. The beta with a maximum energy of 0.317 MeV (5+ to 4+) is observed in 99.9% of all decays of ^{60}Co, while the beta with a maximum energy of 1.49 MeV (5+ to 2+) is observed in only 0.1% of all decays of ^{60}Co. No beta is observed leading directly to the ground state of ^{60}Ni (5+ to 0+). As with isomeric transitions, larger changes in total angular momentum mean a lower probability for that particular pathway.

The same behavior is observed in alpha decay. For example, 94.4% of the time, ^{226}Ra (0+) emits a 4.78 MeV alpha particle when decaying to the ground state of its daughter (^{222}Rn, also 0+). The remaining 5.6% of the time it emits a 4.60 MeV alpha particle, producing an excited state of ^{222}Rn with a J^{π} of 2+. There are a few other pathways observed, but their branch ratios are less than 0.00007 and involve greater changes in spin. Note that both parent and daughter in this decay are *ee*, meaning they will both be 0+. Alpha decay of *ee* nuclides will generally feature few accompanying gamma photons as the ground state to ground state alpha transition will be strongly favored.

Quantum mechanics can also help us understand *how* alpha and cluster decay occur. As suggested in Section 5.8.2, there is an energy barrier to these decay modes. That barrier can be thought of as overcoming the strong force attraction between the emitted particle and the rest of the nucleus (the

---

4   Other isomeric transitions are also observed, but they are also not drawn because they also occur so infrequently.

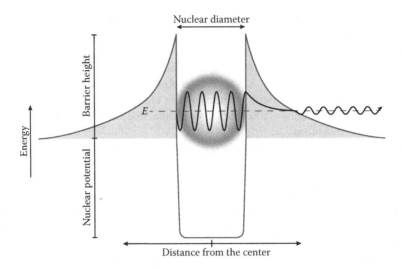

FIGURE 6.10 Alpha particle penetration of the Coulomb barrier.

daughter). So long as the particle (alpha or cluster) is inside the nucleus, it feels the strong force of all the surrounding nucleons—in order to break free, it needs to overcome all of these attractions. It's a bit like throwing a ball against a wall and expecting it to travel through rather than just bounce off. In our macro world, it always bounces. Fortunately, alpha particles (and other emitted clusters of nucleons) have quantum mechanics on their side.

Imagine an alpha particle in a nucleus, bouncing back and forth off the inside of the nuclear surface. It's traveling at about one-tenth the speed of light and hits the barrier about 10^{20} times every second (it's a very small space!). The barrier is significantly higher than the energy of the alpha particle, so classical physics would tell us it's never getting through. Quantum mechanics tells us that it can spend very short periods of time penetrating the barrier, and if the alpha particle has enough energy and the barrier isn't too "thick," it can then tunnel through and escape the nucleus. This is represented in one-dimension in Figure 6.10. When the alpha particle is in the nucleus, it is in the potential well in the center of the diagram. If it has energy E (dashed line), it is part-way up the walls of the barrier (the light gray areas under the curves), but not enough to get over. Its only option is to tunnel, which has a very low probability of success. Fortunately, the alpha particle is bouncing up against the barrier like crazy, eventually it'll get through.

Figure 6.10 represents the alpha particle as a wave because, as a quantum mechanical object, it is best represented as a wave function. Notice that as the wave function penetrates the barrier it diminishes (loses amplitude). When it emerges, it still has the same wave characteristics, except for intensity. In quantum mechanical terms, this simply says that the probability of finding the alpha particle separated from the nucleus is much lower than finding it within. As the energy of the alpha particle increases, the dashed line in Figure 6.10 moves up and there's less barrier to get through. Decay then becomes more probable and the half-life is decreased. While this is not always observed, many groups of nuclides follow this trend. Higher-energy alpha particles are typically observed emitted from nuclides with shorter half-lives.

The thicker the barrier, the greater the loss of amplitude, the lower the probability of alpha decay occurring. If there are more protons in the nucleus, the barrier will be higher. Again, it's not a really solid trend, but, other things being equal, a higher Z will mean a longer half-life. It turns out there's an approximate minimum energy for alpha decay of about 3 MeV. At that point the barrier is so thick, the alpha particle has virtually no chance of making it through. A great example is ^{209}Bi, which emits a 3.1 MeV alpha particle and has a half-life of 2×10^{19} years. That's low probability!

We've looked at this barrier as overcoming the strong force, but that is not strictly correct. If we imagine the opposite process, *adding* an alpha particle *to* a nucleus, we can better understand the barrier. Since an alpha particle and a nucleus both have positive charges, they will repel each other as the alpha particle approaches. We'll see in Chapter 10 that the alpha particle will require a certain amount of energy to overcome this Coulomb barrier. Decay is the opposite process, so we'll get that energy out as the alpha particle separates itself from the nucleus, no longer feels the strong force, and is pushed away by Coulomb repulsion. We'll also have to wait until Chapters 10 and 11 to see how the liquid drop and shell models help us understand nuclear reactions.

We can also further answer a question posed about alpha decay in Section 5.8.2—why is spitting out an alpha particle so popular? As suggested in Section 5.8.2, an alpha particle is a relatively small chunk, so it has less strong force to overcome than cluster decay. We also learned that alpha was favorable because it was doubly magic. Let's think about this in terms of the total energy of decay—having a more stable chunk ejected should provide for greater release of energy. Energies for spitting out various small bits from ^{210}Po are given in Table 6.3. Notice that the only alpha decay leads to a positive total energy of decay. All other emissions are endoergic, and therefore are not possible decay modes.

TABLE 6.3
Total Energy of Decay for ^{210}Po

Decay Mode	alpha	neutron	proton	^{2}H	^{3}H	^{3}He
E (MeV)	5.41	−7.66	−4.98	−10.22	−10.85	−8.43

QUESTIONS

6.1 Define the following: oblate spheroid, the semi-empirical mass equation, liquid drop model, shell model, nuclear potential well, total angular momentum, halo nucleus, and drip line.

6.2 Which image more correctly represents a nucleus? Briefly explain.

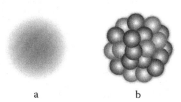

a   b

6.3 List and briefly define the terms that make up the semi-empirical mass equation.

6.4 What is calculated by the semi-empirical mass equation?

6.5 What are the degenerate sublevels in the next higher energy level in Figure 6.5a (where $2(n-1)+l=7$ for the parabolic nuclear potential well)? How many total nucleons will fit into this energy level?

6.6 Why is the rounded square potential well (Figure 6.4b) a more realistic model than the parabolic well (Figure 6.4a)? Why is the rounded square potential still inadequate?

6.7 When spin-orbit coupling is applied to the 4p sublevel, how does it split? Indicate which ends up at lower energy. Do the same with the 2h sublevel. Which splits more (wider energy gap), the 4p or the 2h?

6.8 What does the shell model help predict?

6.9 Where (in the chart of the nuclides) does the shell model not work so well? Briefly explain.

6.10 Using Figure 6.5c (or neighboring nuclides if oo), guess the spin/parity (J^π) for the nuclides listed here. Look up their measured values using an online chart of the nuclides (like nndc.bnl.gov or www-nds.iaea.org) or equivalent app and explain any discrepancies.

$$^{12}\text{N} \; ^{45}\text{V} \; ^{76}\text{Kr} \; ^{95}\text{Zr} \; ^{125}\text{Sn} \; ^{238}\text{Np}$$

6.11 ^{85}Kr (ground state) has a spin and parity of 9/2+ and its metastable state is 1/2–. Which nucleon states have the unpaired neutron for these two nuclides? Is the metastable state likely the lowest-lying excited state for ^{85}Kr?

6.12 ^{207}Tl (ground state) has a spin and parity of 1/2+ and its metastable state is 11/2–. Which nucleon states have the unpaired proton for these two nuclides? Is the metastable state likely the lowest-lying excited state for ^{207}Tl?

6.13 ^{112}In (ground state) has a spin and parity of 1+ and its metastable state is 4+. Which nucleon states have the unpaired nucleons for these two nuclides? Is the metastable state likely the lowest-lying excited state for ^{112}In?

6.14 After the one illustrated in Figure 6.7, the next higher excited state for ^{209}Bi is 13/2+. Sketch its proton configuration (similar to Figure 6.7).

6.15 Use Equation 6.2 to calculate the total energy of decay for ^{86}Br. How does it compare to the value given in Appendix A?

6.16 What is the optimal N/Z ratio for $A = 99$? How does this compare to the estimate made in Section 5.6?

6.17 Draw the nucleon energy levels for ^{14}C, ^{14}N, and ^{14}O and indicate their decay modes like Figure 6.8. Compare your drawing with Figure 6.8.

6.18 Draw a single decay diagram for both ^{130}Cs and ^{130}Ba using information in Appendix A. Obtain spin/parity values for all nuclidic states from a nuclide chart. Do changes in J^π figure into any of these transitions?

6.19 A partial decay diagram for ^{90}Sr is shown in Figure 6.11. The ground states are indicated with a nuclide symbols. Spin and parity are given next to each nuclidic state. Complete the decay diagram by drawing arrows where the most likely decay pathway(s) will occur, indicating the decay mode(s).

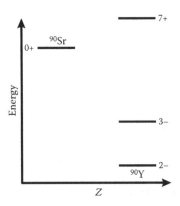

FIGURE 6.11 A partial decay diagram for ^{90}Sr.

6.20 A partial decay diagram for some of the $A = 82$ isobars is shown in Figure 6.12. The ground states are indicated with nuclide symbols. Spin and parity, and relative energy are given to the left and right (respectively) of each nuclidic state. Complete the decay diagram by drawing arrows where the most likely decay pathways will occur, indicating each decay mode. Calculate the maximum energies for all emitted particles and energies of all photons making sure they are clearly associated with a particular transition.

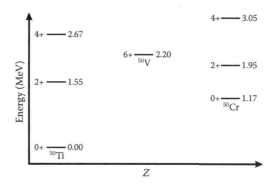

FIGURE 6.12   A partial decay diagram for some of the $A = 82$ isobars.

6.21 A partial decay diagram for ^{209}Po is shown in Figure 6.13. The ground states are indicated with nuclide symbols. Spin/parity and relative energy are given to the left and right (respectively) of each nuclidic state. Alpha decay to all daughter isomers is possible. Draw in the arrows and indicate which path is most likely and which is least likely. Calculate the energies for all three alpha emissions and the corresponding daughter recoil energies. Note that

FIGURE 6.13   A partial decay diagram for ^{209}Po.

the positions of the daughter excited states are not properly drawn to scale.

6.22 The most commonly observed alpha particle is almost always the most energetic alpha possible for that decay. In other words, the likeliest alpha decay pathway is typically from ground state parent to ground state daughter. Briefly explain.

6.23 What are the observed trends between alpha energy and half-life in alpha decay? Between atomic number and half-life? These trends are most consistently observed for *ee* nuclides. Briefly explain.

6.24 The most energetic alpha (MeV) emissions for some of the *ee* isotopes of uranium are given here. List these isotopes in order of increasing half-life.

^{226}U 7.56, ^{228}U 6.68, ^{230}U 5.89, ^{232}U 5.32, ^{234}U 4.78, ^{236}U 4.49, ^{238}U 4.20

6.25 List the following nuclides in order of increasing half-life for alpha decay: ^{230}Th, ^{242}Pu, ^{234}U, and ^{248}Cm. They all decay via alpha and have roughly the same total energy of decay (~4.9 MeV).

6.26 Calculate the total energies of decay for ^{235}U for the emission of various small chunks, like was done for ^{210}Po in Table 6.3. Which decay modes are possible for ^{235}U?

7 Interactions of Ionizing Radiation with Matter

Most of the previous chapters have established a firm understanding of radioactive decay. This chapter now turns to the ramifications of ionizing radiation. Understanding how ionizing radiation interacts with matter is essential to understanding how it can be detected, shielded, and how it affects living systems. We'll start with a more detailed definition of ionizing radiation, and then look at the various ways energetic particles and photons can interact with matter.

7.1 IONIZING RADIATION

Ionizing radiation refers to radiation with enough energy to knock an electron loose from an atom. Alpha particles, beta particles, positrons, gamma rays, and X-rays are all forms of ionizing radiation. Remember that X-rays and gamma rays are located on the high-energy end of the electromagnetic spectrum (Section 1.1) and that the difference between the two is how they are generated. As explained in Section 5.4, X-rays come from electrons (slowing down or transitioning between certain energy levels), and gamma rays from nuclei (IT).

Just below X-rays and gamma rays on the electromagnetic spectrum is UV light. The line between non-ionizing UV and ionizing X- and γ-rays is somewhat fuzzy but is generally drawn at 100 eV. A threshold of 100 eV for ionization may seem artificially high, especially since the electron binding energy of the outermost electrons (valence) for most atoms is less than 15 eV and most chemical bonds (sharing valence electrons) have energies of 1–5 eV. It seems like it should require much less than 100 eV to knock electrons loose from atoms and molecules, but we know that UV is not ionizing. As we will see in this chapter, only a fraction of the photon's (or particle's) energy can be transferred to an electron in an atom or molecule, and a minimum of 100 eV is necessary to consistently knock 'em loose.

7.2 CHARGED PARTICLES

Interactions between charged particles (e.g., alpha or beta particles) and matter can be thought of as the transfer of energy from the particle to the matter. As a particle travels through matter, energy is transferred from it until it has about the same energy as the surrounding matter. It might be strange to think of, but all matter has energy, unless it is at absolute zero! As the particle transfers its energy, it slows down—its energy is kinetic energy and kinetic energy is equal to one half times mass times velocity squared (Equation 7.1).

$$KE = \tfrac{1}{2}mv^2 \tag{7.1}$$

As kinetic energy decreases, then velocity must also decrease, since mass remains constant. These interactions can also be thought of as collisions between the energetic particles and atoms, although

DOI: 10.1201/9781003303787-7

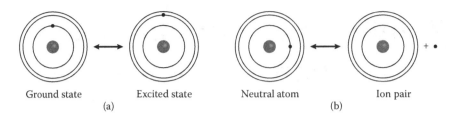

FIGURE 7.1   (a) Electron excitation and (b) atomic ionization.

it is not strictly correct. Even so, collisions are often used to describe the transfer of energy from particle to atom. Perhaps it is because that is how we often observe energy transfer in our macroscopic world (e.g., a bowling ball hitting the pins, a billiard ball hitting other billiard balls, etc.). Please keep in mind that the subatomic world is a very different place. Solids and liquids may appear to completely occupy their space, but on the atomic level they are mostly empty space. An alpha particle traveling through matter is a bit like coasting on a bicycle into a flock of small birds. Each impact on a bird causes the bike rider to slow down a little bit as kinetic energy is transferred from the bicyclist to a bird. If the bicycle and rider represent an alpha particle, then the birds represent electrons. Interactions with electrons are much more likely than an interaction with the nucleus, as the electrons roam most of the space inside an atom. If the bike rider is especially unlucky, it will strike an object inside the flock with a mass greater than or equal to it (the nucleus). If we continue to let the bike and rider represent an alpha particle, then the mass of two SUVs would represent a gold nucleus (ouchies!).

An **elastic** collision is one where the kinetic energy of the particle is cleanly transferred as kinetic energy to the electron. The total kinetic energy of the particle and the electron it hits remains the same. An **inelastic** collision is one where some of the particle's kinetic energy is converted to some other form of energy (such as an electron moving to a higher shell). Inelastic collisions are very common between a large charged particle (proton, alpha, etc.) and an electron. When the electron is promoted to a higher energy level the interaction is an **excitation**. As shown in Figure 7.1a, this process is reversible—the electron will eventually (usually right away) de-excite, emitting electromagnetic radiation such as visible light or X-rays.

If the amount of energy transferred by the particle is greater than the electron's binding energy, the electron is removed from the atom (Figure 7.1b). This is called **ionization** because it results in the formation of a cation and a free electron (an **ion pair**). The creation of an ion pair by ionizing radiation is called a primary ionization. This distinguishes it from **secondary ionizations**, which can subsequently be caused by the freed electron. Because the freed electron can have a significant amount of kinetic energy, it can cause its own ionizations. The freed electron is often called a **secondary electron**, or, more collectively, as **delta rays**. The latter term is somewhat archaic, and somewhat misleading, as the term "rays" refers to photons, not particles. As indicated in Figure 7.1b, ionizations are also reversible, although it is rarely the same electron that returns to the cation reforming the neutral atom.

Excitations and ionizations are two very common results of the interaction of ionizing radiation with matter. What are their relative odds? Electron excitations are roughly two times more common than ionizations in air.

Example 7.1 A 7.6869 MeV α particle emitted during the decay of ^{214}Po undergoes a collision with an M shell electron in an atom of holmium. If the α particle now has only 7.6830 MeV, will the electron be removed from the atom, and if so, how much kinetic energy will it have?

Remember from Section 5.4 that the electron binding energy is the amount of energy required to completely remove an electron from an atom. Thankfully, the energy of an M shell electron in holmium is given in Figure 5.10 (1.71 keV). The energy of the electron booted out of Ho will be the energy transferred minus the binding energy.

The energy transferred is:

$$7.6869 \text{ MeV} - 7.6830 \text{ MeV} = 0.0039 \text{ MeV} = 3.9 \text{ keV}$$

This is greater than the binding energy of an M shell electron in Ho, so the atom will be ionized. The energy of the freed (secondary) electron is:

$$3.9 \text{ keV} - 1.71 \text{ keV} = 2.2 \text{ keV}$$

Note that the electron has very little energy, because it appears that very little energy is transferred. In fact, the amount transferred in this problem represents the maximum amount of energy that could be transferred. Remember that the alpha particle is very massive (you and your bike) compared to an electron (a small bird). At most, an alpha particle can transfer only ~0.05% of its total energy to an electron. This is also true with you and the bird—colliding with a bird isn't going to slow you down very much.

Let's take a closer look at the relationship between kinetic energy and velocity. Earlier in this chapter the following facile formula was quoted (Equation 7.1):

$$KE = \tfrac{1}{2}mv^2$$

Applying this relationship to the alpha particle in Example 7.1 *should* allow calculation of its velocity. A couple of conversion factors are necessary to get conventional velocity units. Kinetic energy needs to be in units of joules (J). Remember that a joule is the same as a kilogram meter squared per second squared.

$$J = \frac{kg \cdot m^2}{s^2}$$

Converting the energy of the alpha particle emitted by ^{214}Po from MeV to J:

$$7.6869 \text{ MeV} \times \frac{10^6 \text{ eV}}{\text{MeV}} \times \frac{1.602 \times 10^{-19} \text{ J}}{\text{eV}} = 1.231 \times 10^{-12} \text{ J}$$

Since energy has units of joules, the mass of the alpha particle (electrons not included!) needs to be expressed in kilograms:

$$4.0015 \text{ u} \times \frac{1.66054 \times 10^{-27} \text{kg}}{\text{u}} = 6.6446 \times 10^{-27} \text{ kg}$$

Plugging energy and mass into Equation 7.1:

$$1.231 \times 10^{-12} \text{ J} = \frac{1}{2}(6.6446 \times 10^{-27} \text{ kg}) \times v^2$$
$$v = 1.925 \times 10^7 \text{ m/s}$$

That looks like a pretty zippy little particle! The speed of light is 2.998×10^8 m/s, and no particle (matter) may travel at or above the speed of light. Our alpha particle isn't all that far from this ultimate speed limit. According to Einstein's theory of relativity, increasing amounts of energy are required to continue accelerating a particle whose velocity is approaching the speed of light. More and more energy is required for smaller and smaller increases in velocity, and the simple relationship between kinetic energy and velocity (Equation 7.1—often referred to as the "classical equation") no longer holds true. In this case, it might be a good idea to calculate velocity using the more complex "relativistic equation" (Equation 7.2).

$$\text{KE} = \left(\frac{1}{\sqrt{1 - \dfrac{v^2}{c^2}}} - 1 \right) mc^2 \tag{7.2}$$

where c is the speed of light. This equation is only necessary for particles moving near light-speed. As written, it can be a little awkward to solve for velocity. Solving for velocity (not trivial!),[1] we get:

$$v = c\sqrt{1 - \left(\frac{mc^2}{\text{KE} + mc^2} \right)^2} \tag{7.3}$$

If our alpha particle's velocity is any different using this equation, we'll know that relativity needs to be taken into account. If it ends up with the same velocity as the classical equation, we'll know that relativity doesn't apply here, and that we've been wasting our time. Let's see!

$$v = 2.998 \times 10^8 \, \frac{\text{m}}{\text{s}} \times \sqrt{1 - \left(\frac{(6.6445 \times 10^{-27} \, \text{kg}) \times \left(2.998 \times 10^8 \, \frac{\text{m}}{\text{s}} \right)^2}{1.231 \times 10^{-12} \, \text{J} + (6.6445 \times 10^{-27} \, \text{kg}) \times \left(2.998 \times 10^8 \, \frac{\text{m}}{\text{s}} \right)^2} \right)^2}$$

$$v = 1.922 \times 10^7 \, \frac{\text{m}}{\text{s}}$$

Once again, it is important to use compatible units. Try the math yourself—there are a lot of buttons to press on this one and it's important to get them in the right order. The result for velocity is nearly the same as with the classical equation. Had we not carried everything out to at least four significant figures, we might not have seen any difference. Despite the small difference, using the relativistic equation is the correct approach for this problem—because of the high level of precision. If the velocity were higher, or if the problem involved a less massive particle (like a proton or an electron), the difference in velocities calculated by the classical and relativistic equations would be more dramatic.

Velocities near the speed of light are often expressed as a fraction of the speed of light (v/c). For the alpha particle from ^{214}Po, this fraction would be:

---

1   My thanks to Dr. Craig Tainter for this derivation—JCB.

$$\frac{v}{c} = \frac{1.922 \times 10^7 \, \frac{m}{s}}{2.998 \times 10^8 \, \frac{m}{s}} = 0.06412$$

Therefore, the alpha particle could be said to be traveling at 6.412% of the speed of light. Despite our earlier amazement at its velocity, this is not really all that fast for a subatomic particle. Notice also that calculating v/c marginally simplifies the relativistic equation (Eq. 7.3).

$$\frac{v}{c} = \sqrt{1 - \left(\frac{mc^2}{KE + mc^2}\right)^2} \qquad (7.4)$$

The faster a particle is moving, the more energy it has to deposit in the matter it travels through. If the particle is traveling through air, an average of 33.85 eV is required for each ionization of an air molecule. Remember that air is ~78% nitrogen (N_2) and ~21% oxygen (O_2), so 33.85 eV/IP is reflective of this particular gas mixture. The average energy required to produce an ion pair in any medium is called its **W-quantity**. This energy cost includes the electron binding energy and is therefore the total energy lost by the ionizing radiation as it travels through the medium and is not the energy of the secondary electron.

As a charged particle travels through matter, it will create a certain number of ion pairs per unit length. This is its **specific ionization** (SI), which depends on the energy of the particle and the nature (density) of the matter it is traveling through. The specific ionization of alpha particles traveling through air typically varies from three to seven million ion pairs per meter (IP/m)—that's a lot of ions!

If we know how many ion pairs a charged particle creates per unit length (SI), and we know the average cost of creating each ion pair (W-quantity), we can also determine the average energy lost by the charged particle per unit length. This is called the particle's **linear energy transfer** (LET) and is obtained by multiplying the specific ionization by the W-quantity.

$$LET = SI \times W \qquad (7.5)$$

Example 7.2 Alpha particles of a certain energy have a specific ionization of 5.05×10^6 IP/m. Calculate the LET value for these particles in air.

Since $W = 33.85$ eV/IP in air:

$$LET = \frac{5.05 \times 10^6 \, IP}{m} \times \frac{33.85 \, ev}{IP} = 1.71 \times 10^8 \, \frac{eV}{m} = 171 \frac{MeV}{m}$$

Wow! That's a lot of energy for every meter traveled in air. The alpha particle will need to be very energetic to travel a full meter, or it will only travel some fraction of a meter before it runs out of gas. **Range** is defined as the average distance a charged particle will travel before being stopped. For alpha and other "heavy," charged particles (nuclei), range can be calculated by dividing the energy of the alpha particle by its average linear energy transfer (LET). Caution: this equation only applies to alpha particles and other nuclei (not electrons!).

$$Range = R = \frac{E}{LET} \qquad (7.6)$$

Remember that specific ionization varies with the energy of the particle, so the amount of energy transferred per unit length (LET) will also vary. Since particles slow down (lose energy) as they travel through matter, their SI and LET values will also vary. Therefore, many calculations, such as Equation 7.6, need to be done with SI and LET values that are based on averages for the particle's entire journey through matter.

Alpha particle range (cm) in air can also be estimated solely from its energy using the equation below.

$$R_{\alpha\text{-air}} \cong 0.31 E_\alpha^{3/2} \tag{7.7}$$

Example 7.3 ^{241}Am is used in smoke detectors because the LET of air is significantly lower than the LET of smoky air. Estimate the range and the average SI and LET in air for the 5.4857 MeV alpha particles emitted by ^{241}Am.

$$R_{\alpha\text{-air}} \cong 0.31 E_\alpha^{3/2} = 0.31 \times (5.4857 \text{ MeV})^{3/2} = 4.0 \text{ cm}$$

$$\text{LET} = \frac{E}{R} = \frac{5.4857 \text{ MeV}}{4.0 \text{ cm}} = 1.4 \frac{\text{MeV}}{\text{cm}}$$

$$\text{SI} = \frac{\text{LET}}{W} = \frac{1.4 \dfrac{\text{MeV}}{\text{cm}}}{33.85 \dfrac{\text{eV}}{\text{IP}} \times \dfrac{\text{MeV}}{10^6 \text{ eV}}} = 4.1 \times 10^4 \frac{\text{IP}}{\text{cm}}$$

These alpha particles are completely stopped by only four centimeters of air! The smoke detector works by detecting the alpha particles after they've passed through ~1–2 cm of air. If smoke gets between the ^{241}Am source and the detector, they absorb more of the alpha particles; the detector does not see as many alpha particles per second and sounds the alarm. ^{241}Am is a pretty good source for this application as it has a 433-year half-life, so its activity is unlikely to change over the lifetime of the detector, yet it is short enough to only require a small amount (~1 μCi).

As previously mentioned, specific ionization and linear energy transfer vary with the energy of the particle. As a particle moves through matter it loses energy (slows down). As it loses energy, it transfers increasing amounts of energy and creates more ion pairs per centimeter traveled. Figure 7.2 generically illustrates this phenomenon for a charged particle moving through matter.

Notice that the specific ionization does not change much initially, and then increases dramatically after it has penetrated some distance into the matter. This means that the particle transfers energy to the matter at a relatively low and constant rate, until it reaches some low-energy threshold. After that point it begins to transfer energy to the matter at ever-increasing rates until, suddenly, it is out of gas and stops.

The behavior illustrated in Figure 7.2 is common to all charged particles (alpha, beta, protons, …) in all media (air, water, humans, …). The peak observed in Figure 7.2 is called the **Bragg peak** and represents the fact that charged particles deposit much of their energy after penetrating matter some distance. One way to think about this is in terms of the probability of interaction. While the particle is moving quickly (higher energy), it spends only a very short time by each atom it passes, decreasing the probability of interaction. At some point it is traveling slowly enough (lower energy) that the time spent near an atom increases the probability of interaction. From that point forward,

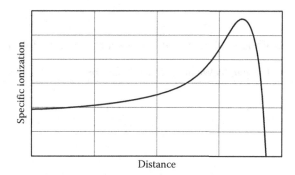

FIGURE 7.2 Generic representation of the change in specific ionization as a charged particle penetrates matter.

the probability of interaction increases dramatically as the particle continues to decelerate. It's a bit like skipping a flat rock over a still lake. The first couple of skips are far apart, and relatively little energy is transferred per unit distance. When the rock slows, it starts to skip often over short distances, transferring a fair bit of energy over a relatively short distance before coming to a stop and sinking into the lake.

What happens when the particle stops? If it's an alpha particle it will steal a couple electrons and form a helium-4 atom, if it's a positron, it'll annihilate with an electron, etc. Keep in mind that the particle's energy does not go to zero, so it doesn't really "stop"—this is only possible at 0 K. The particle will have translational and rotational energy that reflect the temperature of the matter it is now part of. Instead of thinking that the particle comes to a full stop; it is better to understand that it transfers energy to its surroundings until its own energy matches that of its environment.

So far we've focused on the interactions of charged particles with air. What about their interactions with denser media, such as liquids and solids? Solids and liquids have a lot more matter packed into less space. Therefore, they have greater densities and the probability of interaction per centimeter traveled will increase dramatically. Specific ionization will increase (more energy deposited per cm) and range will decrease. One way to compare the shielding effectiveness of different materials is to determine relative stopping power (RSP).

$$RSP = \frac{R_{air}}{R_{abs}} \tag{7.8}$$

where R_{air} is the range in air and R_{abs} is the range in another *abs*orbing material. Larger values for RSP mean that that material is more strongly absorbing (smaller R_{abs}). Table 7.1 provides some RSP and range values for a 7.0 MeV alpha particle in different media.

TABLE 7.1

RSP and Range Values for a 7.0 MeV Alpha Particle in Various Media

Absorber	Density (g/cm³)	RSP	Range (cm)
Air	0.0012	1	6.1
Water	1.0	970	0.0063
Aluminum	2.7	1700	0.0035

TABLE 7.2
Comparisons of LET and Range Values for an Alpha Particle, a Proton, and an Electron

Particle	LET (water, MeV/cm)	Range (air, cm)	Range (water, cm)
10 MeV alpha	920	10	0.011
10 MeV proton	83	120	0.12
10 MeV electron	1.9	4200	5.2

As the density increases, so does relative stopping power, and the range drops dramatically. The bottom line here is that alpha particles are not very penetrating—something we already knew to be true, but now we've got some serious numbers to back up the claim. Since we are made up mostly of water, the numbers in the water row can be approximated for us. Skin averages between 1 and 2 mm thick (that's 0.1–0.2 cm), so we can feel pretty safe around alpha radiation. So long as it remains external to our bodies, the outer layers of our skin will stop alpha radiation.

Much of the discussion in this section has focused on "large" charged particles, such as alpha particles, protons, and other nuclei. What about smaller charged particles like electrons? They could be beta particles, Auger, or conversion electrons emitted during nuclear decay, or they could be from an artificially produced electron beam. We should also consider positrons here as they have the same mass and unit charge. As a result, they tend to interact with matter in much the same way as electrons.

Like the heavier charged particles, electrons can cause excitations and ionizations as they pass through matter. Additionally, the probability of interaction increases as the velocity (KE) of the **incident electron** decreases. The incident electron is the electron that is transferring the radiation to the matter, i.e., the radiation, the one that's causing the whole *incident* (ionizations) to take place. The nature of the interaction and the energy of any secondary electrons produced from these interactions will depend on how much energy is transferred from the incident electron.

The obvious difference between electrons and larger, charged particles is mass. Electrons are thousands of times less massive than even the smallest nuclei. This has two major implications. First, if an electron and proton have the same energy, the electron will be moving a whole lot faster. Electrons will therefore require use of the relativistic kinetic energy equation at much lower energy values (> 3 keV) than protons (> 6 MeV) or alpha particles (> 25 MeV).[2] Since they are moving faster, they will spend less time passing atoms, and therefore have a lower probability of interaction. This means that electrons will be more penetrating than alpha particles or protons with equivalent energy values. Lower probabilities of interaction mean fewer ion pairs created per centimeter and longer ranges, as illustrated in Table 7.2.

Specific ionization values for *electrons* in air can be calculated if their velocity (energy!) is known using Equation 7.9.

$$\text{SI} = \frac{4500 \text{ IP/m}}{v^2/c^2} \tag{7.9}$$

where v is the electron's velocity, and c is the speed of light. Note that this equation works only for electrons and positrons in air.

---

2   The energy thresholds given here are when relativistic values begin to differ from classical by more than 0.5%. Threshold values will vary if a different level of precision is used.

Example 7.4 Calculate the SI and LET in air for the most probable energy beta particle emitted by ^{32}P.

Remember that beta particles share their energy with the antineutrino, and that the most probable energy is $0.3E_{max}$ (see Figure 5.3). For ^{32}P, $E_{max} = 1.709$ MeV, and therefore $0.3E_{max} = 0.51$ MeV (assuming two significant figures). Using the relativistic equation (since beta energy > 3 keV), $v/c = 0.87$ for a 0.51 MeV electron (check it yourself!).

$$SI = \frac{4500 \text{ IP/m}}{(0.87)^2} = 6000 \text{ IP/m} = 60 \text{ IP/cm}$$

$$LET = SI \times W = \frac{6000 \text{ IP}}{m} \times \frac{33.85 \text{ eV}}{IP} \times \frac{MeV}{10^6 \text{ eV}} = 0.20 \text{ MeV/m}$$

Note that the same average energy per ion pair (W) is used here as was used for alpha particles. The source of energy doesn't matter; the average energy it takes to ionize air molecules remains the same. Note also that a lower energy beta particle would have a higher SI and LET. Lower energy means the particle is moving slower, has a higher probability for interaction, is less penetrating, and creates more ion pairs per meter.

The second major difference between electrons and nuclei as projectiles is that electrons are much more likely to be deflected (scattered) as they pass through matter. If most of the interactions between a charged particle and matter involve the electrons, this makes perfect sense. If alpha particles are like a human cyclist coasting into a flock of birds, then beta particles are simply a (high-speed) bird flying into a flock of equally sized birds. Both are slowed by collisions with the flock, but the cyclist's direction is unlikely to change much, and the bird will likely bounce around like a cue ball slamming hard into a bunch of other billiard balls. Because of scattering, proper calculation of the average range for a particular energy beta particle is complex, however, electron range in human tissue (R_e) can be estimated by dividing the electron energy (in MeV) by 2.0 MeV/cm (Equation 7.10). This relationship will become more important when we discuss electron beam therapy in Chapter 15.

$$R_{e^-} = \frac{E}{2.0 \text{ MeV/cm}} \tag{7.10}$$

Scattering of incident electrons doesn't really change the interactions from the perspective of the matter the particle is traveling through. Electrons that are part of the matter will still experience excitations or ionizations as energy is transferred from the incident electron to the matter. The big difference is that electron projectiles can also be scattered by flying near a nucleus. The most important such interaction is called **inelastic scattering by nuclei**. If the projectile electron passes close enough to a nucleus, it will feel an attractive (positrons will be repelled) Coulomb force and its trajectory changed. Classical electrodynamics tells us that the electron will be accelerated as it nears the nucleus and emit electromagnetic radiation as shown in Figure 7.3.

As the electron passes by the nucleus and begins to move away it decelerates. The net effect is that the electron changes direction and slows down (loses kinetic energy). The energy lost in slowing is equal to the energy of the photon (the squiggly line in Figure 7.3), which is typically an X-ray. The level of interaction will vary depending on how close the electron travels to the nucleus, and how many protons (Z) are in the nucleus. If the electron passes very close to the nucleus, it will be scattered through a higher angle, slow more, and emit a higher energy photon than if it were to pass

FIGURE 7.3   Inelastic scattering of an electron by a nucleus.

further away. An infinite number of these kinds of interactions are possible, as the distance between the electron and nucleus can be varied by infinitesimally small increments. The result is that a wide variety of scattering angles can occur, corresponding to a wide variety of energies for the emitted X-rays. An energy spectrum of X-rays produced from the beta particles emitted during the decay of ^{90}Sr and ^{90}Y interacting with aluminum is illustrated in Figure 7.4.

X-rays produced by inelastic scattering of electrons by nuclei are often referred to as **bremsstrahlung**, which is a German word roughly translated as "braking radiation." Because they have a broad energy distribution, they are also referred to as **continuous X-rays**. Notice in Figure 7.4 that not all energies have equal probabilities. In general, the probability decreases as the energy increases. This is because lower energy X-rays are produced when the electron is further from the nucleus, and there are more points in space at a greater distance. Prove this to yourself by drawing two concentric circles. If the nucleus is at the center and each point on the larger circle represents the probability of the lower energy interaction, and each point on the smaller circle represents the probability of the higher energy interaction, which is more likely? In other words, which circle has more points? The larger circle, of course! This means that there are more points further away from the nucleus that will produce a lower energy X-ray.

Now, let's mess with the matter the incident electron is traversing. The effect of increasing the atomic number (Z) of the matter the electron is traveling through can be thought of in two ways. As mentioned above, the intensity of the interaction will increase with Z at the same distance from the nucleus. This is simply due to the increased positive charge of the nucleus. The other effect of increasing Z is that the nucleus will have greater reach. A larger positive charge means that the electron can interact with it from further away. Overall, as Z increases the probability of inelastic scattering also increases. This should make intuitive sense; as atomic number increases, so

FIGURE 7.4   X-ray spectrum produced by beta emissions from ^{90}Sr and ^{90}Y interacting with Al.

FIGURE 7.5   Elastic scattering of an electron by a nucleus: (a) prior to scattering; (b) after scattering.

(generally) does density. Denser matter makes better shielding because ionizing radiation has more opportunities to interact as it is trying to pass through.

Elastic scattering by nuclei happens when the projectile electron hits the nucleus dead on, and kinetic energy is conserved. No excitations or ionizations occur, at least not initially, and no photons are created. This type of scattering is much less probable than inelastic scattering. The nucleus is really small, and Coulomb attraction can act over reasonably long distances. As illustrated in Figure 7.5a, we can think of the electron initially in motion, heading right toward a nucleus, which is not moving. The collision itself is too violent to be depicted here, but Figure 7.5b shows the aftermath. The electron has reversed direction and has lost some velocity (the arrow is smaller), and the nucleus is now moving very slowly in the opposite direction. Only some of the electron's kinetic energy is transferred to the nucleus, and, because the nucleus is so much more massive, it has very little velocity. Elastic scattering of electrons by nuclei is also known as **backscattering** because the electron ends up going back along the path it came from.

Čerenkov radiation is an obscure form of radiation that results from particles (typically electrons) traveling from one medium into another. The speed of light depends (slightly!) on the medium the particle or photon is traveling through. For example, light travels 1.33 times slower in water than in a vacuum. This means that beta particles traveling near the speed of light in one medium that pass into another medium, where light travels more slowly, could now be exceeding the ultimate speed limit. Since no object may move faster than light, the electron has to slow down. In doing so, it has to release energy. This effect isn't really important, but it explains why nuclear reactor cores (Chapter 12) have a cool blue glow. High-energy beta particles are emitted by some of the products formed by the fission reactions and travel from the solid fuel pellet, through the solid fuel rod, and into the liquid (usually water) coolant. The energy released by the speeding electrons into the water is converted into blue light. Čerenkov radiation can also be observed by placing higher energy beta-emitting nuclides like ^{90}Y in water and turning out the lights.

7.3   PHOTONS

The transfer of energy from X- and gamma-ray photons to the matter they traverse is a bit more complicated than it is from particles. Nine different interactions are possible, but we will only look at the three that are generally the most probable and have the most significance for medical applications: Compton scattering, the photoelectric effect, and pair production. These three interactions all transfer energy from high-energy photons to matter through fairly complex mechanisms. We'll discuss them in fairly simple terms, but please keep in mind that they are not as simple as they are described here.

7.3.1   COMPTON SCATTERING

Compton scattering can be imagined as a collision between an incident (incoming) photon and an unsuspecting outer shell (valence) electron in an atom of the matter the photon is traveling through.

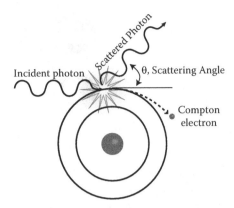

FIGURE 7.6   Compton scattering.

This is represented in Figure 7.6. In reality, it is better to think of it as a transfer of some energy from the incident photon to a valence electron. There's not much electron binding energy to overcome here (valence electrons are typically bound by less than 50 eV), so the electron gets booted away from its atom. The ejected electron is called the Compton (or recoil)[3] electron. Energy is conserved in this process, therefore the energy of the incident photon (E_0) equals the energy of the scattered photon (E_{SC}) plus the energy of the Compton electron (E_{CE}) plus the binding energy of the electron (B.E.).

$$E_0 = E_{SC} + E_{CE} + \text{B.E.}$$

Since the binding energy of a valence electron is really small compared to the other energies, we can neglect it. So, this equation simplifies to:

$$E_0 = E_{SC} + E_{CE} \tag{7.11}$$

The photon is scattered, meaning its direction has changed. The angle through which the photon is scattered is called the scattering angle (θ). The energy of the scattered photon is related to its scattering angle and its original energy.

$$E_{SC} = \frac{E_0}{1 + \left(\dfrac{E_0}{0.511}\right) \times (1 - \cos\theta)} \tag{7.12}$$

All energies in Equation 7.12 *must* be in units of MeV. Basically, this equation tells us that as scattering angle increases, the energy of the scattered photon (E_{SC}) decreases. This makes some sense—if the incident photon "hits" the valence electron dead center, maximum energy is transferred and the scattered photon returns along the same path of the incident photon, traveling in the opposite direction (**backscattering**, $\theta = 180°$). Anything other than dead center will transfer less

---

3   An unfortunate term, as it has nothing to do with daughter recoil—the momentum of the daughter following nuclear (particularly alpha) decay.

TABLE 7.3
Energies of Scattered Photons and Compton Electrons when θ = 180°

Nuclide	E_0 (keV)	E_{SC} (keV)	E_{CE} (keV)	% Transferred to Electron
^{125}I	35.5	31.2	4.3	12
^{99m}Tc	140.5	90.6	49.8	35
^{60}Co	1330	214	1116	84

energy and will deflect the photon less. At the other end of the spectrum, a glancing blow will give a very small scattering angle and transfer only a tiny amount of energy.

Example 7.5 Calculate the energies of the 140.5 keV γ photons that are produced during the decay of ^{99m}Tc when they are scattered through angles of 25° and 180°. Also determine the energies of the Compton electrons that are produced.

Apply Equations 7.12 and 7.11 to each.

$$E_{SC} = \frac{0.1405 \text{ MeV}}{1 + \left(\dfrac{0.1405 \text{ MeV}}{0.511} \right) \times \left(1 - \cos 25° \right)} = 0.1370 \text{ MeV or } 137.0 \text{ keV}$$

$$E_{CE} = 140.5 \text{ keV} - 137.0 \text{ keV} = 3.5 \text{ keV}$$

$$E_{SC} = \frac{0.1405 \text{ MeV}}{1 + \left(\dfrac{0.1405 \text{ MeV}}{0.511} \right) \times \left(1 - \cos 180° \right)} = 0.0906 \text{ MeV or } 90.6 \text{ keV}$$

$$E_{CE} = 140.5 \text{ keV} - 90.6 \text{ keV} = 49.8 \text{ keV}$$

Note that higher angle scattering transfers more energy from the incident photon to the Compton electron.

Table 7.3 lists three nuclides that emit gamma photons of different energies. Notice that the percentage of energy that can be transferred from the incident photon to a Compton electron increases with the energy of the photon. All calculated values are for θ = 180° (backscattering). Higher energy photons transfer a higher percentage of their energy to the Compton electron. Note also that a photon can never transfer all its energy to an electron via Compton scattering. The maximum amount is transferred when the photon undergoes backscattering. In Example 7.5 (and in the second row of Table 7.3), it is when the incident photon transfers 49.8 keV to the Compton electron but retains 90.6 keV.

7.3.2 THE PHOTOELECTRIC EFFECT

When a high-energy photon interacts with an atom via the **photoelectric effect,** we can imagine that the energy of the incident photon is completely absorbed by an *inner* shell electron as illustrated in Figure 7.7. So long as the incident photon has enough energy (greater than the electron binding

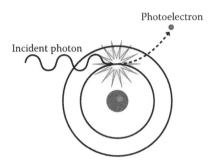

FIGURE 7.7 The photoelectric effect.

energy), odds are good it will be a K shell electron that gets booted. This time the ejected electron is called a **photoelectron** and there is no scattered photon. The energy of the incident photon therefore equals the energy of the photoelectron (E_{PE}) plus the electron binding energy.

$$E_0 = E_{PE} + \text{B.E.} \tag{7.13}$$

Since the binding energy of a K shell electron can be significant compared to the energy of a gamma photon, we can't neglect it. Kicking out an inner shell electron creates an electron vacancy or hole in that shell. This is exactly the same situation that follows electron capture decay (Section 5.4). An electron in a higher energy shell drops down to fill that vacancy and a characteristic X-ray or an Auger electron is then emitted.

7.3.3 PAIR PRODUCTION

The least important of the three photon–matter interactions we will study is **pair production**. Like the photoelectric effect, the atom absorbs the entire energy of the incident photon. The difference is that the incident photon is absorbed by the nucleus and an electron and a positron are created as shown in Figure 7.8. They are the "pair" that is produced.

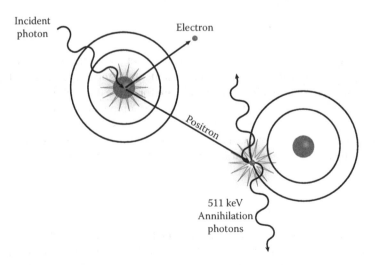

FIGURE 7.8 Pair production.

The incident photon needs to be at least 1.022 MeV in order for pair production to take place. This is the amount of energy needed to create the matter contained in an electron and a positron. If the incident photon has more than 1.022 MeV of energy, the remainder goes into the kinetic energy of the electron and positron. In this case, the energy of the incident photon is equal to the energy of the emitted particles plus 1.022 MeV.

$$E_0 = E_{\beta^+} + E_{\beta^-} + 1.022 \text{ MeV} \tag{7.14}$$

This extra energy is not evenly divided between the pair, anywhere from 20% to 80% of the excess energy can be deposited in either particle. The remainder goes into the other one.

The positron will transfer its kinetic energy to the nearby matter, likely via ionizations and excitations. When it runs out of gas, it will annihilate with an electron to form two 0.511 MeV annihilation photons (Figure 7.8). Note that, theoretically, we could start with a 1.022 MeV incident photon and end up with two 0.511 MeV photons!

Pair production is the least important of the three photon–matter interactions discussed here because it is the least commonly observed for incident gamma photons with low to moderate energies. So, what are the relative probabilities of these three interactions? The relative probability of each depends on both the energy of the photon and the atomic number (Z) of the atom it interacts with.

Figure 7.9 illustrates how the relative probabilities of Compton scattering (solid black lines), the photoelectric effect (dashed lines), and pair production (solid gray line) vary with photon energy.

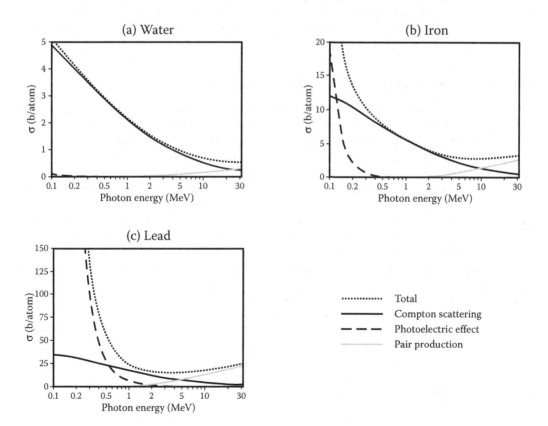

FIGURE 7.9 Relative probabilities (b/atom) of photon–matter interactions with (a) H_2O, (b) Fe, and (c) Pb. Produced using data from XCOM (http://physics.nist.gov/).

The probabilities of all three are added together and graphed as "total" (dotted line). Photon energy increases from left to right, and probability increases from bottom to top on each graph. Graph (a) is for photon interactions with water, (b) with iron (Fe), and (c) with lead (Pb). For water (Figure 7.9a), Compton scattering is most likely for all photon energies graphed, except at the highest energies where pair production becomes the dominant interaction. For Fe (Figure 7.9b), Compton scattering still dominates for most energy values, but the relative odds of the photoelectric effect for low-energy photons, and the relative odds of pair production at higher energies, are both increased compared to water. This trend continues as we move further down the periodic table. The photoelectric effect is dominant for low-energy photons (less than 500 keV) traveling through lead (Figure 7.9c), and pair production becomes the most significant interaction above 5 MeV.

Notice also that the values for the probability (y-axis) scale increase dramatically with atomic number for the three graphs in Figure 7.9. It is generally true that the overall probability of photon interaction with matter increases as the atomic number increases. This can be understood in terms of density. The more matter you pack into the same amount of space, the more likely you'll get high-energy photons to interact within that space. Density generally increases with atomic number.

Finally, it is worth noting that the overall probability of interaction generally decreases with increasing energy of the photon. The overall probability that a photon will interact is the sum of the probabilities for Compton scattering, photoelectric effect, and pair production. For all three graphs in Figure 7.9, the overall probability of interaction (dotted lines) generally decreases as the energy of the photon increases. This is normally true of photon–matter interactions. Like particles, higher energy photons generally have a lower probability of interacting with matter. In this case, we *cannot* rationalize this trend in terms of the time spent near the atom because all photons travel at the same velocity. As you might guess, it's complicated, and is related to the wavelength of the photon, which does change with photon energy. A shorter wavelength (higher energy) generally makes it more difficult for a photon to interact with matter.

7.4 ATTENUATION OF GAMMA- AND X-RADIATION

When gamma- and X-ray photons run through matter, three things can happen. They can be absorbed (photoelectric effect or pair production), scattered (Compton scattering), or they can travel straight through without interacting. If a photon is scattered or absorbed, we say it has been attenuated. For example, if 1000 photons hit a slab of matter (a half-pound cheeseburger? mmm, delicious!) and 200 are scattered, 100 absorbed, and 700 transmitted, then we'd say that 300 (30%) were attenuated.

What if a photon in the middle of a broad beam is scattered through a small angle? Would we notice? Only if our detector is far enough away—far enough that the scattered photon would get itself out of the beam and miss the detector. What if the beam is made up of photons of various energies (polyenergetic)? As we can see in Figure 7.9, low-energy photons are always more likely to interact with matter than their high-energy buddies. It can get complicated.

Let's keep it simple for now and assume that we're working only with a narrow beam of monoenergetic (all having the same energy) photons, all traveling in the same direction. This situation is often referred to as "good geometry," and anything else would be "bad geometry." When we have good geometry, the following equation holds.

$$I = I_0 e^{-\mu x} \tag{7.15}$$

which can be re-written as:

$$\ln \frac{I}{I_0} = -\mu x \tag{7.16}$$

where I is the number of photons making it through the matter (transmitted), I_0 is the original number of photons in the beam, x is the thickness of the matter, and μ is the linear attenuation coefficient. The ratio of I/I_0 is also known as the **transmission factor**—which makes sense; it is the fraction of photons that make it straight through the matter. Conversely, the attenuation coefficient is a measure of how much scattering and absorption takes place, i.e., it is based on how many photons don't make it straight through. The attenuation coefficient depends on the material and the energy of the photon beam. It should always be the same for a particular material and a specific photon energy. The units on the linear attenuation coefficient are usually reciprocal centimeters (cm^{-1}) requiring x to have units of cm. Other units of length can be used but should cancel when used in the same calculation. With this equation, it should be apparent that we could never fully attenuate a photon beam. There is always a finite probability that some photons in any beam will travel through any matter. Test this for yourself with Equation 7.15 by making the thickness (x) bigger and bigger. Use arbitrary, but consistent, values for I_0 (like 1 million) and μ (1.0 cm^{-1}).

Example 7.6 A narrow beam of 2000 monoenergetic X-ray photons is reduced to 1000 photons by a 1.0 cm thick piece of copper foil. What is the attenuation coefficient of the Cu foil?

$$\ln \frac{1000}{2000} = -\mu \times 1.0 \text{ cm}$$

$$\mu = 0.69 \text{ cm}^{-1}$$

In Example 7.6, the beam is attenuated to half its original intensity; therefore 1.0 cm is the **half-value layer** (HVL)—sometimes called the half-value thickness (HVT). Since the half-value layer will always have an I/I_0 of 1/2, HVL can be mathematically defined with Equation 7.17.

$$HVL = \frac{\ln 2}{\mu} \tag{7.17}$$

Some half-value layer values in water for various photon energies are listed in Table 7.4. Notice that HVL increases with the photon energy. As the photon's energy increases, it takes more water to attenuate half of the beam, and the photons become more penetrating.

TABLE 7.4
Half-Value Layers in Water for Different Photon Energies

Photon Energy (MeV)	HVL in Water (cm)
0.10	4.0
0.50	7.2
1.0	9.8
5.0	23
10	31
20	38

Example 7.7 What would the beam intensity be in Example 7.6 if the foil's thickness is tripled?

$$I = I_0 e^{-\mu x} = 2000 \text{ photons} \times e^{-0.69 \text{ cm}^{-1} \times 3.0 \text{ cm}} = 250 \text{ photons}$$

This makes sense! For every HVL, half of the beam is cut out. We now have three HVLs in place. The first centimeter cuts the beam from 2000 to 1000, the second from 1000 to 500, and the third from 500 to 250. This is exponential change, just like radioactive decay!

A **tenth-value layer** (TVL) is the thickness required to attenuate a photon beam to one-tenth of its original value. It is calculated in a manner very similar to HVL.

$$\text{TVL} = \frac{\ln 10}{\mu} \tag{7.18}$$

The attenuation coefficient can take other forms and units. Some look pretty crazy. The mass attenuation coefficient (μ_m) is defined as the linear attenuation coefficient divided by the density (ρ) of the absorbing material.

$$\mu_m = \frac{\mu}{\rho} \tag{7.19}$$

If density has units of grams per cubic centimeter and the linear attenuation coefficient (μ) is per centimeter, then μ_m would have units of cm²/g.

The atomic attenuation coefficient (μ_a) is defined as the linear attenuation coefficient multiplied by the average atomic mass of the absorbing material, divided by its density times Avogadro's number.

$$\mu_a = \frac{\mu \times \text{ atomic mass}}{\rho \times \text{ Avogadro's number}} \tag{7.20}$$

Believe it or not, μ_a has units of square centimeters per atom (cm²/atom).

Example 7.8 Calculate the mass and atomic attenuation coefficients for Example 7.6.

The density of copper is 8.96 g/cm³ and its average atomic mass is 63.55 g/mol.

$$\mu_m = \frac{0.69 \text{ cm}^{-1}}{8.96 \dfrac{\text{g}}{\text{cm}^3}} = 0.077 \frac{\text{cm}^2}{\text{g}}$$

$$\mu_a = \frac{0.69 \text{ cm}^{-1} \times 63.55 \dfrac{\text{g}}{\text{mol}}}{8.96 \dfrac{\text{g}}{\text{cm}^3} \times \left(6.02 \times 10^{23} \dfrac{\text{atom}}{\text{mol}}\right)} = 8.2 \times 10^{-24} \frac{\text{cm}^2}{\text{atom}}$$

Remember, all of this is for good geometry (monoenergetic, narrow beam). With a broad beam, percent transmittance doesn't drop off as rapidly, because some photons can be scattered through small angles, and they might not escape the beam by the time we detect them.

Polyenergetic beams are also problematic because the attenuation coefficient varies with the energy of the photon. As implied by Figure 7.9, low-energy gamma- and X-ray photons are more likely to interact with all types of matter. Therefore, a polyenergetic beam will have a higher energy distribution after it passes through some attenuating matter. This is called **filtration**. Higher energy photons are called "hard," and lower energy photons "soft." A beam that has been filtered would then be "harder." Filtration is commonly done with medical X-rays. Most medical X-ray machines produce a polyenergetic beam that is filtered before it hits the patient (usually with an aluminum disk—take a look the next time you go to the dentist!). This is a good thing, as the soft X-rays are more likely to be absorbed by the patient, and unnecessarily increase their dose during the procedure.

For polyenergetic beams, an *effective* attenuation coefficient can be determined using Equation 7.21.

$$\mu_{eff} = \frac{\ln \frac{I_0}{I}}{x} = \frac{\ln 2}{HVL} \tag{7.21}$$

It is based on measurements, and is specific to the X-ray source, as different sources will produce different spectra (energy distributions). Calculation of other flavors (analogous to μ_m and μ_a) of the effective attenuation coefficient are performed in the same way as Equations 7.19 and 7.20.

Example 7.9 An X-ray beam has an HVL of 1.5 mm for Cu. What are the effective linear and mass attenuation coefficients?

$$\mu_{eff} = \frac{\ln 2}{0.15 \text{ cm}} = 4.6 \text{ cm}^{-1}$$

$$\mu_{meff} = \frac{\mu_{eff}}{\rho} = \frac{4.6 \text{ cm}^{-1}}{8.96 \frac{g}{cm^3}} = 0.52 \frac{cm^2}{g}$$

QUESTIONS

7.1 Calculate the wavelength and frequency of a photon with 100 eV of energy.

7.2 Why is a γ-ray considered ionizing radiation, but photons of purple light are not?

7.3 Define the following: ionizing radiation, linear energy transfer, bremsstrahlung, Compton scattering, photoelectric effect, pair production, and half-value layer.

7.4 An alpha particle emitted during the decay of ^{222}Rn (5.4895 MeV) transfers 0.0010% of its energy to an argon atom. Is this enough energy to ionize the argon atom by booting any of its electrons? Table 7.5 provides electron binding energies for all occupied subshells in an argon atom. If it can eject electrons from any of the subshells listed, calculate the energy of the secondary electron that is produced.

TABLE 7.5
Electron Binding Energies for Argon

K	L_1	L_2	L_3	M_1	M_2	M_3
3.20 keV	327 eV	250 eV	248 eV	27 eV	14 eV	14 eV

7.5   How many ion pairs will be formed by a single 5.00 MeV α particle traveling through air? Assume all of the particle's energy goes into ion pair formation.

7.6   Would two beta particles, with different energies, have different ranges in air? Briefly explain.

7.7   If the range of a charged particle in dry air (at standard temperature and pressure) is known, what other data must be obtained to calculate the range of the same particle in titanium?

7.8   Calculate v/c for a 7.6868 MeV proton. Repeat for an electron (beta particle) of the same energy. Use both the classical and relativistic equations for each particle. Which gives more accurate results?

7.9   Calculate the kinetic energy (MeV) of an electron traveling at a velocity exactly equal to one-half the speed of light. Repeat the calculation for a proton, and an alpha particle.

7.10 What is the maximum velocity for a beta particle emitted during the decay of ^{6}He? Why must this be considered a "maximum" velocity?

7.11 Which of the following are true?
   a.   Compton scattering is the most likely interaction for any X-ray photon energy.
   b.   Pair production is the least likely of the three interactions studied.
   c.   The probability of all interactions generally increases with increasing Z.
   d.   Pair production is never the dominant form of interaction.
   e.   The greater the energy of the photon correlates to a greater probability of interaction.

7.12 Calculate the mass of 1.0 μCi of ^{241}Am. AmO_2 is its chemical form in smoke detectors. What is its specific activity? Do ^{241}Am-containing smoke detectors pose a radiation hazard for people who have them in their homes?

7.13 Former KGB agent Alexander Litvinenko was poisoned with ^{210}Po in 2006. ^{210}Po emits an alpha particle with an energy of 5.3044 MeV. What is the specific ionization of this alpha particle in air? On average, how many ion pairs will be created in air by each alpha particle emitted by ^{210}Po? Why is ^{210}Po a particularly clever poison?

7.14 How would you expect protons to interact with matter? Compare to alpha particles and electrons of equal energy.

7.15 How many times more massive are alpha particles compared to electrons?

7.16 Define the following: tenth-value layer, attenuation, photoelectron, Compton electron, and Čerenkov radiation.

7.17 An elastic collision takes place between a 2.51 MeV electron and a stationary ^{12}C nucleus. Assuming that exactly 25% of the electron's energy is transferred to the ^{12}C nucleus, calculate the resulting velocities of both.

7.18 An electron initially traveling at 2.997×10^8 m/s slows as it transfers to another medium, emitting a photon of blue light (wavelength = 455 nm). Calculate the electron's new velocity. What is the blue light known as?

7.19 Calculate the energies of the backscatter photon and the Compton electron for the 662 keV gamma ray observed in the decay of ^{137}Cs.

7.20 Calculate the energies of the scattered photon and the Compton electron when incident gamma radiation of 167 keV (from ^{201}Tl) is scattered through an angle of 23°. Assume the Compton electron originated from the L_1 subshell of oxygen (binding energy = 37.3 eV).

7.21 What is the linear energy transfer (LET) of beta particles emitted by ^{90}Y in air? Use the most probable energy of a beta particle emitted by ^{90}Y. Estimate the range in air of an alpha particle with the same energy and compare this to the average range of the ^{90}Y beta particle (11 mm) in human tissue.

7.22 Humans can be considered as mostly water. If a 25 cm thick human is exposed to gamma radiation from ^{137}Cs, estimate the percentage of the gamma ray intensity attenuated. The mass attenuation coefficient for water exposed to ^{137}Cs is 0.088 cm²/g. What is the most likely interaction between photon and water molecule?

7.23 Copper has a density of 8.9 g/cm^3, and its total atomic attenuation coefficient is 8.8×10^{-24} cm^2/atom for 500 keV photons. What thickness of copper is required to attenuate a 500 keV photon beam to half of its original intensity?

7.24 The mass attenuation coefficient for iron exposed to 1.5 MeV X-rays is 0.047 cm^2/g. Calculate the half- and tenth-value layers.

7.25 Using information from Question 7.24, calculate the atomic attenuation coefficient. Also calculate the intensity of the original beam if the measured intensity is 5800 photons/s on the other side of a 70 mm thick piece of iron.

7.26 Calculate the kinetic energy of a photoelectron released from the K shell (binding energy = 36.0 keV) as a photon with a frequency of 9.66×10^{18} s^{-1} interacts with a chunk of cesium. What part of the electromagnetic spectrum does this photon belong to?

8 Detection of Ionizing Radiation

Perhaps the main reason that most people fear ionizing radiation is that humans have no way to directly sense it. We need instruments to convert ionizing radiation into signals that we can observe and understand. Remember that ionizing radiation can interact with matter by ionization (creation of an ion pair) or excitation. We can take advantage of both in our quest to detect ionizing radiation. This chapter is all about how these instruments work. Special emphasis will be placed on Geiger-Müller tubes (also known as Geiger counters) as they are so commonly used, and on gamma spectroscopy, because of its importance to nuclear medicine.

8.1 GAS-FILLED DETECTORS

The basic design and function of any gas-filled detector is illustrated in Figure 8.1. Ionizing radiation passes through some gas in a container, creating ion pairs from the gas molecules. The negatively charged electrons then begin to migrate toward a positively charged piece of metal (anode) and the positively charged cations move toward a negatively charged cathode. When the ions hit the metal, they create an electrical signal, which can be read. Many different electrode configurations are possible, but the most common is a wire anode running down the center of a cylindrical cathode. Remember from Chapters 1 and 7 that X-rays and gamma rays have a relatively low probability of interacting with matter. Since gases have very low densities, gas-filled detectors don't have a lot of matter available to interact and are therefore quite inefficient in detecting photons. Typically, less than 1% of all photons entering the tube will interact with the gas inside the tube. That means that more than 99% fly right on through, undetected. However, gas-filled detectors are very efficient at detecting beta and alpha radiation as they have much higher probabilities of interacting with matter.

The major difference between the three flavors of gas-filled detectors discussed in this section is the voltage applied to the electrodes. We will look at them in order of increasing voltage.

8.1.1 IONIZATION CHAMBERS

Ionization chambers are the lowest voltage form of gas-filled detector, typically running at 50–300 V. The idea is to have just enough voltage to keep the ions moving toward the electrodes without accelerating them. If they are accelerated, there's a chance they may cause other ionizations as they travel through the chamber. If we can keep their velocity more or less constant, then the only things detected are the ions created directly by the ionizing radiation.

Ionization chambers are usually filled with air. Remember that, on average, it only takes about 34 eV to create an ion pair in air; so many ion pairs *can* be expected from every particle or photon of ionizing radiation that enters the chamber. Even so, ionization chambers cannot always detect

DOI: 10.1201/9781003303787-8

FIGURE 8.1 Schematic of a gas-filled detector.

FIGURE 8.2 Schematic drawing of a pocket dosimeter.

individual events, because the electrical signal from these events can be too weak. Fortunately, most radioactive sources emit radiation at a pretty high rate, and the chamber can measure the overall current created, which can be calibrated to a count rate.

One application of ionization chambers is the pocket dosimeter. A schematic of one is pictured in Figure 8.2. The inner wire (anode) and the outer case (cathode) are given a charge, which is stored in a capacitor. The user can then look through a window on one end and read the level of charge on a gauge. As ionizing radiation interacts with the gas inside the dosimeter, the charge is gradually dissipated. The extent of charge dissipation is a measure of the amount of radiation the user was exposed to. These dosimeters are commonly used when the possibility of a significant dose in a relatively short time exists. The accuracy of these devices is poor, only about 20%.

8.1.2 PROPORTIONAL COUNTERS

One disadvantage of ionization chambers is that they are not very sensitive. They are designed to detect radiation in bulk, not as individual photons and particles. They can have trouble detecting a single ionizing particle because the signal generated is sometimes too weak. Fortunately, we can make a couple of simple changes to amplify the signal.

First, we can crank up the voltage. A higher potential means that the electrons will now be accelerated as they approach the anode. If they accelerate enough, they can begin to cause additional ionizations, which will generate more electrons, which also accelerate, causing more ionizations … and, when all these electrons hit the anode, a stronger signal is generated. This process of accelerating the electrons to create more ion pairs is called a **cascade** or **avalanche**. Typically, 10^6 times as many ion pairs are generated in the cascade as were originally generated by the ionizing radiation.

Secondly, we can change the gas. Air contains oxygen, which has a bad tendency to react with free electrons, preventing them from getting to the anode, dampening the signal in ionization chambers.

Proportional counters typically use Ar or Xe gas inside the detector, in addition to running at higher voltages than ionization chambers.

The signal is now strong enough that individual counts can be made. Another bonus is that the total charge produced is *proportional* to the energy deposited by the ionizing radiation. Therefore, both count rate and energy information on the incident radiation can be obtained. This can be helpful in discerning alpha from beta radiation, and in identifying alpha-emitting nuclides.

8.1.3 GEIGER-MÜLLER TUBES

Geiger-Müller (G-M) tubes are designed for maximum signal amplification (Figure 8.3). To do this the voltage is usually cranked up to around 800 V. The detector gas is almost always Ar. As a result, electrons are accelerated like crazy toward the anode. Typically, 10^{10} times more ion pairs are created. There's no missing that signal!

Lots of atomic excitations also happen in the mad rush of electrons to the anode. When these atoms de-excite, electromagnetic radiation (visible light, UV, and X-rays) is given off—*lots* of it. Some of these photons have enough energy to knock electrons loose from the metal electrodes (Einstein's photoelectric effect). Some of the photons also cause additional ionizations in the detector gas.

This cascade could easily get out of hand, so what shuts it down? The slow-moving cations! The cations are much more massive than the electrons, and therefore move much slower under the same conditions. So many ionizations are created near the anode by all the electrons accelerating toward it that a concentric cylinder (much smaller than detector itself) of cations forms around the anode. Eventually, all gas inside this smaller cylinder is ionized, and the cascade shuts down.

As the cations finally approach the cathode (~300 μs later), electrons come off the wall (the cathode) to combine with them. As these electrons accelerate toward the vast cation cloud, they excite electrons in the Ar atoms they meet along the way, and more electromagnetic radiation is generated. This new burst of photons threatens to start another cascade. To shut this down (**quench**), chlorine is added to the Ar gas inside the G-M tube.

$$Cl_2 \xrightarrow{\text{photons}} 2Cl\bullet \rightarrow Cl_2$$

Chlorine efficiently absorbs the higher energy photons, which symmetrically break the Cl–Cl bond, forming two chlorine radicals. These radicals then recombine, dissipating the released energy as heat.

Because of the huge gas amplification (acceleration toward the electrodes), G-M tubes detect individual events quite well. However, the strength of the observed signal is constant, regardless of the amount of energy deposited in the tube by the ionizing radiation. Therefore, only count information can be recorded with G-M tubes, not energy.

For a certain amount of time after ionization takes place in a G-M tube it is unable to distinguish another ionization of the argon gas. Even after the cascade has been quenched, a little more time

FIGURE 8.3 Schematic of a Geiger-Müller detector.

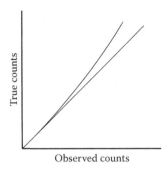

FIGURE 8.4   Dead time correction graph for a G-M tube.

must pass before we can be sure another event will be recorded. A number of terms are applied to this time interval:

Dead time is the time interval after a pulse has occurred during which the counter is insensitive to further ionizing events.

Resolving time is the minimum time interval by which two pulses must be separated to be detected as two individual pulses.

Recovery time is the time interval that must elapse after a pulse has occurred before another full-sized pulse can occur.

The differences between these three terms are fairly subtle, but the implication of this required delay between counts is clear. At higher count rates, G-M tubes will begin to miss some counts. For example, if two beta particles enter a G-M tube at about the same time (**coincidence**), only one count will be recorded. What kind of time interval are we talking about here? It depends a bit on the tube, but typically it is a few hundred microseconds. Because of the possibility of additional ionizations occurring within this dead time increases with the count rate, higher count rates means more counts will be missed!

Fortunately, it is easy to estimate the true number of counts by experimentally determining how many counts are being missed at various count rates and plotting the results in a graph such as that shown in Figure 8.4. Correcting for the missing counts is usually called a **dead time correction**. The straight line shows the entirely theoretical case when true counts are always equal to observed counts. The curved line shows the reality of observed counts being less than true counts at higher count rates.

In the real world, count rates can also be affected by variances in line voltage, background, sample geometry (how the sample is positioned relative to the detector), atmospheric conditions, movement of wires, etc. As mentioned in Chapter 2, the percentage of decays detected relative to all the source's decays in the same time interval is called the percent efficiency. As is clear here, the percent efficiency can be affected by a number of factors and would therefore be expected to vary somewhat from detector to detector.

8.2   SCINTILLATION DETECTORS

Gas-filled detectors work by converting ionizing radiation into an electrical signal via the formation of ion pairs. Another way to detect ionizing radiation is to convert it to visible light. This is exactly what scintillation detectors do, and it is accomplished through electron excitations (albeit indirectly). The visible light is observed by converting it to an electrical signal. This section begins by

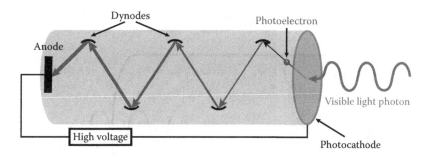

FIGURE 8.5 Schematic of a photomultiplier tube (PMT).

looking at the conversion of light to electronic signal (an electronic eyeball), then describes the two most common types of scintillators, inorganic and organic.

8.2.1 PHOTOMULTIPLIER TUBES

When ion pairs recombine, or excited electrons drop back to the ground state, visible light can be given off. **Scintillators** are materials that do this particularly well. Only a few thousand photons are usually produced when ionizing radiation travels through a scintillator. This is not very bright, and, therefore, very difficult to detect, so we will need a way to amplify the visible light signal. **Photomultiplier tubes** (PMTs) do this by converting the visible light photons to electrons then amplifying the signal. This process is illustrated in Figure 8.5.

When the visible light photons leave the scintillator, they hit the photocathode. The photocathode converts the photons into electrons (photoelectrons). These electrons now represent the signal generated by the ionizing radiation. The electrons are accelerated using electrical charge, toward a series of electrodes (dynodes) arranged in a zigzag pattern. The dynode is made from a material that ejects more electrons than hit it at any given moment. By the time the electrons are ejected by the last dynode, there are a whole lot more electrons than started out. This is how the signal is amplified. Finally, the electrons hit the anode on the far end of the PMT and register an electrical signal. The energy of the original signal is preserved through the PMT, as it is proportional to the number of photoelectrons produced at the photocathode. This means that we can plot an energy spectrum (count rate vs. energy) to see how many counts are observed at specific energy values.

8.2.2 INORGANIC SCINTILLATORS

Inorganic scintillators are salts (inorganic compounds) that light up when hit by ionizing radiation. The most commonly used are crystals of NaI (sodium iodide), CsI (cesium iodide), and $LaBr_3$ (lanthanum bromide). In all three cases, the crystals light up a lot better if they are doped with a tiny amount of either thallium (Tl) or cerium (Ce). The doped crystalline materials are represented as NaI(Tl), CsI(Tl), and $LaBr_3$(Ce).

Notice that these materials all contain a fair amount of high atomic number (Z) elements. Higher Z means higher probabilities of interacting with the elusive X-ray and gamma ray photons, especially through the photoelectric effect (see Figure 7.9). These materials are also very dense when compared to the gases used in gas-filled detectors. In other words, they're packing a lot more matter, increasing the probability of interacting with ionizing radiation. As a result, inorganic scintillators are much more efficient at detecting high-energy photons.

Figure 8.6 demonstrates how a typical NaI(Tl) detector works. The crystal is cylindrically shaped and sealed inside a shiny metal case. The case is reasonably thin, so that X-rays and gamma rays

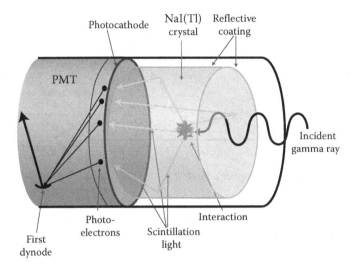

FIGURE 8.6 A NaI(Tl) detector at work.

can easily penetrate it. Once inside the crystal, the photon interacts, most likely with the iodide inside the crystal,[1] causing ionizations and excitations, eventually creating some visible (scintillation) light. The nature of these interactions is discussed in more detail in Section 8.4. The light then goes directly toward the photocathode, or it is reflected off the shiny metal enclosure into the photocathode.

After the PMT amplifies the signal, it is run through a multichannel analyzer (MCA), which sorts the signal by energy (actually, it is by channel, with each channel covering a short energy range). Both count and energy data are obtained, allowing the production of a spectral plot (photon energy vs. count rate). Such spectra will be indicative of its source.

NaI(Tl) detectors are relatively inexpensive and are commonly used with gamma-emitting nuclides. However, they are shock-sensitive (crystals are easy to crack or break if dropped), rather bulky (because they require a PMT), and suffer from poor energy resolution. Gamma photons are monoenergetic, and therefore should be observed in a spectrum as having a single energy value. Instead, a distribution of energies is observed centered on the actual value. The wider this distribution, the poorer the energy resolution, i.e., the harder it would be to see (resolve) two different gamma energies that just happen to have energy values close to each other.

Inorganic scintillators suffer from poor energy resolution because the signal generated at the photocathode is limited. First, not all of the incident photon energy is converted to visible light photons, some is converted to heat (displaced electrons traveling through matter will generate some heat!). Second, not all the visible light photons will make it to the photocathode *and* generate photoelectrons. This is when the signal is the weakest. As with any measurement, greater relative error is experienced with smaller numbers being measured.

There is also random error associated with the number of photoelectrons produced, which are then amplified with the signal in the PMT. The poor energy resolution of inorganic scintillators also results from the fact that it takes so many steps (conversions of energy) from the incident photon to the final electrical signal. There are random errors introduced at each step, leading to significant random error in the final signal. Fortunately, most gamma energies in a single spectrum are well

---

1 There's not much Tl and the Na is much lower Z.

FIGURE 8.7 Organic scintillation gives off photons of visible light.

enough separated to be observed using inorganic scintillators. Their resolution may be poor, but they are still useful.

8.2.3 ORGANIC SCINTILLATORS

All of the detection methods mentioned so far suffer from the fact that they can only detect radiation that enters the detector (and interacts). Since radiation is typically emitted in all directions equally, less than 50% of all emitted radiation can be expected to enter the types of detectors we've seen so far. If we have a low activity sample, especially if it emits low-energy radiation, it would be better if we could place the sample inside our detector, so that we could detect all decays. This can be done with an **organic scintillator**, as shown in Figure 8.7.

Organic scintillation works by dissolving the radioactive material in a "scintillation cocktail." This cocktail is made up of organic (low Z) materials: a solvent, a scintillator, and sometimes other stuff to help make it glow brighter or with the right wavelength. Low Z materials are used to minimize bremsstrahlung that can occur when beta particles are emitted during decay. This solution is placed in a glass or plastic vial, which is put in a dark place (inside an instrument!) with PMTs all around it to watch for visible light photons and record data.

How does this work? First the decay takes place. Since the cocktail is mostly solvent, it is most likely the solvent that absorbs energy from the ionizing radiation. The solvent then (eventually) transfers this energy to a scintillator molecule, which becomes excited. The scintillator then de-excites, emitting a photon. The photon hits a PMT, which amplifies the signal and records it. The more PMTs we can pack around the scintillation vial, the more photons we'll see, and the higher the counting efficiency.

As mentioned above, this is a great way to measure samples emitting low-energy radiation, such as weak beta-emitting nuclides. It is not very efficient for hard X-rays or gamma rays, as these photons are unlikely to interact with the low Z materials inside the vial.

Another drawback is **quenching**. Quenching here is a different issue than for gas-filled detectors. In terms of scintillation counting, it refers to anything that reduces light output. It can be a fingerprint or other smudge on the outside of the vial. A highly colored sample (like blood) could also be a problem. If the sample volume is significant compared to the volume of the cocktail, the photon output will be reduced. Finally, O_2 can inhibit the transfer of energy from the solvent to the scintillator and is best purged from the solution.

8.3 OTHER DETECTORS

8.3.1 Semiconductor Detectors

Given the poor energy resolution of inorganic scintillation detectors, it might be nice to have a detector with higher resolution when it is important to resolve signals from two gamma photons with similar energy values. As you remember from Section 8.2.2, the Achilles heel of a scintillation detector is the low number of photoelectrons that are produced at the photocathode. We need the detector made out of a solid so it has a reasonable chance of interacting with gamma photons, but it'd be nice if it worked like the gas-filled detectors, which convert ionizing radiation more directly to an electronic signal. Finally, we need a stronger signal, i.e., ion-pair formation from the incident radiation needs to be more efficient.

The answer would be a detector made from semiconductors. Semiconductors are those elements on the periodic table that have chemical and physical properties that are intermediate between the metals and the non-metals (like Si, Ge, and Te). These elements are sometimes called metalloids, but most important for us is the fact that they are neither conductors of electricity, nor are they insulators—they are something in between, semiconductors. This means that sandwiching a crystal between two electrodes can allow electrons knocked loose by ionizing radiation to migrate to an electrode and produce an electrical signal.

If you take a slab of silicon (Si) or germanium (Ge) and sandwich it between a couple of electrodes, it makes a **semiconductor detector**. These detectors are very similar to gas-filled detectors as depicted in Figure 8.1, except they are filled with a Si or Ge crystal instead of a gas. The main disadvantages of semiconductor detectors is that they are expensive to make (very high purities are normally necessary, although they can also be doped with certain elements), and typically need to be cooled to ~100 K (liquid nitrogen!) in order to work. The main advantage is that their energy resolution is excellent, as illustrated in Figure 8.8.

The improvement in energy resolution is largely a reflection of the greater number of ionizations created by the incident radiation. Incident gamma photons interact with semiconductors by knocking loose an electron and creating an electron "hole." The hole can be imagined as simply the place where the electron is supposed to be. Instead, the electron has been given extra energy and now resides in what is called the conduction band and is free to move about the semiconductor (toward the anode). Previously, it had been lodged in the valence band, where it can't move. The hole also "moves," but toward the cathode. This really isn't physical movement of the hole so much as an electron between the hole and the cathode being pushed into the hole resulting in a new hole closer to the cathode.

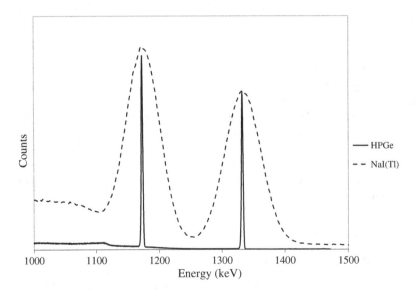

FIGURE 8.8 Comparison of the gamma spectrum (^{60}Co) energy resolution of an NaI(Tl) (dashed line) and high-purity germanium (solid line) detectors.

FIGURE 8.9 Thermoluminescence.

As mentioned above, two drawbacks of Si- and Ge-based detectors are their expense and their need to be very cold to work properly. A third drawback is that they have relatively low atomic numbers ($Z = 14$ and 32, respectively). Being made of lower Z material means that detection efficiency will suffer. More recently, new materials have been found that do not share all of these drawbacks. In particular, CdTe and $Cd_{1-x}Zn_xTe$ ($0.04 \leq x \leq 0.20$) have shown significant promise. The latter is more commonly known as "CZT." Neither require special cooling, in fact, both function well at normal room temperatures. They also incorporate higher Z material ($Zn = 30$, $Cd = 48$, and $Te = 52$), and therefore are more efficient in detecting gamma and X-ray photons. Detectors made from these materials are still expensive relative to NaI(Tl)-based detectors, but they are finding application in medicine, research, and national security where their compact size, good energy resolution, higher counting efficiencies, and convenience (no, or less, cooling necessary) are valued.

8.3.2 THERMOLUMINESCENT DOSIMETERS

Certain salts (inorganic compounds) can be used to store information on exposure to ionizing radiation. Good materials for this are lithium fluoride, LiF, and calcium fluoride doped with a little manganese, $CaF_2(Mn)$. When electrons in crystals of these salts are excited by interaction with ionizing radiation, they get moved up from the valence band to the conduction band, but then get stuck (trapped) in intermediate energy levels. They can stay trapped for months or even years! When the crystal is heated up, the electrons are freed from their traps, and return to the ground state. Along the way they emit visible light, which can be recorded by a PMT. The intensity of the light emitted is proportional to the amount of radiation the crystal was exposed to. This process (**thermoluminescence**) is illustrated in Figure 8.9.

These crystals are typically built into badges worn by personnel working with radioactive material or radiation-producing devices to estimate their occupational dose. These badges are called **thermoluminescent dosimeters** (TLDs) and can be collected on a monthly or quarterly basis to monitor exposure. Typically, three or four crystals will be placed inside a single badge, with different levels of shielding. This way the penetrating ability of the radiation the worker is exposed to can also be estimated.

8.4 GAMMA SPECTROSCOPY

Gamma spectroscopy is the detection of both count and energy information from gamma rays. By plotting counts vs. energy, we can see different energy gamma rays at the same time. It is sometimes referred to as pulse-height or energy spectrometry. Regardless of the name, it is a wonderful tool for detecting and discerning various radioactive nuclides. Unfortunately, the information is sometimes complicated by what appear to be odd features in the spectrum. Fortunately, these features can all be understood in terms of the interactions of high-energy photons with matter discussed in Chapter 7.

As you already know, gamma rays are very penetrating, i.e., they have a low probability of interacting with matter. We can boost this probability by making them try to pass through relatively dense matter. That's why lead (Pb) is commonly used for shielding high-energy photons. If we

FIGURE 8.10   Gamma spectrum for ^{137}Cs using a NaI(Tl) detector.

want to detect them, we'll also have to use fairly high Z material. As previously discussed, that's why inorganic scintillators or semiconductor detectors are commonly used. Remember that these detection methods give count *and* energy information from gamma-emitting sources! Instead of just getting count data, we get count data spread out over the gamma energy spectrum, like Figure 8.10 for ^{137}Cs.

But wait a minute! We know that the decay diagram for ^{137}Cs (Figure 8.11) only shows one gamma ray being emitted, with an energy of 662 keV. Since nuclear energy states are quantized, we should *only* see counts for 662 keV. Why is the peak at 662 keV so broad, and what's all the other stuff in the spectrum? Why is there a broad continuum of counts below ~500 keV, what is the bump around 200 keV, and the really big peak at relatively low energy? The answers to these important questions (and maybe some others!) have to do with how gamma rays interact with matter, because that's how they are detected.

Figure 7.9 shows us that if we build a detector out of low or medium Z matter, most of the gamma photons that interact with it will do so via the Compton effect. If we make it with high Z materials, the photoelectric effect will dominate, at least for low- to medium-energy gamma rays. Figure 8.12 shows the relative probabilities of the photoelectric effect, Compton scattering and pair production with iodine. Notice that the photoelectric effect is dominant for photon energies below ~300 keV. Above 300 keV, Compton scattering is most likely.

Does it make a difference whether the photon interacts with the iodide in the detector via the photo-electric or Compton effect? You betcha. When a photon interacts via the photoelectric effect, all its energy is dumped into the detector, giving us a signal corresponding to the energy of that photon. In the gamma ray spectrum this is called the **photopeak** (or the total energy peak, or the gamma peak). This is what we want to happen; therefore, we prefer to make gamma ray detectors out of high Z materials. As noted in

FIGURE 8.11   Decay diagram for ^{137}Cs.

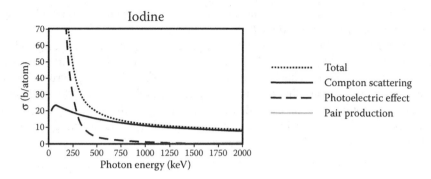

FIGURE 8.12 Relative interaction probabilities of high-energy photons with iodine.

Chapter 7, high Z also means a higher overall probability of interaction, meaning fewer gamma photons will pass completely through the detector without interacting.

If the photoelectric effect gives the energy of the gamma photon, why is the photopeak in Figure 8.10 so broad? As explained in Section 8.2.2, it is because of the weakness of the signal and because the energy of the incident photon undergoes so many conversions that these types of detectors have larger random error.

When the gamma photon interacts with the detector via the Compton effect, the photon *could* deposit only some of its energy in the detector then leave, taking with it some of its original energy as the scattered photon. Since the amount of energy transferred by the photon undergoing Compton scattering is variable, then different amounts of energy will be deposited in the detector for each photon undergoing the Compton effect then escaping the detector. This will give rise to a broad continuum of counts called the **Compton continuum**. This is the broad plateau of counts below (lower in energy) the photopeak in Figure 8.10. One way to decrease the number of counts under the Compton continuum relative to the photopeak is to increase the size of the detector. A larger detector will give the scattered photon more opportunities to interact on its way out, leaving all of the incident photon's energy inside the detector.

What's the maximum energy Compton scattering can leave in the detector? Assuming the gamma photon only undergoes one Compton scattering interaction inside the detector, it would be when it is scattered through 180°. This energy can be calculated based on the original energy of the photon using Equation 8.1:

$$E_{ED} = \frac{E_0^2}{E_0 + 0.2555} \tag{8.1}$$

where E_0 is the energy of the incident photon, and E_{ED} is the "maximum energy" that can be deposited in the detector. Both energy values must have units of MeV for this equation to be true. E_{ED} is also known as the **Compton edge**. The Compton edge for the 662 keV photopeak observed in the ^{137}Cs spectrum is 478 keV.

$$E_{ED} = \frac{0.662^2}{0.662 + 0.2555} = 0.478 \text{ MeV} = 478 \text{ keV}$$

Careful examination of Figure 8.10 shows that the Compton edge does not define a strict upper limit to the continuum, but rather it is the point where counts begin to decrease with increasing energy. Why are some Compton continuum counts still observed above the Compton edge? This is partly due to the poor energy resolution of NaI(Tl) detectors. It can also result from a gamma photon having more than one Compton scattering interaction inside the detector and exiting leaving slightly more energy behind than is calculated for E_{ED}.

Compton scattering can also take place *outside* the detector. A gamma photon originally traveling away from the detector could be scattered off an air molecule, you, your lab partner, or some other matter. If it is scattered back into the detector, it shows up in the gamma spectrum with a specific energy. The energy has a certain value because the scattering angle has to be 180° (or pretty close to it) for the scattered photon to end up in the detector. The energy of this small peak (E_{BS}, also known as the **backscatter** peak) can be calculated using Equation 8.2, where all energy values are again given in MeV:

$$E_{BS} = \frac{E_0}{1 + \left(3.91 \times E_0\right)} \tag{8.2}$$

Despite the fact that they originate from different phenomena, there's a simple mathematical relationship between E_{BS} and E_{ED}, they add up to energy of the incident photon!

$$E_0 = E_{BS} + E_{ED} \tag{8.3}$$

This equation is really the same as Equation 7.11, repeated here for your convenience:

$$E_0 = E_{SC} + E_{CE}$$

where E_{SC} is the energy of a scattered photon and E_{CE} is the energy of the Compton electron. Equation 8.3 represents a special case of Equation 7.11—when backscattering takes place, then E_{SC} is equal to E_{BS}, and E_{CE} is equal to E_{ED}.

^{137}Cs emits a 0.662 MeV gamma ray, therefore its backscatter peak should be found at 184 keV.

$$E_{BS} = \frac{0.662}{1 + 3.91(0.662)} = 0.184 \text{ MeV} = 184 \text{ keV}$$

We now understand most of what we see in the gamma spectrum for ^{137}Cs. These features are labeled in Figure 8.13.

So what's up with the low-energy peak, and why is it labeled "Ba X-ray"? The ^{137m}Ba isomer (a metastable state) that is formed after ^{137}Cs emits a beta particle (see Figure 8.11) can also decay to the ground state by spitting out a conversion electron. The conversion coefficient (α_K) is 0.093, which means that ~1 in 10 decays produces a conversion electron instead of the 662 keV gamma photon. The resulting hole in the K-shell of the ^{137}Ba daughter can be filled by an electron from an L shell, which could generate an X-ray with an energy of 32 keV. These X-ray photons are also detected by the NaI crystal, producing the low-energy peak observed in the gamma spectrum of ^{137}Cs.

Equipped with a solid understanding of the interactions of gamma rays with matter (Section 7.3), let's see if we can understand what other odd features we might come across in gamma spectra.

The relative intensities of gamma peaks can depend on a few factors. As we already know, higher energy photons have a lower probability of interacting with matter. Therefore, we would expect to see smaller peaks for higher energy photons even if the source is emitting equal numbers of low- and high-energy photons. Intensities of two different peaks may also vary if they are part of two different decay branches, or if they have significantly different conversion

FIGURE 8.13   Labeled gamma spectrum for ^{137}Cs.

coefficients. Finally, relative intensities can vary due to the relative probability that each photon will be emitted as part of the decay.

A great example is ^{57}Co. As illustrated in Figure 8.14, it decays via electron capture to an excited state of ^{57}Fe with a 5/2 spin state. This excited state has two options. It can decay directly to the ground state (spin 1/2), emitting a 136 keV gamma photon (gray), or it can decay to a lower energy excited state (spin 3/2) emitting a 122 keV gamma photon (black). The 122 keV gamma photon is more likely to be observed (more intense in the spectrum) than the 136 keV gamma photon because the change in nuclear spin states is less dramatic. We would also expect the 136 keV peak to be smaller because it is higher in energy than the 122 keV peak. The observed relative intensities are 85.6:10.7.

Using NaI(Tl) detectors it is not possible to resolve the two ^{57}Co gamma peaks—their energy values are too close to each other. Since the 122 keV peak dominates intensity, the 136 keV peak tends to be buried underneath it. Figure 8.15 shows the gamma spectrum for ^{57}Co using a NaI(Tl) detector. The 122 keV peak is the largest feature on the spectrum, and it appears to have a low shoulder on the right (higher energy) side. This is probably a part of the 136 keV photopeak.

There is another feature in the ^{57}Co spectrum (Figure 8.15) that is often observed in the spectra of low-energy gamma photons. It shows up as a small peak that is 28 keV lower in energy than the photopeak (at ~94 keV in this case). When a low-energy gamma photon enters a NaI(Tl) detector, it will most likely interact with the iodide (I^-) via the photoelectric effect. This creates an electron vacancy in the K shell of the iodide. When an electron from a higher energy shell fills this vacancy, an X-ray photon is generated. Normally the crystal absorbs the energy of this photon, i.e., it interacts before it can get out of the crystal. However, if it manages to find its way out of the detector, the recorded energy will be missing 28 keV (the energy of the iodine K_α X-ray). This is called an **iodine**

FIGURE 8.14   Decay diagram for ^{57}Co.

FIGURE 8.15   Gamma spectrum of ^{57}Co.

escape peak. It is only observed with low-energy gamma photons, since they are most likely to interact with I⁻ via the photoelectric effect. Low-energy photons also have a higher probability of interacting with matter, in general. Therefore, odds are better that they will interact close to the surface of the detector. This increases the probability that the K_{α} X-ray will escape the detector, since it (potentially) has less matter to travel through to get out.

Figure 8.16 shows the decay diagram for ^{60}Co. After beta emission, an excited state of ^{60}Ni with a spin state of 4 is formed. Like the ^{57}Co example above, this excited state has two choices; decay to another excited state (spin 2) or go all the way to the ground state (spin 0). In this case, the change in spin state is too dramatic to allow direct decay to the ground state. As a result, photopeaks are observed at 1.17 and 1.33 MeV.

If a sample of ^{60}Co is reasonably toasty (high activity), a small peak may also be observed at 2.50 MeV. However, this is not due to a 4+ → 0+ transition but is a **sum peak**. This happens when two different gamma photons hit the detector at the same time (or within the detector's resolving time) and are recorded as a single count. Sum peaks become more probable when a well detector is used. A cutaway diagram of a well detector is shown in Figure 8.17. Basically, it's a NaI(Tl) detector with a hole drilled into it so the source can be placed in the center of the crystal. This dramatically increases the counting efficiency as well as the odds that two photons will enter the scintillation crystal at more or less the same time.

There's also a gamma spectral feature that indicates positron decay. Figure 8.18 shows the decay diagram for ^{22}Na, which decays by positron emission and electron capture. Both decay modes form

FIGURE 8.16   Decay diagram for ^{60}Co.

FIGURE 8.17   Cutaway view of a well detector.

the same excited state of the ^{22}Ne daughter, which emits a 1.275 MeV gamma photon when it decays to the ground state. As expected, its gamma spectrum has a photopeak at 1.275 MeV, but a large peak is also observed at 0.511 MeV. This peak is due to positron annihilation (with an electron) and is called an **annihilation peak**. The positron will most likely annihilate outside of the detector, generating two 511 keV photons. A number of these photons enter the detector and those that deposit all of their energy are recorded at 0.511 MeV. The ^{22}Na spectrum can also exhibit sum peaks at 1.02 MeV (two annihilation photons entering the detector at the same time), and at 1.786 MeV (1.275 + 0.511 MeV).

If the energy of the gamma photon is greater than 1.022 MeV (as in the last two examples), a peak at 0.511 MeV may be observed as a result of pair production *outside* the detector crystal. Remember that pair production creates an electron and a positron. If the positron annihilates outside the detector and sends one of the annihilation photons into the detector and it deposits all of its energy in the detector, a count will be recorded at 0.511 MeV. Remember also that the odds of pair production are pretty low or zero at energies we're likely to see gamma photons (below ~1.5 MeV, Figure 8.12), i.e., it's pretty unlikely you'll ever observe this from a gamma source. It would require a relatively high activity source emitting pretty high-energy gamma photons—something you wouldn't want to be near.

What if a high-energy photon gets inside the detector and interacts via pair production? Then the positron and the annihilation photons are created inside the detector. If one or both of the annihilation photons get out of the detector without interacting, then counts will be recorded with an energy of 0.511 (one escapes) or 1.022 MeV (both escape) below the photopeak. The peak that is 511 keV

FIGURE 8.18   Decay diagram for ^{22}Na.

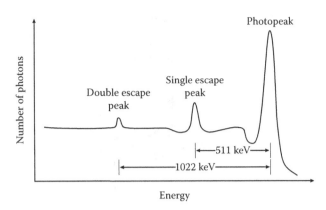

FIGURE 8.19 A gamma spectrum with escape peaks.

below the photopeak is called a **single escape peak**. The peak 1.022 MeV below the photopeak is called a **double escape peak**. An example of a spectrum exhibiting these escape peaks is shown in Figure 8.19. To observe these peaks, the incident photon must have an energy well above the 1.022 MeV threshold for pair production and have a fairly high activity. Again, the odds of most humans seeing this phenomenon in their lifetime is pretty low. Notice also that the intensity of the double escape peak is much lower than the single escape peak. The odds of both annihilation photons escaping is much lower than just one.

If lead is used for shielding or as a collimator, or is just lying around nearby, a peak at ~75 keV could be observed. If a gamma photon from the radioactive source interacts with Pb via the photoelectric effect, it can create a vacancy in the lead K electron shell. When this vacancy is filled with an electron from a higher energy shell, a lead characteristic X-ray can be produced. The X-ray can then enter the detector and record a count of its energy, which is called a **lead X-ray** peak.

When *a lot* of beta particles, relative to the number of gamma photons, are emitted from a radioactive source, **bremsstrahlung** is observed. Remember that bremsstrahlung is the production of X-rays as a high-speed electron slows down while interacting with matter. These interactions are highly variable and can produce a broad energy spectrum of X-ray photons. Bremsstrahlung can occur either inside or outside the detector. All that matters to us is whether the detector records the X-ray. If it does, it'll show up as a count at the low-energy end of the gamma spectrum. Collectively, it is manifested in the far left of the gamma spectrum as a slow rise in the count rate as photon energy decreases. Figure 8.20 gives an example spectrum and the corresponding decay diagram.

The decay diagram for [86]Rb (Figure 8.20b) shows that the beta decay branch ratio leading to gamma emission is small. This nuclide is spitting out mostly beta particles, and only a few gamma photons. As a result, some bremsstrahlung is also observed in its gamma spectrum (Figure 8.20a).

If something is between the source and the detector, the gamma photons *can* undergo Compton scattering while traveling through the object on their way into the detector. This is known (logically!) as **object scattering**. Since Compton scattering lowers the energy of the photon, object scattering leads to relatively more counts observed under the Compton continuum and relatively fewer counts under the photopeak.

The **peak-to-total ratio** is a measure of detector quality. It is determined by summing the counts for all the channels under the photopeak and dividing by all counts observed in the entire spectrum (Figure 8.21). The peak-to-total ratio will depend on the energy of the gamma ray. The ratio generally decreases with increasing energy of the gamma ray. This makes sense as the *relative* probability of Compton scattering increases with increasing gamma ray energy. Overall, the probability

FIGURE 8.20 (a) Gamma spectrum (adapted from Heath, R. L. *Scintillation Spectrometry: Gamma-Ray Spectrum Catalogue*, 2nd ed., 1997, Idaho National Laboratory, Idaho Falls, ID). (b) Decay diagram for ^{86}Rb.

FIGURE 8.21 Peak-to-total ratio.

of interaction goes down as gamma energy increases, but the odds of undergoing the photoelectric effect go down a lot faster than the odds of Compton scattering (Figure 8.12), therefore the relative odds of Compton scattering increase for higher energy photons. This means that the peak-to-total ratio is expected to be lower for a higher energy peak because more of its photons will interact with the detector via Compton scattering rather than by the photoelectric effect. If the detector is damaged, this ratio will increase.

Another measure of detector quality is its ability to resolve two peaks close in energy to each other. As mentioned previously in this chapter, resolution is a measure of how much random error is introduced by the detector and is manifested in the width of the photopeak. A wider peak means more error. Resolution is quantified by measuring the width of the peak halfway to the top (full width at half-maximum, or FWHM) and dividing by the energy of the photopeak (E_0).

$$\% \text{ resolution} = \frac{\text{FWHM}}{E_0} \times 100\%$$

Full width at half-maximum is illustrated in Figure 8.22. NaI(Tl) detectors typically have a resolution of 7–8% at 662 keV (^{137}Cs) while lanthanum bromide detectors are ~3% and germanium detectors are a phenomenal 0.3% at the same energy.

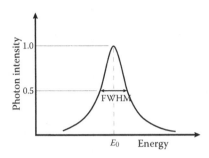

FIGURE 8.22   Resolution for a photopeak.

QUESTIONS

8.1   How could a proportional counter discern between alpha and beta radiation?

8.2   List the three types of gas-filled detectors in order of increasing voltage. Very briefly give a distinguishing characteristic, or typical use for each.

8.3   Using information from Question 7.4, determine which electronic transitions in argon atoms produce X-rays, UV, and visible light.

8.4   Would a Geiger-Müller counter be efficient at detecting neutrinos? Briefly explain.

8.5   A high-energy (~1.3 MeV) gamma-emitting nuclide is placed about 2 cm below a G-M tube, which records 360 counts over 30 s. A thin (~0.035 mm) sheet of aluminum is then placed directly over the source and the detector records 460 counts over 30 s. Such an increase in counts is often observed under these circumstances. Can you explain why?

8.6   What would be the best detector to use if we only wanted approximate count information for an intense, polyenergetic photon beam and we didn't want to significantly attenuate the beam?

8.7   What would be the best detector to obtain count and energy information from a beta-emitting nuclide? A gamma-emitting nuclide? Assume only one energy (particle or photon) is emitted in each case.

8.8   A nuclear medicine technologist working in a radiopharmacy performs swipe tests weekly to look for areas of radioactive contamination. The swipe test is performed by wiping a small piece of paper (like filter paper) across the surface to be tested, placing it in a glass test tube, then placing the assembly into a gamma spectrometer. If the only radionuclide in use at this pharmacy is ^{99m}Tc, comment on why this might not be an especially good quality control (QC) procedure.

8.9   How is bremsstrahlung avoided in organic scintillators?

8.10 Radiation therapists work with equipment that generates high-energy photons (X-rays). If they wanted to determine the monthly dose from small amounts of radiation "leaking" from their machines (not the main beam) into the treatment room, what would be the best way to go about it?

8.11 A 0.100 MeV gamma ray interacts inside a NaI(Tl) detector by the photoelectric effect. What element is most likely involved in this interaction? What is the atomic shell most likely to lose an electron from this element? Briefly explain how an iodine escape peak might be formed from this gamma ray. Calculate the energy of the freed electron.

8.12 In terms of detecting ionizing radiation what is the fundamental difference between gas-filled and scintillation detectors?

8.13 Calculate the energy of the backscatter peak and the Compton edge for the gamma ray most commonly observed in the decay for ^{41}Ar.

8.14 The energy of a backscatter peak is observed at 213 keV. What is the energy of the corresponding Compton edge?

8.15 Backscatter peaks are often observed at slightly higher energies than calculated. Why might this be?

8.16 Sketch the gamma ray spectrum for ^{99m}Tc labeling the important features.

8.17 During a routine survey of the Crab Nebula, the science staff of the Stargrazer discovers an abundance of one element. Sensor analysis shows it to be Al, P, Cu, Br, or Tm. You obtain a small sample and place it in a neutron beam generated by one of the EPS conduits. Upon removal, you detect a 0.51 MeV photon using a flat-faced NaI(Tl) detector. The same sample shows peaks at 0.51 and 1.02 MeV when placed in a well detector. Using either detector, at least one other peak is observed in the gamma spectrum, but interference from the nebula is preventing you from observing their energies. Your report on this element is due to the ship's science officer in one hour. What can you contribute to the solution of this mystery? What experiment would you suggest be done next?

8.18 Explain the origin of the Compton continuum often observed in gamma spectra.

8.19 Starting with the relevant type of decay, outline the process by which an annihilation peak is observed in a gamma spectrum.

8.20 What two types of decay are likely to eventually produce an X-ray peak in a gamma spectrum.

8.21 Label the features observed in the gamma ray spectrum of ^{65}Zn (Figure 8.23).

8.22 When a lead collimator is used with a gamma camera, a peak at 75 keV is observed. When the collimator is absent, no peak is seen at 75 keV. Briefly explain the origin of this peak.

FIGURE 8.23 The gamma spectrum of ^{65}Zn.

8.23 Why are Compton scattering and pair production undesirable interactions inside a gamma detector? How are detectors designed to minimize these interactions?

8.24 A gamma photon enters a NaI(Tl) detector, but a count is not registered under the photopeak. Give four reasons why this might happen.

8.25 The dashed line in the Figure 8.24 represents that gamma spectrum of a radionuclide in air, while the solid line is the spectrum for the same source inside a patient. Briefly explain the differences between the two spectra. What happens to the peak-to-total ratio as this source is put inside a patient?

FIGURE 8.24 Gamma spectra taken of a nuclide outside (dashed line) and inside (solid line) a patient.

9 Applications of Nuclear Science II

Medicine and Food

Radioactivity was a newly discovered phenomenon early in the 20th century. During the early decades of the century, it was commonly believed that deliberate exposure to ionizing radiation had health benefits. There were numerous radiation "treatments." Radium salts (inorganic compounds) were sold in health elixirs. People paid for mine tours to breathe air containing relatively high concentrations of radon, or to soak in waters containing radioactive ions. Some of these practices continue today, and a health benefit from low-level exposures to ionizing radiation may be real.

There was also some excitement over consumer products containing radioactive materials at that time. Radium-containing cosmetics and radium-painted clock and watch dials were popular in the early part of the 20th century because they glowed all night long. For much of the 20th century, uranium compounds were used to provide colors in the glazes of ceramic products, most famously in orange and ivory Fiesta®[1] dinnerware from 1936 until 1973. Uranium was also used to provide a yellow or yellow green color to transparent glass. Called "uranium," "Vaseline," or "gemstone" glass, they were also manufactured until the early 1970s.

The excitement faded then turned to horror as the adverse health effects from significant doses of ionizing radiation became public. First were the "radium girls." Some of the workers (almost all young women) that applied radium-containing paint to clocks, watches, and aircraft instruments, began to get sick and die. As they painted, they used their lips to sharpen the tips of their brushes, ingesting a significant amount of radium. In the middle of the century, images of the adverse health effects experienced by Japanese exposed to significant amounts of ionizing radiation from the Hiroshima and Nagasaki bombings were made public.[2] Shortly after World War II, numerous fictional representations of the possible horrors resulting from exposure to ionizing radiation became a staple of popular fiction that unfortunately persists today.

While fiction took considerable advantage of the possible health effects of exposure to ionizing radiation, scientists sought to quantify it. As discussed in Section 1.6, it is clear that above certain levels, adverse health effects are observed, and the risk increases in a roughly linear fashion with increased dose. At lower doses, it remains less clear whether there is still some risk, no risk, or a health benefit. While much of the scientific evidence supports either no risk or a health benefit to low doses of ionizing radiation, and there is very little evidence to support the linear non-threshold hypothesis, radiation workers still follow the ALARA (keeping doses as low as reasonably achievable) rule. Health physicists also work to lower patient doses whenever possible, assuming that additional exposure *could* present a risk to the patient. What if the additional exposure is also a health

---

1　Fiesta is a registered trademark of the Homer Laughlin China Co., Newell, West Virginia.
2　Less publicized were long-term studies of the radium girls and Japanese atomic bomb survivors that received lower doses showing they generally lived long, healthy lives.

DOI: 10.1201/9781003303787-9

benefit? If evidence mounts to support this notion, we may discover that *some* of our horrors over the radiation treatments of the early 20th century are misplaced.

This chapter deals mostly with the applications of ionizing radiation to health care, primarily in the treatment of cancer (radiation therapy), and the diagnosis of a variety of diseases (radiography and nuclear medicine). These health professions are normally considered part of the medical specialty of radiology, so it will be briefly discussed first. Because of its relationship to radiation therapy, food irradiation is also briefly discussed in this chapter. It may sound like an odd relationship now but hang in there, it's a pretty good one.

9.1 RADIOLOGY

Radiology is a medical specialty that uses both ionizing (like X-rays) and non-ionizing (e.g., ultrasound) radiation for the diagnosis and treatment of disease. We'll only be concerned with those modalities that use ionizing radiation. One that is often thought of as involving ionizing radiation is magnetic resonance imaging (MRI). Perhaps it is because it was formerly known as nuclear magnetic resonance (NMR). MRI takes advantage of the fact that the nuclei in the molecules in our body are all little (very little!) magnets. If placed in a strong magnetic field, they will align with the field. Once aligned, they can be tipped over using a radio wave, and, as they realign with the field, they give off their own radio signals, the exact energies of which depend on their local environment. Thus, muscle looks different from the blood vessels running through it. Radio signals are transmitted by photons, but they are low-energy (see Figure 1.1) and are non-ionizing. MRI simply tickles nuclei rather than violently transforming them.

Radiology departments are very high-tech and include a wide variety of imaging and therapeutic techniques. Perhaps the most familiar part of radiology is radiography—using X-rays to image the body's innards. The technique is pretty simple, shine a beam of X-rays on part of a patient and collect an image on the other side. As discussed in Section 7.4, X-rays are more likely to be absorbed by higher density materials; therefore, they are generally absorbed by bones and transmitted through soft tissues. Radiography is therefore well suited to bone imaging, as bone shows up with the greatest contrast. Soft tissues have slight variations in density, which then appear as subtle differences on the radiograph (X-ray image). When necessary, contrast media can be administered to help light up particular areas. For example, a patient could swallow a $BaSO_4$ suspension,[3] which would then absorb X-rays trying to pass through the gastrointestinal system.

Conventional radiography is two-dimensional and cannot provide cross-sectional views. The X-ray beam travels through the body in one direction and produces an image of everything it travels through. If there are different organs stacked along the beam's path, they may be difficult to discern. Compton scattering will also take place as the beam passes through the (mostly low Z) patient, which will cause some fogging of the image overall, making it difficult to see some of the more subtle differences in the image. If these limitations are important, they can be overcome using computed tomography (CT, also known as a CAT scan). The key difference from conventional radiography is that CT imaging involves the movement of both the X-ray source and detector around the patient, as shown in Figure 9.1. Several X-ray images taken at different angles can be put together to form a three-dimensional image of the patient's anatomy.

Computed tomography scans allow cross-sectional views, without having to slice the patient in two (messy!). It also allows separate and complete viewing of organs that could otherwise be in the way of each other on a conventional X-ray. Finally, the errors produced by scattered photons can be minimized by the use of collimators both before the beam enters the patient and after it exits. The collimators are basically honeycombs of lead that will absorb any photons not traveling on a straight line from the X-ray source to the detector.

---

3   Barium sulfate has an extremely low solubility in water and will therefore pass through the patient as a solid.

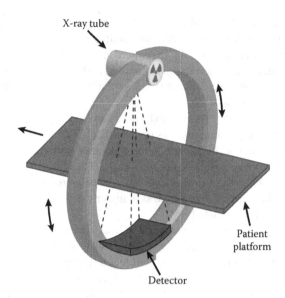

FIGURE 9.1 A schematic of a CT scanner.

CT scans provide fairly detailed anatomical information but come with a cost. As you might guess, taking lots of X-rays to get enough for a clear 3-D image means higher doses to the patient. In fact, CT scans typically involve the highest patient doses for any diagnostic procedure. Manufacturers of CT machines, physicians, and health physicists have worked diligently to find ways to lower CT doses while still obtaining the information needed. The focus has been on limiting the amount of patient scanned while using the lowest intensity of photons needed to obtain a quality image. Their efforts have largely paid off, and patient doses have been lowered for many CT protocols.

Various detectors are used for CT and 2-D radiography, and most are based on the detector technologies presented in Chapter 8. An important consideration for any X-ray detector used in medicine is its efficiency. The more efficient it is, the lower the dose needs to be for the patient. As you might guess, higher Z solids tend to dominate, because they can place more matter in the beam path and are more likely to interact.

Storage phosphor detectors work in a manner similar to thermoluminescent detectors (Section 7.3.2). When the X-rays interact with the phosphor, electrons are promoted and trapped in higher energy states. Instead of heat, laser light is used to release the electron, which then gives off a photon of visible light (at a different wavelength than the laser), which can be read. If made properly, phosphors can be quite efficient in detecting X-rays. However, getting the image is a two-step process—exposing the phosphor, and then reading it.

Also quite efficient are detectors based on inorganic scintillators (Section 7.2.2). These detectors use a scintillator, like CsI(Tl), to convert the X-ray to visible light. The light then can be read several different ways, but one of the more interesting is by a charge-coupled device (CCD). CCDs are commonly used in digital cameras to snap pictures of you and your friends. These types of detectors produce the image in a single step, instantly displaying on a computer screen after the patient is zapped.

9.2 RADIATION THERAPY

Radiation therapy (RT) is the medical application of ionizing radiation to treat disease. Today, most radiation therapy procedures involve the bombardment of tumors with high-energy X-rays. The remaining RT procedures involve the placement of a sealed radioactive source inside or near diseased tissue with the intent to kill it. The former are generally termed external beam therapy and the latter implant or brachytherapy. The goal in performing radiation therapy is to deliver the greatest dose to the tumor while minimizing dose to the surrounding (healthy) tissue. The ratio of abnormal cells killed to normal cells killed is called the **therapeutic ratio**. These ratios are generally high for tumor treatment because tumor cells are rapidly dividing, and therefore are much more susceptible to radiation damage from a high dose delivered over a relatively short time. The history of RT is all about increasing the therapeutic ratio.

The first fifty years (1895–1945, or so) of external beam therapy were dominated by treatment using machines that could only generate photons up to 500 keV. Because these instruments generated X-rays in much the same fashion as diagnostic X-ray machines (used to look for a broken bone …), they are generally called "conventional" X-ray machines.[4] These treatments were often limited by skin reactions because the maximum dosage was always delivered to the skin. In fact, treatment time was often determined by observation of the skin under the X-ray beam—when the skin started getting too red, treatment was terminated.

Implant therapy was fairly popular in those days. Back then, radioactive materials were pretty much limited to the members of the naturally occurring decay series. ^{226}Ra was commonly used as it could be separated from uranium ores. As you will recall (Chapter 1), ^{226}Ra is in the middle of the ^{238}U decay series—there are still a dozen or so radioactive nuclides in the series before it stops at ^{206}Pb. This is bad news, as it means that most of these other nuclides will be present in any sample of ^{226}Ra once it has sat around for a while (about a month). All kinds of different ionizing radiation will be emitted. Enough high-energy gamma radiation is emitted that those working with these sources on a daily basis couldn't help but be exposed to significant doses.

Alpha decay is a big part of the decay of ^{226}Ra to ^{206}Pb. Every one of these alpha particles becomes a helium atom inside the sealed source. Pressure will eventually build up from the helium (and the ^{222}Rn produced in the decay chain) and burst the container. This was a big problem with radium implant sources because the radium was in the form of a fine powder—when a sealed source burst, it created quite a mess to clean up.

Finally, it was tough to get these sealed sources to stay put. Our bodies move—even at rest we breathe, circulate blood, and digest. It is very difficult, even for the most disciplined person, to remain motionless for long. When we move, something placed in our tissues will also move—not always in the same way. As a result, it was often difficult to ensure adequate dose distributions from sealed sources placed inside humans. For all the reasons mentioned above, implant therapy fell out of favor during the first half of the 20th century.

After World War II, nuclides like ^{60}Co (emitting 1.17 and 1.33 MeV gamma rays during decay) became available, and radiation therapy instruments using ^{60}Co sources became the mainstay for about 30 years. As a result, some aspects of being a radiation therapist became more complicated. Higher energy photons meant that the maximum dose was now received below the skin, improving the therapeutic ratio, but eliminating skin damage as a way to determine treatment times.

It was also during this time that radiation therapists started repositioning the gamma ray beam during or between treatments—firing at a tumor from different angles as shown in Figure 9.2. This is known as **isocentric** therapy because the tumor is located at the intersection (isocenter) of the various beam paths. Isocentric therapy minimizes the dose to the healthy tissue between the tumor and the source by changing the path the photons take to zap the tumor. This technique greatly increases the therapeutic ratio. Determining variables such as treatment time, optimal beam shape, and angles of entry became quite complex when using isocentric techniques. Fortunately, computers were also developed in this time period, which greatly facilitated treatment planning.

In the late 1960s, researchers at Los Alamos National Laboratory developed compact linear accelerators, which could generate X-rays with maximum energies greater than 1 MeV in a

---

4   When generating X-rays from a tube potential of 150–500 kV they are known as "orthovoltage" or "deep therapy" machines.

FIGURE 9.2 Isocentric radiation therapy.

relatively small space (like a hospital room). Once again, higher energy beams meant the ability to deliver more dose deeper inside the patient, improving the therapeutic ratio. As a result, these **megavoltage** machines (commonly called "**linacs**") started replacing ^{60}Co instruments in the 1970s. It took a while, but they are now pretty much the only external beam therapy machines used by radiation therapists (Figure 9.3). With the dawning of the megavoltage era, still greater complexity was introduced. These are pretty high-tech instruments that produce very intense, very high-energy, very sharp beams. They can also be moved with greater flexibility relative to the tumor, and some can even change the shape of the X-ray beam as the beam is moved around the patient (**intensity-modulated radiation therapy**, or IMRT). Imaging of the tumor can also be done at the same time as treatment. After all, you're running a whole lot of X-rays through a patient; might as well see if you're hitting the target. This technique is called **image-guided radiation therapy** (IGRT) and allows for modification of the beam shape in real-time with the therapy. Finally, treatment can be

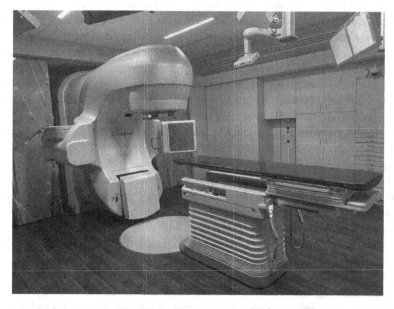

FIGURE 9.3 A modern-day linear accelerator radiation therapy machine. Photo courtesy of Zak Baudry.

synchronized with the patient's breathing (**respiratory gated therapy**) when the treatment volume moves with respiration. Treatment planning is now heavily dependent on computers, but therapeutic ratios are higher than ever.

A few ^{60}Co radiation therapy treatment machines have survived for specialty applications. A machine known as a **gamma knife** arranges a couple of hundred ^{60}Co sources in a dome-shaped device. Each source is carefully collimated so only a narrow beam of photons is not absorbed by shielding (these machines are heavy!). The beams are oriented so they all converge at a point. The patient's tumor (typically brain) is placed at the convergence point, and the patient is moved so the entire tumor gets some quality time at this point. A more flexible version of the gamma knife is the **cyberknife**. A cyberknife machine has a single ^{60}Co source in a shielded container that produces a narrow beam. This container is mounted on a robotic arm that can then position the beam hundreds of different ways for treatment. By maximizing the number of pathways taken by the high-energy photons, the therapeutic ratio can be improved. Treatment with the cyberknife is best for relatively small tumors with lots of angles of approach. As a type of IGRT, cyberknife machines collect diagnostic images during therapy to optimize the location of the beam.

Implant therapy has experienced a bit of a renaissance in recent decades, thanks in part to the use of new nuclides such as ^{192}Ir and ^{103}Pd. Iridium and palladium are chemically inert, and therefore require minimal, or no encapsulation. Other implant sources were typically encapsulated in stainless steel, while ^{192}Ir and ^{103}Pd can be encased in thin plastic. This allows more ionizing radiation (low-energy gamma photons and beta particles) to penetrate the casing. ^{192}Ir and ^{103}Pd both emit lower-energy gamma photons, and ^{192}Ir also emits beta particles. Because of their high specific activities and minimal encapsulation requirements, these sources can be quite small. In fact, they are often referred to as "seeds." Implant sources can be placed permanently inside a patient, but they are usually only temporarily put inside someone. To accomplish this, they can be placed on a wire or plastic line and inserted into the patient through a catheter. To minimize dose to the therapist, these sources can be inserted into and removed from the patient by remote control. This is known as **remote afterloading**.

Electron and proton beams are sometimes used for radiation therapy. Electron beams are readily available from the same X-ray machines used by radiation therapists to generate beams of high-energy photons. As explained in more detail in Chapter 14, X-rays are generated from beams of high-energy electrons hitting a metal target. Simply removing the target allows these electron beams to hit the patient instead. Proton beams are a little harder to come by, as they need a cyclotron to accelerate them, and rather large and powerful magnets are required to bend the beam. While a typical therapy room may not look very big (Figure 9.4), the rest of the equipment for the room is at least as large as the room itself. As a result, proton therapy centers are relatively expensive to construct and maintain and are much less common than those using X-ray and electron beams.

Since both electrons and protons are charged particles, they tend to deposit a large proportion of their energy at a specific depth below the surface of the matter (the Bragg peak—see Section 7.2). This depth depends on the energy of the particle, the greater the energy, the greater the penetration. This means that electron and proton beams tend to have great therapeutic ratios. The downside for electron beams is that they can undergo scattering by the nuclei of the matter they penetrate (more from Section 7.2) causing the beam to spread a bit inside the patient. Both electrons and protons also have a much greater probability of interacting with matter than high-energy photons, and therefore cannot always penetrate as deeply as needed. Despite their drawbacks, both have specific applications where they outperform high-energy photon beams or implants, so both experience some use.

Beams of other particles have been examined for radiation therapy, but none have found a practical application. Beams of neutrons, pions (also known as π-muons), alpha particles, and nuclei of ^{20}Ne and ^{40}Ar have all been examined. None are currently in common use.

As mentioned later in this chapter, nuclear medicine is currently researching molecular radiotherapy—designer molecules with a very specific biodistribution and a radioactive payload.

FIGURE 9.4   A proton therapy gantry. Photo courtesy of the National Association for Proton Therapy.

If successful, these therapies might be well suited to conditions that cannot be effectively treated with current beam or implant techniques. One example would be a cancer that has metastasized to several different locations within a patient. Beam and implant therapies work best on tumors that are well localized—if one has spread throughout a vital organ or the body, it is difficult to treat with these techniques without destroying the organ or body.[5] Ionizing radiation delivered on the cellular or molecular level could be much more effective.

9.3   FOOD IRRADIATION

Food irradiation is the use of ionizing radiation to kill biological organisms that are deleterious to foods. It may seem strange to place this topic in the same chapter with radiation therapy, but they have a lot in common. Both typically use high-energy photons to kill the bad stuff (diseased cells or bacteria) while trying to leave the bulk (healthy tissue or edible food) unaffected. Before we get into food irradiation, let's take a quick look at some other uses of high-energy photons.

The U.S. Postal Service uses ionizing radiation to kill anthrax spores, or other potentially harmful organisms that could be sent through the mail. High-energy photons are typically used because of their ability to penetrate packaging. Ionizing radiation is also widely used to sterilize medical supplies like bandages and consumer products such as tampons and condoms, which are difficult to sterilize by more conventional means (steam). A common public misconception is that such irradiation will make their mail, bandage, or condom become radioactive. We will learn in Chapter 10 that this can't be true—only projectiles such as neutrons, protons, and other nuclei could cause residual radioactivity in the matter they pass through. Most likely, ionizations will be caused, and chemical bonds will be broken—hopefully killing the bacteria and/or bugs while leaving the rest of the irradiated object alone. Just like radiation therapy, we want to kill the bad stuff, while minimizing damage to the rest.

---

5   Some studies suggest that low half-body or full-body X-ray doses (~150 cGy over 5 weeks) are effective against more distributed cancers and are helpful in preventing recurrence. Similar studies during the COVID-19 epidemic showed low doses helped alleviate symptoms in the lungs, easing breathing.

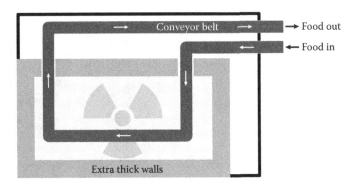

FIGURE 9.5   A schematic for a food irradiation facility.

Food irradiation uses X-rays, gamma rays, or electron beams, to rid food of pests or bacteria so that the food will not be damaged or spoil as quickly. In some cases, it can serve as an effective substitute for chemical pesticides and preservatives. It only works well with certain foods, and is typically used on fruits, vegetables, and spices. It can dramatically increase the shelf life of some of these products. Food irradiation can be used with meats and is an effective method to prevent *E. coli* contamination in ground beef. Unfortunately, irradiation of meats can leave it with a bad taste—some say it is like burnt hair, disgusting!

Because very high doses are used, we can't expect that the ionizing radiation will only target the bugs. Some chemical damage is also done to the food. In the case of meat, it clearly alters some of the chemicals that give it flavor. Opponents of food irradiation suggest that irradiation affects the chemical makeup of any food and should not be used. Attempts to detect these chemicals have largely proved fruitless, which is expected since these chemicals could only be formed in very low quantities. The important question to ask is which has the most benefit and the least risk: treatment with chemicals, treatment with ionizing radiation, or no treatment at all. There are risks and benefits to each.

A typical layout for an irradiation facility is shown in Figure 9.5. The stuff to be zapped is loaded on a conveyer belt that takes it into the irradiation room then out again to be distributed. The irradiation room has extra-thick walls and contains a high-activity source (typically ^{60}Co or ^{137}Cs) or a radiation-generating device (for an X-ray or electron beam). The energy and type of the ionizing radiation as well as the size and materials of the packaging must all be considered to ensure that all of the package contents receive the required dose. When not in use, the radioactive source can be lowered into a shielded storage area or switched off.

9.4   NUCLEAR MEDICINE

Nuclear medicine is a health-care specialty that uses radioactive nuclides to diagnose and treat disease. Nuclear medicine technologists (NMTs) prepare radiolabeled drugs (**radiopharmaceuticals**), which localize in a specific organ, then obtain images that provide information on the structure and function of the target organ. A wide variety of scans are possible, covering every major organ system in the human body. Nuclear medicine scans are even possible for some animals other than humans. This is a growing field, and entire textbooks are devoted to it. This section is an introduction to nuclear medicine from the perspective of the radionuclide.

9.4.1   RADIONUCLIDE PRODUCTION

Radioactive nuclides can be prepared using a variety of nuclear reactions (which we'll explore in depth in the next few chapters) using accelerators or nuclear reactors. Traditionally, most nuclides for use in nuclear medicine came from nuclear reactors. There are two ways reactors can prepare these nuclides: (1) as a product in a fission reaction, and (2) from neutron bombardment of a material placed in or near a reactor. To use a fission product, the nuclear fuel could be removed, and the desired nuclide separated from the various other components in the fuel pellet. Fission products make up only about 3% of the spent fuel, and hundreds of different fission product nuclides are formed during fission. Additionally, the spent fuel is rather toasty (hot and radioactive!), making it difficult to handle. For all of these reasons, obtaining medical radionuclides from spent nuclear fuel is usually undesirable.

A more attractive option for obtaining fission products would be to place a relatively pure ^{235}U target in the reactor. As the neutrons bombard it, some of its nuclei will fission, providing a much higher percentage of fission products when compared to the starting mass (fuel pellets are mostly ^{238}U). ^{99}Mo, ^{131}I, and ^{133}Xe were all commonly obtained this way.

$$^{235}U + {}^1n \rightarrow {}^{135}Sn + {}^{99}Mo + 2{}^1n$$

$$^{235}U + {}^1n \rightarrow {}^{131}I + {}^{102}Y + 3{}^1n$$

$$^{235}U + {}^1n \rightarrow {}^{133}Xe + {}^{100}Sr + 3{}^1n$$

With nuclear fission, hundreds of fission reactions are possible; the three shown here are simply examples that happen to form desired products. Other reactions that produce more neutron-rich isobars can also decay to produce the desired products. For example, if ^{99}Nb is formed by a fission reaction, it will quickly decay ($t_{1/2}$ = 15.0 s) to ^{99}Mo.

$$^{99}Nb \rightarrow {}^{99}Mo + {}_{-1}e$$

Using highly enriched uranium (HEU) as a target to make nuclides for radiopharmaceuticals has become a concern in recent years. If intercepted while being shipped, the HEU could be used to make a nuclear weapon. Alternative methods, most still using fission, but with lower enrichments of ^{235}U, are now used to allay proliferation concerns.

Placing other elements/nuclides as targets in a nuclear reactor to undergo a nuclear reaction is also an attractive option. For example, ^{50}Cr can absorb one of the many neutrons zipping around inside a nuclear reactor and form ^{51}Cr along with a gamma photon.

$$^{50}Cr + {}^1n \rightarrow {}^{51}Cr + \gamma$$

Many such reactions are possible, but the main point is that if the target is a pure element or nuclide, then isolating the product should be a lot easier than fishing it out from among all the other stuff present in spent fuel or irradiated ^{235}U. Regardless of how they are prepared, reactor-produced nuclides *tend* to be neutron-rich and undergo beta (minus) decay (note that ^{51}Cr is an exception!).

Increasing numbers of medical radionuclides are being prepared with cyclotrons. Cyclotrons are a type of accelerator, which are machines used to get particles (electrons, protons, etc.) moving fast enough to slam them into nuclei to make nuclear reactions happen. Cyclotrons are the accelerator of choice for clinical applications because they are relatively compact (about the size of a large fridge or a really small car) and affordable when compared to larger, more powerful accelerators. Protons, deuterons, and alpha particles are typically the particles accelerated in a clinical environment. For example, ^{18}F can be produced from firing a proton into ^{18}O. This reaction also spits out a neutron.

$$^{18}O + {}^1H \rightarrow {}^{18}F + {}^1n$$

Accelerators are necessary for nuclear reactions because the projectile (the ^1H in the reaction above) usually needs to be moving fast enough to hit the target nucleus (^{18}O). Since both have positive charges, they naturally repel each other; so, if the proton doesn't have enough velocity, it'll simply be deflected by a ^{18}O nucleus. For some reactions, the required velocity might be too high for a cyclotron—an inherent limitation that must be considered. Other practical limitations are that the target nuclide be readily available, and that it should be easy to separate the product. Efficient separation of the product can sometimes be of paramount importance because of the short half-lives of some of the cyclotron-produced nuclides. Some are so short that they are administered directly to the patient from the cyclotron facility (e.g., ^{13}N, $t_{1/2}$ = 9.97 min). In contrast to reactor-produced nuclides, cyclotron-produced nuclides tend to be proton-rich, typically decaying via positron emission or electron capture.

How can a nuclear medicine department in a hospital, clinic, or private practice, be assured of a consistent supply of the radionuclides it needs? Both cyclotrons and nuclear reactors are expensive facilities to have around, so most medical radionuclides are purchased and shipped to where they are needed. The farther your practice is from the production facilities, the more limited the variety of radionuclides that can be shipped. Those with very short half-lives would not survive the journey if it takes too long.

A short half-life is desirable for nuclear medicine (*vida infra*). How can a radiopharmacy assure a continuous supply of short half-life nuclides in the absence of a reactor or a cyclotron? If a longer-lived parent (to the desired nuclide) can be produced at the reactor or cyclotron, it may be packaged and shipped to a remote location. As the parent decays, the daughter is gradually produced. If the two can easily be separated, then the daughter can be used for radiopharmaceutical preparations. Such a system is called a **radionuclide generator**.

The parent needs to stay inside the generator while the daughter flows out. This is usually accomplished with an absorbent or ion exchange material. After the daughter is removed from the generator, time must pass for its activity inside the generator to build back up. Since separation can be repeated many times at regular intervals, the generator is often called a "cow" and the separation procedure "milking." Radionuclide generators are generally examples of transient equilibria and the formulas discussed in Section 2.6.2 can be used to determine how often to milk the cow.

The most common generator in use today is the $^{99}Mo/^{99m}Tc$ ("moly-tech") generator. The ^{99}Mo is typically produced in nuclear reactors as described earlier in this section. It is not produced for direct use in a radiopharmaceutical, but to generate ^{99m}Tc. Figure 9.6 shows the effects on activity

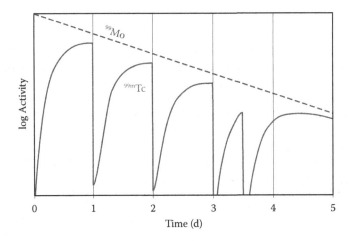

FIGURE 9.6 ^{99}Mo (parent, dashed) and ^{99m}Tc (daughter, solid) activities when ^{99m}Tc is removed at occasional Intervals.

by typical milking of this generator. If enough time is allowed to elapse between milking, daughter (^{99m}Tc) activity can build up to close to parent activity. Remember from Section 2.6.2 (Equation 2.15) that maximum ^{99m}Tc (daughter) activity can be calculated from the half-lives of ^{99}Mo (parent, 2.7476 d = 65.942 h) and ^{99m}Tc (6.008 h).

$$
\begin{aligned}
t_{max} &= \left[\frac{1.44 t_{1/2(^{99}\text{Mo})} t_{1/2(^{99m}\text{Tc})}}{\left(t_{1/2(^{99}\text{Mo})} - t_{1/2(^{99m}\text{Tc})} \right)} \right] \times \ln \frac{t_{1/2(^{99}\text{Mo})}}{t_{1/2(^{99m}\text{Tc})}} \\
&= \left[\frac{1.44 \times 65.942 \text{ h} \times 6.008 \text{ h}}{65.942 \text{ h} - 6.008 \text{ h}} \right] \times \ln \frac{65.942 \text{ h}}{6.008 \text{ h}} \\
&= 22.8 \text{ h}
\end{aligned}
$$

This time interval (22.8 h) is quite convenient for us humans, as it is very close to one day. In Figure 9.6, ^{99m}Tc is removed every 24 hours for the first three days; shortly after the ^{99m}Tc activity has peaked. Later in the third day, additional doses were necessary, so the generator was milked again. If insufficient time between milking elapses, daughter activity will still be well below parent activity. The generator was not milked on the fourth day, and transient equilibrium will finally be reached late on the fifth day since the generator is not milked again (for more than seven daughter half-lives).

Note also that parent (^{99}Mo) activity also decreases with time in Figure 9.6, as expected for transient equilibrium. Eventually (typically after a week or two), the generator will no longer be able to crank out sufficient ^{99m}Tc activity to provide the doses necessary and a new generator will be purchased. Finally, notice that daughter (^{99m}Tc) activity never exceeds parent (^{99}Mo) activity in Figure 9.6. This might be a bit puzzling, because we learned that it should, back in Section 2.6.2. Daughter activity is a little lower in Figure 9.6 because ^{99m}Tc is only produced in 86% of all ^{99}Mo decays. Those pesky branch ratios keep popping up.

A cut-away view of the ^{99}Mo/^{99m}Tc generator is shown in Figure 9.7. Overall, the generator is a cylinder, about 25 cm tall. Sterile saline solution is added through a port on the top. This solution flows through an Al_2O_3 (alumina) column with the ^{99}Mo packed in at the top. Before loading it on the alumina, it is oxidized to MoO_4^{2-} (molybdate, probably using H_2O_2). TcO_4^- (pertechnetate) is produced when molybdate undergoes beta decay and does not bind to the alumina as strongly as molybdate. The relative affinity for alumina can be related to ionic charge. Molybdate has a 2– charge, while pertechnetate only has only a 1– charge. Generally speaking, the higher the charge, the more likely it'll stick to alumina. This allows elution of pertechnetate using the saline solution, without bringing much, if any, parent along. The solution is usually eluted directly into a **kit**—a bottle containing chemicals that will react with pertechnetate to form a radiopharmaceutical. Other radionuclide generators used in nuclear medicine include ^{68}Ge/^{68}Ga, ^{82}Sr/^{82}Rb, and ^{113}Sn/^{113m}In.

Nuclides that are produced without the presence of isotopes or isomers (carrier) of the desired nuclide are called **carrier-free**. In reality, most radionuclide preparations will result in the presence of some carrier. Therefore, most radionuclide preparations are termed as **no carrier added** (NCA) to indicate that no isotopes or isomers of the product were deliberately added. The ^{99m}Tc preparation described above may initially seem carrier-free, but as soon as it is produced, ^{99m}Tc begins to decay to ^{99}Tc. Some ^{99}Tc (carrier) will be present in any sample of ^{99m}Tc. The presence of carrier is undesirable because isotopes or isomers of the desired nuclide will have different nuclear properties, but nearly identical chemical properties. They will therefore be unlikely to help in imaging the target organ and may even interfere. The term "carrier" generally refers to a stable isotope or isomer, but a radioactive isomer like ^{99}Tc is also considered a carrier because it is (very nearly) a purely beta-emitting nuclide with a very long half-life (213,000 years!).

FIGURE 9.7   Cut-away view of a $^{99}Mo/^{99m}Tc$ generator.

9.4.2 RADIOPHARMACEUTICALS

As already mentioned, a radiopharmaceutical is a drug containing a radioactive nuclide used for diagnostic or therapeutic purposes. While the vast majority of nuclear medical procedures are diagnostic, therapeutics are increasing. Radiopharmaceuticals can be as simple as an atom (^{133}Xe), ion ($^{125}I^-$ or $^{82}Rb^+$), or a small organic (^{18}F-glucose) or inorganic ($^{99m}Tc(CNR)_6^+$) compound; or it can be as complex as a large bio-molecule (radio-labeled protein, antibody, etc.). The exact nature of the radiopharmaceutical depends on how it will be used.

What are the qualities of a good diagnostic radiopharmaceutical (a.k.a. a radiodiagnostic agent)?

1. It should have a good biodistribution. In other words, it should localize primarily in the target organ. It is also helpful if it does not localize in other organs nearby, as this may interfere with obtaining a good image of the target.
2. The radionuclide should emit only gamma rays. Gamma emissions are important, as a nuclide emitting solely alpha or beta particles will not be visible outside of the patient and will give the patient a significant internal dose. If it decays via IT, it should have a low conversion coefficient (α_T). If it decays via EC, it should have a high fluorescence yield (ω_T). Additionally, the gamma rays should be between 50 and 500 keV in energy (roughly speaking). Any lower, and the patient will absorb too many, and much higher will make detection too inefficient (see Section 7.3). The amount of radionuclide administered should be enough to provide a good image, but no more—minimizing the dose to the patient.
3. The radioactive nuclide should also have a short half-life. The half-life needs to be long enough for the radiopharmaceutical to be prepared, injected into the patient, localize in the target organs, and be imaged. As a practical matter, it should be at least a few minutes. A short half-life will maximize the number of decays that take place during imaging, ensuring good pictures. It will also minimize the dose to the patient (and others nearby!) and possible contamination (due to excretion) following the procedure.
4. Finally, the nuclide should be readily available, relatively pure, and decay to a stable, benign daughter. As discussed earlier, the radionuclide needs to be produced on-site, or shipped to

the site. A radioactive daughter will mean additional patient dose and may interfere with the scan. Benign simply means that it should not be harmful to the patient and the environment— the radionuclide should not decay to an element with high toxicity for humans, or that is particularly harmful to the environment. These are not great concerns, as very small amounts of radioactive materials are used in diagnostic procedures.

5. The radiopharmaceutical should be relatively inexpensive and easy to make. If the materials (radionuclide + chemicals) to make a radiopharmaceutical are too costly, it is less likely to be approved by insurance companies (the kiss of death!). If its synthesis is too complex, radiopharmacies are less likely to be able to make it.

6. Once imaging is complete, the radiopharmaceutical should be metabolized and excreted by the patient in a reasonable amount of time. This also minimizes the patient dose. If the patient will excrete a significant amount of the dose soon after the procedure, then a special bathroom should be used. The patient should also be warned if their urine and/or feces represent any radiological hazard.

^{99m}Tc is the most commonly used radionuclide for diagnostics. Its generator is convenient in that ^{99}Mo is easily prepared in a reactor and the half-lives of ^{99}Mo and ^{99m}Tc lend themselves well to human schedules (as shown earlier in this section). Also previously noted is that the Mo-Tc generator has a useful lifetime of about one week, also convenient for institutions working a five-day week. The only significant disadvantage to ^{99m}Tc is that it does not decay to a stable daughter.

^{99m}Tc generators are also relatively inexpensive,[6] and ^{99m}Tc emits a 140 keV gamma photon in 89% of its decays (low α_T) with a half-life of six hours. Technetium is an element with a great deal of chemical flexibility. It can form a wide variety of chemical compounds—with coordination numbers (number of things attached to the metal) around Tc from 4 to 9, and oxidation states (relative number of electrons localized on the metal) from –1 to +7. ^{99m}Tc is typically produced in the generator as TcO_4^- (pertechnetate ion, coordination number = 4, oxidation state = +7), and reduction (lowering of the oxidation state) is almost always required. This is accomplished by the tin(II) (Sn^{2+}) ion, which is oxidized to tin(IV) (Sn^{4+}). When the technetium is reduced, other chemicals (**ligands**) must bind to it. For example, the preparation of the myocardial perfusion agent, ^{99m}Tc-Sestamibi, can be characterized by the following chemical reaction.

$$TcO_4^- + 6CNR + 3Sn^{2+} + 8H^+ \rightarrow Tc(CNR)_6^+ + 4H_2O + 3Sn^{4+}$$

where CNR is 2-methoxy isobutyl isonitrile, the ligand. Technetium in $Tc(CNR)_6^+$ has a coordination number of 6 and oxidation state of +1.

The chemistry is often easy to write down on paper, but there are usually challenges when preparing radiopharmaceuticals. First, the solutions are very dilute—$^{99m}TcO_4^-$ has a concentration of about 10^{-9} moles/liter (nanomolar!). The reactions must also proceed in high chemical yield, as injection of other radioactive chemical species will likely interfere with imaging of the target organ.

$Tc(CNR)_6^+$ has a rather symmetric structure and appears to the body as a large positively charged sphere. The body confuses it for a potassium ion, so a fair bit of it ends up in the heart, making it a good myocardial imaging agent. Other radiopharmaceuticals used for heart imaging look even more like the potassium ion (K^+); they are $^{82}Rb^+$, $^{13}NH_4^+$, and $^{201}Tl^+$. ^{82}Rb and ^{13}N both decay via positron emission. The positron annihilates in nearby tissue, generating two 511 keV annihilation photons. Because of the higher photon energy (relative to ^{99m}Tc), ^{82}Rb or ^{13}N are sometimes preferred for

---

6   Atomic Energy of Canada, Ltd. (AECL) produced a significant portion of the ^{99}Mo used for Mo-Tc generators worldwide and did not recover its full production costs. Like the element itself, the cost for a Mo-Tc generator was artificial. This changed when AECL shut down the reactor responsible for ~40% of the world's ^{99}Mo production in 2018.

larger patients. Because they have more tissue for the photons to penetrate, more of ^{99m}Tc's 140 keV photons will get absorbed before they can leave a larger body. Note also that two photons are produced for every positron decay.

Sometimes the physical form of the radiopharmaceutical is more important that its chemical form. Several ^{99m}Tc colloids are used for different imaging applications. A colloid is a bunch of molecules that stick together or a really large molecule. Colloids are still small enough to disperse, but not dissolve, in a solvent, and they are not big enough to precipitate out. Generally speaking, a colloid is the dispersal of one phase of material in another. Smoke is a solid dispersed in gaseous air and is therefore a colloid. Milk is also a colloid because it is a suspension of liquid fats in liquid water—the fat is a colloid. The subsequent mixture is called an emulsion because its two components are immiscible. The key to a ^{99m}Tc colloid is the size of the multimolecular aggregate that forms, as the size of the colloid will determine its application.

Most ^{99m}Tc radiopharmaceuticals in current use can be related to ^{99m}Tc-Sestamibi in that they started out as a chemical compound that was easily prepared from nanomolar solutions of pertechnetate, was stable under physiological conditions, and had some (typically unanticipated) favorable biodistribution. Most were discovered through trial and error, which can be a time-consuming and inefficient way to discover new radiopharmaceuticals. More recently, radiopharmaceutical research has started with biodistribution, and then looked for a way to attach a radioactive nuclide. Two successful approaches are discussed next.

The first involves forming a chemical linkage between a molecule with a known biodistribution, such as an antibody, carbohydrate, enzyme, protein, or peptide. In the example shown in Figure 9.8, a diethylenetriamine pentaacetate (DTPA) molecule is wrapped around an indium(III) ion (^{111}In^{3+}). When a single ligand binds to a single metal ion with more than one group, it is called a **chelate**. Chelating ligands like DTPA can also be used to help sequester radioactive materials that have been inadvertently ingested (chelation therapy). The DTPA wraps itself around the radioactive ion and facilitates its clearance from the body. Back to the reactions in Figure 9.8, notice that one of the five acetate groups (–COO$^-$) remains unbound after the DTPA binds the indium ion, and is used to form a chemical bond to the biomolecule. This approach could be called a "tetherball," where the tether is the biomolecule, and the ball is the chelated metal ion. More formally these types of radiopharmaceuticals are called "radioconjugates."

This approach can fail if the presence of the ball affects the tether's ability to localize in the desirable organ. It could be that the ball binds to a critical part of the tether, or that the addition

FIGURE 9.8 Preparation of a "tetherball" radiopharmaceutical using indium and DTPA.

FIGURE 9.9   (a) Glucose and (b) FDG.

of the ball changes the overall shape of the biomolecule to a significant extent. Synthesis of the radiopharmaceutical can also be more complex if the two steps need to be separated.

The most elegant approach is to incorporate the radioactive nuclide in the biomolecule—replacing one of the atoms with a radioactive isotope. For example, replacing the nitrogen in an amino acid (a basic biochemical building block!) with a ^{13}N. The synthetic challenges here are daunting, especially since most of the radioactive nuclides that could be used have rather short half-lives and are cyclotron-produced. ^{18}F-fluorodeoxyglucose (FDG) is probably the most widely used of this type of radiopharmaceutical, although it represents a minor biochemical compromise.

As can be seen in Figure 9.9, the compromise is that FDG replaces one of the hydroxyl (–OH) groups with a fluorine atom (–F). In terms of size and arrangement of electrons, these two groups are similar enough that the body mistakes FDG for glucose. FDG is used to study metabolism in the brain and heart, and, most significantly, is used to detect tumors at the cellular level. Tumors are sugar hogs, so FDG will collect in tumors (even pretty small ones) if the body isn't doing much else. As a result, oncology patients injected with FDG are asked to sit or lie quietly lest the FDG start collecting in whatever muscles are in use. The Mayo Clinic used to allow FDG patients to watch television while they waited for their scans but found uptake in the right thumb and parts of the neck, especially in male patients. FDG in the right thumb came from continuous channel surfing, and the neck from holding the head in a position to watch TV. They removed the TVs from the injection rooms, and the "anomalous" uptakes disappeared. Just goes to show that watching TV is quite a workout!

Like ^{82}Rb and ^{13}N mentioned earlier, ^{18}F is a positron-emitting nuclide; therefore, FDG is used as a positron emission tomography (PET) agent. As you remember from Section 5.3, positrons annihilate with electrons to produce two 511 keV photons that travel on trajectories that are approximately 180° apart. If a patient is injected with some FDG, it will be emitting positrons, which travel a short distance before annihilating, generating two photons, traveling in opposite directions. To efficiently detect these photons, the patient is placed in the center of a ring of detectors (Figure 9.10).

The chief advantage of PET over using a nuclide like ^{99m}Tc is that PET generates two photons per decay, whereas ^{99m}Tc only produces a single photon to be detected. Both can provide 3-D images of the target organ provided detectors can (more or less) encircle the patient. When this is done with a single-photon nuclide like ^{99m}Tc it is called single-photon emission computed tomography (**SPECT**).

PET imaging is growing in importance within nuclear medicine, in part because radiopharmaceuticals such as FDG are examples of "**molecular imaging**." That is to say, they illuminate biological processes at the molecular or cellular level. They allow understanding of physiology through the actual biochemical reactions that take place inside human beings. Since these processes are somewhat unique to each individual, this has the effect of personalizing the diagnosis and treatment of disease. Molecular imaging is so prominent for nuclear medicine that the U.S. professional organization for nuclear medicine, the Society of Nuclear Medicine and Molecular Imaging (SNMMI), expanded its mission in 2007 to include "advancing molecular imaging and therapy." As our understanding of human beings continues to become more detailed at the biochemical level, our ability to detect and treat disease at this level will become more effective.

FIGURE 9.10   A modern PET/CT camera. Photo courtesy of Kaitlyn Neefe.

Coupled with the increasing abilities of radiopharmaceuticals to detect disease at the cellular and molecular levels are **hybrid** (or fusion) technologies. Not to be confused with high-mileage cars, this type of hybrid involves the coupling of nuclear medical imaging with computed tomography (CT) or magnetic resonance imaging (MRI). The nuclear medicine scan provides physiological information, while CT or MRI provides anatomical information. When the two scans are put together, it provides medical professionals with detailed information within and around the volume of interest. To accomplish both scans, the immobilized patient is typically placed on an exam table that travels through both types of scanners. For example, a PET/CT camera will pass the patient through a CT scanner, then a PET detector like poking your finger through the holes of two doughnuts sitting up on their sides (Figure 9.10). Some of these cameras are made so both PET and MRI occupy the same space—just one doughnut. Others are made deep enough so the whole body can be imaged all at once.

So far, we've only discussed diagnostic agents. As mentioned earlier, nuclear medicine is moving toward "molecular imaging *and therapy*." As molecular imaging advances, it is easy to imagine swapping out a gamma- or positron-emitting nuclide with one that emits alpha or beta particles, Auger or conversion electrons, or low-energy gamma rays. These forms of ionizing radiation will deposit a lot of energy (i.e., do a lot of damage!) in nearby tissue. Only specific tissues (like tumors) are targeted for destruction, so the radiotherapy agent will need an excellent biodistribution. Like radioimaging agents, they should also be readily available and easy to make.

At the time of this writing, molecular radiotherapy remains an active area of research, but only a few agents are currently in use. Part of the problem is that the agents do not localize strongly or quickly enough in the target tissue, leading to a significant dose to healthy tissue.

A common radiotherapy agent in use by nuclear medicine today is a simple ion: $^{131}I^-$. As mentioned earlier in this chapter, ^{131}I can be separated from the other fission products of ^{235}U, but it can also be produced by throwing neutrons at ^{130}Te, producing ^{131}Te, and then letting it decay.

$$^{130}Te + {}^{1}n \rightarrow {}^{131}Te + \gamma$$

$$^{131}Te \rightarrow {}^{131}I + {}_{-1}e$$

Iodide collects in the thyroid, so $^{131}I^-$ is used to treat hyperthyroidism and thyroid cancer. ^{131}I emits gamma rays as part of its beta decay, so it can also be used for imaging. This is also handy when used for therapy because the biodistribution can be determined as well as the dose received by the thyroid. ^{131}I decays to the stable, and chemically inert nuclide, ^{131}Xe.

$$^{131}I \rightarrow {}^{131}Xe + {}_{-1}e$$

Other radioactive nuclides used in therapy are: ^{153}Sm, ^{89}Sr, and ^{32}P used for bone pain palliation for patients with bone cancer; ^{90}Y used in combination with Zevalin to treat non-Hodgkin's lymphoma; and ^{32}P used to treat polycythemia vera (increased red blood cell counts due to bone marrow hyperactivity) and certain cases of leukemia. More recently, ^{177}Lu, ^{225}Ac, and $^{223}Ra^{2+}$ are seeing applications in certain patients with metastasized cancer. The first two are used as part of a radioconjugate, but the radium ion is used as is. Chemically, it mimics the calcium ion, so it is rather efficiently incorporated into parts of the body experiencing rapid bone growth (bone tumors). Because of the different ranges (Section 7.2) of their emitted particles, ^{177}Lu (decays via beta) is preferred for treating larger tumors while alpha-emitting ^{225}Ac and ^{223}Ra are used for smaller ones. ^{177}Lu also emits gamma photons allowing diagnostic scanning simultaneously with therapy. Nuclides like this are called "theranostics"[7] and are an active area of research.

Radiopharmaceuticals are sometimes referred to as radiotracers, or simply as tracers. These are more general terms that can apply to experiments using small amounts of radioactive nuclides to "follow" any chemical or biological process. The basic idea is that a radioactive nuclide will exhibit the same chemical and physical behavior as its stable isotopes. By mixing a small percentage of radioactive nuclide into a system under study, a lot can be learned about it without changing it in any way. Examples include a wide variety of analyses, such as determining the solubility product constant (how much dissolves, K_{sp}) of relatively insoluble salts, or understanding how CO_2 is assimilated in plants during photosynthesis.

9.4.3 Gamma Cameras

The NMT injects a diagnostic radiopharmaceutical into the patient; it localizes in the target organ and is busy decaying away, sending out gamma photons in all directions. In order to "see" the organ being imaged, this radiation needs to be detected, but also localized as to its point of origin. To accomplish this, a **gamma camera** (also known as a scintillation or Anger camera) is used. Most gamma cameras work like an inorganic scintillator detector, described in Section 8.2.2, that is spread out over two dimensions to allow them to locate the gamma ray interaction within the crystal. In essence, they take a photo of the gamma rays emerging from the patient.

One of the key modifications over a simple detector is the addition of a collimator between the scintillator and the source of the radiation (the patient). Collimators are large slabs of high Z metal[8] with holes drilled through them along the short dimension. These holes can be hexagonal (making it look like a honeycomb), square, triangular, or round—the first two shapes are the most common. As shown in Figure 9.11, this allows photons to travel only on a straight-line path from the patient to the camera. This path is perpendicular to the face of the scintillator crystal.[9]

To better understand how a gamma camera works, let's start with the various ways a gamma photon can emerge from the patient and examine how they may or may not end up generating a signal in the camera, as shown in Figure 9.11. Path a is the desirable path because these photons travel straight up from the patient to the camera, i.e., we know where it came from. Path b is also

---

7   Apparently a better portmanteau than "diapeutics."
8   High atomic number means a higher probability of absorbing the photon. Lead and tungsten are most common.
9   Collimators can have different configurations than the one shown in Figure 8.11.

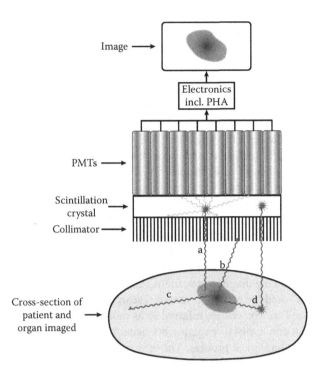

FIGURE 9.11   How a gamma camera works.

straight from the patient to the detector but is at an angle. If this generates a signal, we'll think it came from directly below its point of entry in the camera, which would be bad information. The collimator thankfully absorbs this photon. Path c misses the camera altogether and is not recorded. Path d starts out like it will miss the camera, but then undergoes Compton scattering within the patient and ends up in the camera. This is a potential problem that will be resolved, but first we need to know a little more about how a gamma camera works.

The scintillation crystals used in gamma cameras vary in size and shape. They are often roughly 38×50 cm^2 (15×20 in^2) but are only about 1 cm (~1/2 inch) thick. The thinness is important, as it minimizes multiple Compton scattering within the crystal, which could give erroneous location information. The thinness is also a tradeoff as a thinner crystal will be less efficient in detecting gamma photons, i.e., too many photons will traverse the crystal without interacting. Cameras with thicker crystals are used for nuclides that produce higher energy photons and thinner crystal machines for lower energy photons.

When a gamma photon deposits its energy in an inorganic scintillator via the photoelectric effect (an "event"), it will produce visible light photons. These photons can then take different paths through the crystal, as shown in Figure 9.11, entering a number of different photomultiplier tubes (PMTs) covering the top face of the crystal. Modern cameras have 60–120 PMTs packed on top of their scintillator. The tube directly above the event will likely see the most photons, but those around will also see varying amounts. By weighting the signal intensities from nearby PMTs, the exact position of the event (x, y coordinates) within the crystal (and therefore within the patient) can be calculated.

As part of the electronic processing of the signals, all of the PMT signals for each event are summed (called the "Z pulse") by the pulse height analyzer (PHA) to determine if the total energy of the gamma photon was deposited in the crystal. Because of the relatively large amount of random error associated with inorganic scintillation detectors, a broad energy window is usually set. This insures inclusion of the entire photopeak. The main purpose of the PHA is to reject lower energy signals that may be due to Compton scattering, such as from path d in Figure 9.11. It will also reject X-ray photons that enter the crystal after a gamma photon interacts with the collimator (path b).

The scintillator and PMTs and some of the electronics constitute what is called the **detector head**. Detector heads on a particular camera can be held in place relative to the patient and collect a two-dimensional image. They can also move around the patient collecting multiple images from a variety of angles, like a CT scan, to produce three-dimensional images. Cameras often have two to three heads to facilitate 3-D imaging. In the case of PET imaging, detectors completely encircle the patient (Figure 9.10).

While the head design described here remains the mainstay for gamma cameras, new designs are available. These designs can provide greater counting efficiency (lowering patient dose and/or counting time) and more accurately locate the origin of the radiation within the patient. The detector heads are also considerably smaller than conventional units. These designs use lots of small scintillator crystals, each with a position-sensitive PMT, which allows location of the event within the crystal. Others use semiconducting CZT (see Section 8.3.1).

QUESTIONS

9.1   Why is ALARA followed when working with radioactive materials and X-ray machines? How is ALARA potentially harmful when applied to possible ionizing radiation exposures to the general public following nuclear power plant accidents like Chernobyl and Fukushima?

9.2   Why might some working in radiology departments be expected to have greater longevity than the general population?

9.3   What areas of radiology use ionizing radiation? Briefly describe each.

9.4   TLDs can be placed on a patient to measure X-ray exposure during radiography procedures. The badge is often invisible on the image. How can this be?

9.5   What is the difference between a CT scan and a conventional (2-D) radiograph?

9.6   How might CT scans be useful in radiation therapy? In nuclear medicine?

9.7   What is the therapeutic ratio and how do radiation therapists improve it?

9.8   Why is gas production undesirable for sealed implant sources? Where does this gas come from?

9.9   Calculate the specific activity of ^{192}Ir. Why is high specific activity good for an implant source?

9.10 Why do electrons tend to spread out their dose when used in therapy, when photons do not?

9.11 What is the most likely interaction between high-energy photons (~10 MeV) and the low Z matter that makes up our bodies?

9.12 The dose delivered to radiation therapy patients undergoing external photon beam therapy is due primarily to the photons interacting and what else?

9.13 Describe a simple way that dose at some depth inside a human could be estimated for external beam photon therapy.

9.14 A patient has a small tumor in his neck. It is only 2 cm below his skin, but his spine is not far behind the tumor. The spine is very sensitive to ionizing radiation, so it is important to minimize its dose. Would it be better to treat this tumor with photons or electrons? Briefly explain.

9.15 Calculate the activity for an ^{192}Ir seed that was calibrated 200 days ago at 3.50×10^5 MBq.

9.16 Briefly explain why there might be some convergence of radiation therapy and nuclear medicine in the future.

9.17 How are radiation therapy and food irradiation alike?

9.18 What are the main benefits of food irradiation? What are the main drawbacks?

9.19 Why do reactor-produced nuclides tend to be neutron-rich while accelerator-produced nuclides tend to be proton-rich?

9.20 Why would radioactive nuclides produced by cyclotrons be preferred over the same ones produced in a nuclear reactor? When would those from reactors be preferred?

9.21 A 28 mCi dose of a ^{99m}Tc myocardial agent is injected into a patient. How many atoms of ^{99m}Tc were injected? What mass of ^{99m}Tc is present? This radiopharmaceutical kit also contains 0.050 mg of $SnCl_2 \cdot 2H_2O$. What is the mole ratio of tin to technetium?

9.22 A ^{99}Mo/^{99m}Tc generator is loaded with ^{99}Mo, but currently has no ^{99m}Tc activity. How long will it take for ^{99m}Tc activity to peak?

9.23 What would be a simple, safe, and expeditious way to test for ^{99}Mo impurities in a ^{99m}Tc sample recently eluted into a radiopharmaceutical kit from a ^{99}Mo/^{99m}Tc generator?

9.24 What is the purpose of a radionuclide generator? How does it relate to dairy farming?

9.25 Would ^{60m}Co be a good choice for a nuclear medicine diagnostic procedure? Explain.

9.26 Could ^{34m}Cl be used as a nuclear medicine diagnostic agent? Explain.

9.27 What are the differences between a diagnostic and a therapeutic radiopharmaceutical?

9.28 Why are ^{82}Rb$^+$ and ^{201}Tl$^+$ good myocardial imaging agents?

9.29 Why are PET images inherently fuzzy?

9.30 If ^{131}I is produced exactly as illustrated at the end of Section 9.4.2, would it be carrier-free, no carrier added, or neither? Briefly explain.

9.31 How could a lead X-ray be observed by a gamma camera? Why would it not be used to generate the image?

9.32 Why are the new detector materials mentioned at the end of Section 9.4.3 more efficient at detecting gamma photons?

10 Nuclear Reactions

Radioactive decay is not considered a nuclear reaction. Some readers may consider this strange because of the obvious analogy to chemical reactions. One chemical compound breaking up to form two or more compounds is called a decomposition reaction by chemists, and considered no less a chemical reaction than an acid–base neutralization reaction, or any other type of chemical reaction. So why is nuclear decay not considered a reaction?

Nuclear scientists define nuclear reactions as the collision of two nuclei, or a nucleus and a neutron, resulting in the production of a different nuclide. This definition does not include decay processes, which could be defined as the spontaneous transformation of one nuclide into another, usually accompanied by the nuclear emission of a particle or photon. This may all seem like a relatively pointless semantic debate, but the reader should be aware that it is generally considered improper to refer to nuclear decay as a reaction.

The following is a generic representation of the nuclear reactions we'll study in this chapter:

$$A + x \rightarrow B + y$$

where A is the **target** nuclide, x is the **projectile**, B is the **product** nuclide, and y is an emitted particle or photon.[1] The projectile x and emitted y can be a gamma photon or a wide range of particles, such as an alpha particle (α or ^{4}He), neutron (n), proton (p or ^{1}H), deuterium (d or ^{2}H), or tritium (t or ^{3}H). Curiously enough, something is always spit out; the target and projectile do not simply combine to form a new nuclide. We can however, imagine the temporary formation of C from the addition of projectile to target. Inserting C into our generic representation of a nuclear reaction gives:

$$A + x \rightarrow [C]^* \rightarrow B + y$$

The square brackets around C indicate its transient nature, meaning that it doesn't exist for reasonable amounts of time ($\sim 10^{-17}$ s). The asterisk indicates that this nuclide is in an excited state, typically much higher in energy than the ground state. Taken together, $[C]^*$ represents what is called a **compound nucleus**.

The following shorthand can also be used to represent a generic nuclear reaction:

$$A(x,y)B$$

where the target and product nuclides are placed outside the parentheses, and the projectile and emission are placed inside. As a result, these reactions are often referred to as "**x,y-type reactions**."

---

1 Students have suggested the term "emitron" for y. A more commonly used term is "ejectile."

DOI: 10.1201/9781003303787-10

From this generic point of view this may not seem like much of a shorthand. Hopefully, its utility will be illustrated in the examples to come.

Neutron capture is a fairly common nuclear reaction. An example would be tossing a neutron at ^{51}V to form ^{52}V. This reaction emits a gamma photon, so it would be called an "n,gamma" reaction. ^{51}V is the target, the neutron is the projectile and ^{52}V is the product, while a gamma photon is emitted.

$$^{51}_{23}\text{V}_{28} + {}^{1}_{0}\text{n} \rightarrow {}^{52}_{23}\text{V}_{29} + \gamma$$

$$^{51}\text{V}(n,\gamma)^{52}\text{V}$$

Notice that the numbers of nucleons are conserved in this n,gamma reaction. Just like decay equations, the atomic and mass numbers on both sides of the arrow need to each be equal.

10.1 ENERGETICS

Are mass and energy collectively conserved in nuclear reactions? Let's do the math!

Example 10.1: How much energy is produced, or required, by the n,gamma reaction of ^{52}V?
We can do this by looking at the difference in mass between reactants and products, just like we did with decay.

Mass of reactants	^{51}V	50.943960 u
	^{1}n	+1.008665 u
		51.952625 u
Mass of product	^{52}V	−51.944780 u
Reactants − products		0.007845 u

Note that *atomic* masses are used for ^{51}V and ^{52}V just as we did with decay. Also note that mass is lost in this n,γ reaction; therefore, energy is produced. The variable Q is typically used to designate energy in nuclear reactions, instead of the E that is used for decay. While potentially confusing, it serves to further differentiate these two types of nuclear processes. Let's calculate how much energy is produced by this reaction:

$$Q = \frac{931.5\,\text{MeV}}{\text{u}} \times 0.007845\,\text{u} = 7.308\ \text{MeV}$$

When mass is lost in the course of the reaction, then Q is greater than 0 and the reaction is termed **exoergic** (or exothermic). Decay is always exoergic, but that won't be true for nuclear reactions. Occasionally, we will run into reactions where mass is gained, Q is negative, and the reaction is termed **endoergic** (or endothermic).

The energy released in Example 10.1 can also be calculated using mass excess values (Appendix A). Perhaps it is less intuitive, but it is definitely more efficient.

$$Q = \left(-52.201 + 8.071\ \text{MeV}\right) - \left(-51.437\right) = 7.307\ \text{MeV}$$

Dealing with energy in nuclear reactions is much more complex than with decay processes. For example, in the n,γ reaction above, the energy produced comes from binding of an additional neutron to a ^{51}V nucleus and is released through the gamma photon and the kinetic energy of the product nuclide. There's also energy on the reactant side that we need to be concerned with; namely, the kinetic energy of the projectile, which is also nonzero. In some reactions, the projectile *must* have a certain amount of energy before a reaction can reasonably be expected to proceed. How can we sort all of this out? We'll get to all of this eventually; first, let's consider the projectile.

The projectile in Example 10.1 is a neutron. Neutrons make great projectiles because they have no electrical charge. As a result, they are neither repelled from nor attracted to an atom's protons or electrons. This means a neutron can approach a nucleus without hindrance. Positively charged projectiles do not have the same advantage. As they approach a nucleus they will begin to feel Coulomb repulsion of the positively charged nucleus.

The first example of a charged particle being used in a nuclear reaction was discovered by Ernest Rutherford in 1919! It is the reaction of an alpha particle with ^{14}N, forming ^{17}O and a proton.

$$^{14}_{7}N + ^{4}_{2}He \rightarrow ^{17}_{8}O + ^{1}_{1}H$$

In this reaction ^{14}N is the target, an alpha particle is the projectile, ^{17}O is the product, and a proton is emitted. This is an "alpha,p" reaction and can also be represented as ^{14}N(α,p)^{17}O. We'll take a detailed look at the energetics of this reaction, starting by calculating the energy released (or required!).

Mass of reactants	^{14}N	14.003074 u
	^{4}He	+4.002603 u
		18.005677 u
Mass of products	^{17}O	16.999132 u
	^{1}H	+1.007825 u
		18.006957 u

$$\text{Reactants} - \text{products} = 18.005677 - 18.006957 = -0.001280 \text{ u}$$

Mass is *gained* in this reaction, so energy needs to be put in—this is an endoergic reaction. Nowadays this is usually accomplished by accelerating the projectile using sophisticated equipment. Rutherford was able to accomplish this in 1919 by using an alpha particle projectile that was already packing a punch. The alpha particles were generated from the decay of ^{214}Po, which have an energy of 7.69 MeV. How much energy does this reaction require?

$$Q = \frac{931.5\,\text{MeV}}{u} \times (-0.001280\,\text{u}) = -1.19 \text{ MeV}$$

The Q value is the minimum energy required by this reaction to manufacture the necessary mass. We also need to be concerned with the energy, or momentum, of the products. Both mass and momentum are accounted for in the **threshold energy** (E_{tr}, Equation 10.1):

$$E_{tr} = -Q \times \left(\frac{A_A + A_X}{A_A} \right) \tag{10.1}$$

where A_A is the mass number of the target and A_X is the mass number of the projectile. Note the negative sign in front of Q. This equation only makes sense when Q is negative, i.e., the reaction is endoergic. For Rutherford's reaction:

$$E_{tr} = 1.19 \text{ MeV} \times \left(\frac{14 + 4}{14} \right) = 1.53 \text{ MeV}$$

This tells us the alpha particle needs a minimum of 1.53 MeV of energy for this reaction to produce the required mass and momentum of the products. While this is greater than the 1.19 MeV required solely for matter creation, it is still quite a bit less than the 7.69 MeV of the ^{214}Po alpha particle. Any excess energy brought in by the projectile will end up as kinetic and excitation energy of the products. In other words, the products are movin' and shakin' more than they would otherwise.

Great! We've got Rutherford's reaction under control, right? Not quite. We need to be concerned about the fact that we're bringing together two positively charged particles. They will try to repel each other (Coulomb repulsion) with increasing force as they get closer and closer together. The projectile must have a certain amount of energy to overcome this repulsion. The minimum amount of energy required to overcome this **Coulomb barrier** (E_{cb}, in MeV) can be approximated as shown in Equation 10.2:

$$E_{cb} \approx 1.11 \times \left(\frac{Z_A Z_X}{A_A^{1/3} + A_X^{1/3}} \right) \tag{10.2}$$

where Z_A and Z_X are the atomic numbers for the target and projectile respectively. This energy doesn't get soaked up and doesn't enter into our energy accounting for the reaction; it is simply a barrier to overcome. Just as we did with Q, we have to also be concerned with conservation of momentum here. The minimum energy required to overcome the Coulomb barrier and provide enough energy for the momentum of the products is the **effective Coulomb barrier** (E_{ecb}). Its calculation is shown in Equation 10.3, and it also has units of MeV.

$$E_{ecb} \approx 1.11 \times \left(\frac{A_A + A_X}{A_A} \right) \times \left(\frac{Z_A Z_X}{A_A^{1/3} + A_X^{1/3}} \right) \tag{10.3}$$

Note that Equations 10.2 and 10.3 apply to any nuclear reaction with a positively charged projectile. For Rutherford's reaction, the effective Coulomb barrier is:

$$E_{ecb} \approx 1.11 \times \left(\frac{14+4}{14} \right) \times \left(\frac{7 \times 2}{\sqrt[3]{14} + \sqrt[3]{4}} \right) = 5.00 \, \text{MeV}$$

For endoergic reactions, the minimum amount of energy required to make it proceed in good yield is the greater of the threshold energy (E_{tr}) and the effective Coulomb barrier (E_{ecb}), with the effective Coulomb barrier usually being the larger of the two. The alpha particles in this example have ample energy (7.69 MeV) to make it over this hump.

Finally, we should also be concerned about aim. Projectiles don't always hit their targets dead-on. Glancing blows can sometimes be productive, but problems with conserving angular momentum come into play. Fortunately for us, this concern is beyond the scope of this text.

Let's take a step back and look at this reaction as two separate steps. The first step is the formation of the compound nucleus ([^{18}F]*), followed by its decomposition into the ^{17}O product and a proton.

$$^{14}_{7}\text{N} + ^{4}_{2}\text{He} \rightarrow \left[^{18}_{9}\text{F} \right]^* \rightarrow ^{17}_{8}\text{O} + ^{1}_{1}\text{H}$$

The compound nucleus can have two types of energy, excitation and kinetic. Excitation is placement of nucleons in a configuration other than the ground state, and kinetic in the energy of motion. The compound nucleus is moving because the projectile just slammed into the target. Equation 10.4 provides us with a simple way to estimate the kinetic energy of the compound nucleus:

$$E_{KC} \approx E_{KX} \times \left(\frac{A_X}{A_C} \right) \tag{10.4}$$

where E_{KX} is the kinetic energy of the projectile, A_C is the mass number of the compound nucleus, and A_X remains the mass number of the projectile. Note that this equation applies to *any* nuclear reaction (regardless of the sign of Q). Applying this equation to Rutherford's reaction:

$$E_{KC} \approx 7.69 \text{ MeV} \times \left(\frac{4}{18} \right) = 1.71 \text{ MeV}$$

We can use this information to calculate just how far $[^{18}F]^*$ is above ground state ^{18}F. Consider the (hypothetical) reaction:

$$^{14}N + {}^4He \rightarrow {}^{18}F$$

Let's also assume (even though it is ridiculous!) that all three nuclides are in the ground state. How much energy is produced or consumed? In other words, what is Q?

$$Q = (14.003074 \text{ u} + 4.002603 \text{ u} - 18.000938 \text{ u}) \times \frac{931.5 \text{ MeV}}{\text{u}}$$
$$= +4.41 \text{ MeV}$$

Therefore, when the Rutherford reaction takes its first step and forms a compound nucleus, 4.41 MeV of energy is *produced*. Remember that the alpha particle already has 7.69 MeV of energy. Where does all this energy go? There are only two places it can go—excitation and kinetic energy of ^{18}F. The excitation energy of the $[^{18}F]^*$ in the Rutherford reaction is therefore the energy of the projectile, plus the energy released for the first step of the reaction minus the kinetic energy of the compound nucleus.

$$7.69 \text{ MeV} + 4.41 \text{ MeV} - 1.71 \text{ MeV} = 10.39 \text{ MeV}$$

That's one excited compound nucleus! It's no wonder it can't even hold itself together. If you're clever enough, you can now calculate the energy (excitation plus kinetic) of the products, but this question can be answered without the hassle of calculating the excitation energy of the compound nucleus:

$$7.69 \text{ MeV} - 1.19 \text{ MeV} = 6.50 \text{ MeV}$$

where the 7.69 MeV is still the energy that the alpha brings to the reaction, the 1.19 MeV is the energy needed to make the mass (Q, from a couple pages back), and 6.50 MeV is the energy (excitation plus kinetic) of the products.

With all this information, we can draw a tidy little diagram that shows the energy changes during the course of this reaction (Figure 10.1). We know that going from the ground state of the reactants to the ground state of the products costs 1.19 MeV of energy (Q), so the products should be slightly higher in energy than the reactants. Note that the products of nuclear reactions

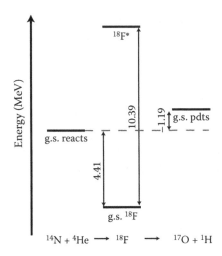

FIGURE 10.1   Energy diagram for Rutherford's reaction.

are rarely generated in their ground state. They typically have some level of excitation and kinetic energy. Therefore, the line representing the products in Figure 10.1 is somewhat misleading. We can also show the ground state, and actual state of the compound nucleus, since 4.41 MeV is released in the formation of the compound nucleus and it is formed in an excited state that is 10.38 MeV above that ground state.

Example 10.2: What is the minimum energy required to make the following reaction proceed in good yield?

$$^{240}_{94}\text{Pu} + {}^{4}_{2}\text{He} \rightarrow {}^{243}_{96}\text{Cm} + {}^{1}\text{n}$$

This is an alpha,n reaction whose shorthand representation would be: ^{240}Pu(α,n)^{243}Cm. First find the difference in mass between reactants and products:

Mass of reactants	^{240}Pu	240.053814 u
	^{4}He	+4.002603 u
		244.056417 u
Mass of products	^{243}Cm	243.061389 u
	^{1}n	+1.008665 u
		244.070054 u

Reactants – products = 244.056417 – 244.070054 = –0.013637 u

Once again, mass is gained, therefore this is an endoergic reaction. The energy required to make the mass needed for this reaction (Q) is calculated next.

$$Q = \frac{931.5\,\text{MeV}}{\text{u}} \times \left(-0.013637\,\text{u}\right) = -12.7\ \text{MeV}$$

Since it is endoergic, we need to calculate threshold energy (mass + momentum):

$$E_{tr} = 12.7 \text{ MeV} \times \left(\frac{240+4}{240} \right) = 12.9 \text{ MeV}$$

Since it involves the collision of two nuclei, we also should calculate the effective Coulomb barrier.

$$E_{ecb} \approx 1.11 \times \left(\frac{240+4}{240} \right) \times \left(\frac{94 \times 2}{\sqrt[3]{240} + \sqrt[3]{4}} \right) = 27.2 \text{ MeV}$$

This is a lot higher than the threshold energy and is therefore the minimum amount of energy required to make this reaction proceed in good yield. Alpha particles will need to be moving pretty fast to make this reaction work.

In both of the examples we've looked at so far, the effective Coulomb barrier was higher than the threshold energy and, therefore, determined the minimum energy required for the reaction. The effective Coulomb barrier will vary somewhat consistently with the masses of the colliding objects as illustrated in Figure 10.2. The two lines show how the effective Coulomb barrier changes for ^{1}H and ^{4}He as the projectile while the atomic number of the target nucleus increases. Only data for the most abundant, stable, target nuclides were used to generate this figure.

The upward sloping lines in Figure 10.2 tell us that the bigger the projectile and target are, the harder it is to get them together. This makes sense as the amount of positive charge generally increases with size. This is nicely illustrated in our two examples. In Rutherford's reaction we slammed an alpha particle (two protons) into a ^{14}N (seven protons) atom and saw an effective Coulomb barrier of ~5 MeV. In the second example, the effective Coulomb barrier jumped to ~27 MeV because we ran an alpha particle into a ^{240}Pu (94 protons!) nucleus.

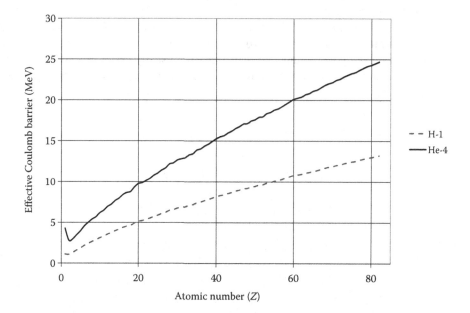

FIGURE 10.2   Variations in the effective Coulomb barrier with projectile and target nucleus.

TABLE 10.1
Some Nuclear Reactions Producing ^{18}F

Reaction #	Reaction	Shorthand
1	$^{15}_{7}N + ^{4}_{2}He \rightarrow ^{18}_{9}F + ^{1}n$	$^{15}N(\alpha,n)^{18}F$
2	$^{16}_{8}O + ^{3}_{1}H \rightarrow ^{18}_{9}F + ^{1}n$	$^{16}O(t,n)^{18}F$
3	$^{17}_{8}O + ^{2}_{1}H \rightarrow ^{18}_{9}F + ^{1}n$	$^{17}O(d,n)^{18}F$
4	$^{18}_{8}O + ^{1}_{1}H \rightarrow ^{18}_{9}F + ^{1}n$	$^{18}O(p,n)^{18}F$
5	$^{19}_{9}F + ^{1}_{1}H \rightarrow ^{18}_{9}F + ^{1}_{1}H + ^{1}n$	$^{19}F(p,pn)^{18}F$
6	$^{20}_{10}Ne + ^{2}_{1}H \rightarrow ^{18}_{9}F + ^{4}_{2}He$	$^{20}Ne(d,\alpha)^{18}F$

Let's approach nuclear reactions from another perspective. Let's say we wanted to make some ^{18}F (commonly used in PET imaging). What possible reactions could be used? A few possibilities are listed in Table 10.1.

That's a lot of options, but wait, there's more! Figure 10.3 illustrates possible "low-energy" nuclear reactions leading to a particular product. Low-energy simply means these reactions are accessible with fairly conventional accelerators (like clinical cyclotrons), not huge, high-power atom smashers. Figure 10.3 is laid out just like a chart of the nuclides with numbers of neutrons increasing from left to right and atomic number increasing from bottom to top. The square in the center (bull's-eye) represents the desired product nuclide, and the reaction types needed to get from a particular starting nuclide to the product are given in the surrounding squares. For example, we know that an n,gamma reaction produces a nuclide that is one square to the right on the chart. Sure enough, "n,γ" is listed as a possible reaction type one square to the left of the desired product. A deuterium-proton (d,p) reaction accomplishes the same result as an n,gamma reaction and therefore appears in the same square. Adding deuterium (^{2}H) increases both N and Z by one; emitting a proton lowers Z by one—the net result is an increase of N by one.

Our list of possible reactions leading to ^{18}F suddenly looks a little short. That's because those in Table 10.1 are limited to a single reaction starting from a naturally occurring nuclide. This is our first criterion in considering potential nuclear reactions—it's usually easier to start with a nuclide that is naturally occurring. Let's see if we can narrow the list down a bit more. Reactions that emit

FIGURE 10.3   Nuclear reactions leading to a particular product (center square).

more than one particle (such as p,pn) generally require more energy than those that only emit one particle. We can cross off reaction #5.

We also want a high yield of product. Just like chemical reactions, sometimes nuclear reactions work well and make lots of product[2] and other times they don't. We'll discuss this more later. Finally, we'd also like to keep the energy requirements within reason. Of the remaining reactions, numbers 4 and 6 have the highest yields and lowest energy requirements. Deciding between these two may depend on what kind of gun is available—in other words, can you easily accelerate protons (reaction #4) or deuterons (reaction #6). It may also depend on how easy it is to separate the product. One advantage of reaction #4 is that the ^{18}O can be incorporated into water molecules, and then the ^{18}F will be produced as the fluoride anion, which can easily be separated from the ^{18}O-water using an ion-exchange column. Both reactions 4 and 6 are commonly used to produce ^{18}F for PET imaging. As an example, let's check out the energetics of reaction #4.

Example 10.3: What is the minimum energy required to make reaction #4 proceed in good yield?

$$^{18}_{8}O + ^{1}_{1}H \rightarrow ^{18}_{9}F + ^{1}n$$

Mass of reactants	^{18}O	17.999161 u
	^{1}H	+1.007825 u
		19.006986 u
Mass of products	^{18}F	18.000938 u
	^{1}n	+1.008665 u
		19.009603 u

Reactants − products = 19.006986 u − 19.009603 u = −0.002617 u

Energy required:

$$Q = \frac{931.5\,\text{MeV}}{\text{u}} \times (-0.002617\,\text{u}) = -2.44\ \text{MeV}$$

Threshold energy:

$$E_{tr} = 2.44 \times \left(\frac{18+1}{18}\right) = 2.57\ \text{MeV}$$

Effective Coulomb barrier:

$$E_{ecb} \approx 1.11 \times \left(\frac{18+1}{18}\right) \times \left(\frac{8 \times 1}{\sqrt[3]{18} + \sqrt[3]{1}}\right) = 2.59\ \text{MeV}$$

The threshold energy and the effective Coulomb barrier are about the same for this reaction, and they are both relatively low. The effective Coulomb barrier is slightly higher, so the minimum energy required for this reaction to proceed in good yield is 2.59 MeV.

10.2   CROSS SECTION

You've probably noticed that we've always qualified the minimum energy requirement with the phrase "in good yield." This suggests that the reaction can proceed at lower energies, but in poor yield. This is true. Nuclear reactions sometimes do proceed at energies lower than they should. When this happens, it is called **tunneling**. It is somewhat analogous to a car driving through a hill rather than over it. Driving through will take a lot less energy than driving over. This is analogous but opposite to how alpha decay takes place (Section 6.4.2). Sadly, tunneling isn't common enough

---

2   This is relative to other nuclear reactions. As we'll soon see overall yields from nuclear reactions are abysmally low compared to chemical reactions.

$$^{63}Cu + {}^1H \longrightarrow [{}^{64}Zn]^* \begin{cases} {}^{63}Zn + {}^1n \\ {}^{62}Cu + {}^1H + {}^1n \\ {}^{62}Zn + 2{}^1n \\ {}^{61}Cu + {}^1H + 2{}^1n \end{cases}$$

FIGURE 10.4 Possible products for the reaction of ^{63}Cu with ^1H.

to run nuclear reactions in good yield on the cheap, so it's still usually a good idea to give those projectiles enough energy to ensure success.

Like chemical reactions, you don't always get what you want with nuclear reactions. Various products can be obtained with the same target/projectile combo. For example, hitting a ^{63}Cu target with a proton projectile can result in all of the products shown in Figure 10.4. The product ratios will depend on the energy of the projectiles and the probabilities (**cross sections**) for the reactions. With higher energy projectiles, a greater product distribution is generally observed.

Every possible nuclear reaction has a cross section, which is symbolized by the Greek letter sigma (σ). When the projectile is a neutron, a subscript following σ indicates what is emitted along with the formation of product. For instance, σ_γ is the cross section for an n,gamma reaction. While it is best to think of cross section as the probability of success, it is more strictly defined as the *apparent* cross-sectional area of the target nucleus as seen by the projectile. Cross section has units of **barns** (b). One barn is equal to 10^{-24} cm^2 (10^{-28} m^2), which is roughly the projected area of the average nucleus. If nuclear size were the only criterion for the chances of a reaction proceeding, then we'd expect cross section to be low for low Z nuclides and high for high Z nuclides. While this is sometimes true, there are other criteria.

Nuclear cross sections depend not only on Z, but also N (and therefore A), the density of the target, and the charge, mass (size), and velocity of the projectile. Figure 10.5 shows how dramatically the probability of all neutron-induced reactions (σ_a) varies with the velocity (energy) of the neutron for the element silver. The general trend is a decreasing probability of reaction as the energy of the neutron increases. This can be rationalized in terms of time (see also Section 7.2). As the neutron moves faster, it spends less time near each nucleus it passes by, thereby decreasing the probability of interaction. Neutrons slowed to that of their surroundings by room temperature water and/or plastic are called **thermal neutrons** and have an energy of about 0.025 eV. More appropriately, it could be understood in quantum mechanical terms of overlap between the deBroglie wavelength of the neutron and the wave functions of the nucleus. As the neutron's wavelength gets shorter, the probability of interaction decreases. Interestingly, this is the same trend observed for gamma photon interactions with matter (Section 7.3).

Cold neutrons have even lower energies as they are passed through materials below room temperature. For example, cold neutrons are generated at Oak Ridge National Laboratory by passing them through supercritical hydrogen at 20 K. Fast neutrons are those with energies significantly higher than room temperature.

The "spikes" observed in Figure 10.5 for fast neutrons indicate that, for neutrons with certain energies, the probability of reaction increases dramatically. These spikes are due to "**resonance capture**," which is the formation of a compound nucleus with an especially favorable amount of excitation energy. The energy where they are observed is called the **resonance energy**. They usually occur for 1–100 eV neutrons. The region of Figure 10.5 where resonance capture is observed is often referred to as the **epithermal region**.

What do the spikes mean? If a particular nuclear reaction is desired, then there may be a specific projectile energy that will greatly increase the probability of reaction. Therefore, it may be desirable to run that reaction with projectiles at the resonance energy. The (enhanced) probability that a reaction will occur at a particular resonance energy is called the **resonance integral**.

FIGURE 10.5 Probability of a neutron-induced reaction with a silver target as the energy of the neutron increases. Produced using data from the evaluated nuclear data file (ENDF) at www.nndc.bnl.gov/.

The bottom line is that variations in nuclear cross sections are not as simple as comparing sizes of target nuclei. They sometimes vary wildly even for closely related target nuclides. For example the thermal neutron cross section for the n,gamma reaction involving ¹H is 0.332 b, but is 0.52 mb (*milli*barns!) for ²H.

The shell model for the nucleus (Section 6.3) can shed some light on the seemingly random variations in cross section. For example, how likely is a neutron capture reaction for a nuclide with a magic number of neutrons? If you thought, "not very," you're right! We know that adding the next neutron (or proton) to a nucleus with a magic number means sticking it in at a much higher energy (Figure 6.5). ⁸⁸Sr has 50 neutrons and a cross section of only 5.8 mb toward an n,gamma reaction with thermal neutrons, yet ⁸⁷Sr (49 neutrons) has a cross section of 17 b for the same reaction. Scan your favorite chart of the nuclides for similar examples.

Likewise, the paring term in the liquid drop model (Section 6.2) suggests that adding a nucleon to a nucleus containing an odd number of that same nucleon might fare a little better than adding it to one that already has an even number. If we extend our previous look at strontium isotopes we observe that ⁸⁶Sr (48 neutrons) has a cross section of 0.82 b toward an n,gamma reaction with thermal neutrons. The pairing term generally has a small effect on binding energy, so its effect is likely to be more or less visible for a group of isotopes. For example, the cross section for the same reaction with ⁸⁹Sr (51 neutrons) is 0.42 b. Higher than ⁸⁸Sr, but not as dramatic as some of the other differences we've seen.

The shell model might also allow us to anticipate whether a proton, neutron, or some other particle might be more likely emitted. For example, if a compound nucleus is formed with a lone proton occupying an energy level much higher than the other nucleons, proton emission is favored. If we hit ¹⁴N with a thermal neutron, the compound nucleus is more likely to emit a proton than a gamma photon.

$$\,^{14}_{7}\text{N} + \,^{1}\text{n} \rightarrow \left[\,^{15}_{7}\text{N}^{*}\right] \rightarrow \,^{14}_{6}\text{C} + \,^{1}_{1}\text{H} \quad \sigma = 1.9\,\text{b}$$

$$\,^{14}_{7}\text{N} + \,^{1}\text{n} \rightarrow \left[\,^{15}_{7}\text{N}^{*}\right] \rightarrow \,^{15}_{7}\text{N} + \gamma \quad \sigma = 0.08\,\text{b}$$

Not every interaction between projectile and target results in a reaction. Sometimes the projectile is simply scattered by the target. The cross section for all possible interactions can be represented as

σ_{tot} while the cross section for all scattering interactions could be σ_s, and those involving absorption σ_a. The total probability of interaction is equal to the sum of the probability of scattering and the probability of absorption.

$$\sigma_{tot} = \sigma_s + \sigma_a$$

The nature of the product makes a difference because it could be stable or radioactive. If it's hot, it'll be decaying as it is produced, making it a little more challenging to calculate yield.

10.3   YIELD

So, how can we tell how much product is made? Reaction yield depends on how much target there is, the projectile flux, the reaction cross section, time, and the nature of the product. **Flux** (Φ, the Greek letter phi) is a measure of how many projectiles pass through a certain area per unit time. Our focus is on neutron flux, so it'll have units of neutrons per square centimeter per second (n/cm²·s).

We'll look at three examples, each with a different product scenario: (1) production of a stable product; (2) production of a radioactive product; and (3) production of a radioactive product with a long half-life. In every scenario we'll assume: (1) the flux does not change with time; (2) the number of target atoms (A) is so large, relative to the number of product atoms (B), it can be considered constant throughout the time of irradiation (t); and (3) the target is thin enough that the flux entering the target is not significantly decreased as it goes through. These assumptions may seem a little perilous, but they are necessary to make the math a bit more tractable, and generally speaking, they are true.

When the product is stable, the amount of product (N_B) formed can be calculated using Equation 10.5:

$$N_B = \sigma \Phi N_A t \tag{10.5}$$

where N_B is the number of atoms of the product nuclide formed after an irradiation time t with units of seconds. σ is the reaction cross section in square centimeters, Φ is the flux of projectile x in particles per square centimeter per second, and N_A is the number of target atoms A (the target doesn't need to be pure nuclide).

Example 10.4: A sample containing 1.00 g of ^{16}O is placed in a nuclear reactor for 10.0 hours. The reactor has a neutron flux of 6.00×10^{10} n/ cm²·s. If $^{16}O(n,\gamma)^{17}O$ is the only reaction that takes place, how many ^{17}O atoms are formed?

The neutron cross section (0.19 mb) can be obtained from a chart of the nuclides or a variety of online sources and converted to cm². Time also needs to be converted to seconds. Finally, the number of target atoms need to be calculated from their mass.

$$0.19 \text{ mb} = 0.19 \times 10^{-27} \text{ cm}^2$$

$$10.0 \text{ h} \times \frac{60 \text{ min}}{\text{h}} \times \frac{60 \text{ s}}{\text{min}} = 36000 \text{ s}$$

$$N_A = \left(1.00\,\text{g} \times \frac{\text{mol}}{15.995\,\text{g}} \times \frac{6.022 \times 10^{23} \text{ atoms}}{\text{mol}}\right) = 3.76 \times 10^{22} \text{ atoms}$$

$$N_B = \sigma \Phi N_A t$$
$$= \left(0.19 \times 10^{-27}\,\text{cm}^2\right) \times \left(6.00 \times 10^{10}\,\frac{\text{n}}{\text{cm}^2 \cdot \text{s}}\right) \times \left(3.76 \times 10^{22}\,\text{atoms}\right) \times \left(36000\,\text{s}\right)$$
$$= 1.55 \times 10^{10} \text{ atoms of } {}^{17}O \text{ formed}$$

If the rate of formation of product is desired, Equation 10.5 can be modified by dividing both sides by time.

$$\text{rate of formation of B} = \frac{N_B}{t} = \sigma \Phi N_A \tag{10.6}$$

If the product is radioactive, we'll have to worry about it decaying as we make it. The amount of hot product formed after irradiating for time t is given by Equation 10.7:

$$N_B = \frac{\sigma \Phi N_A}{\lambda}\left(1 - e^{-\lambda t}\right) \tag{10.7}$$

where λ is the decay constant and should have units of reciprocal seconds. The other variables are the same as we've seen before. Note that N_B can be calculated only for the moment the target is removed from the projectile beam. At any time after that, N_B will be smaller due to its radioactive decay.

Since $A = \lambda N$, Equation 10.7 can also be written as:

$$A_0 = \sigma \Phi N_A \left(1 - e^{-\lambda t}\right) \tag{10.8}$$

where A_0 is the activity of the product at the moment the target is removed from the projectile beam and has the units of decays per second (dps or Bq). Activity is therefore a function of irradiation time (t) and will eventually **saturate**. Saturation occurs when the rate of formation of product is equal to the rate of decay of product. Sound familiar? It's a bit like nuclear equilibria (Section 2.6). For production of radionuclides, irradiation for seven product half-lives gives 99% of saturation, and is a reasonable time to harvest the product nuclide, if maximum yield and minimum time are desired. The path to saturation is illustrated in Figure 10.6. Activity gradually increases as more and more product is made, until it reaches saturation. At saturation, $e^{-\lambda t}$ becomes

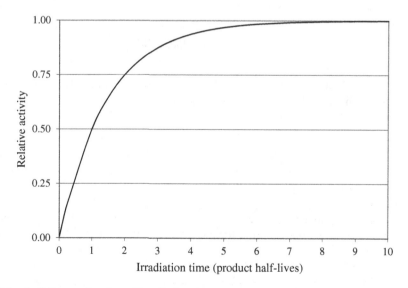

FIGURE 10.6 Activity as a function of irradiation time.

relatively small and $1 - e^{-\lambda t}$ (sometimes called the saturation term) approaches one. Equation 10.8 then simplifies to:

$$A_{sat} = \sigma \Phi N_A \qquad (10.9)$$

at saturation, where A_{sat} is the activity at saturation.

Equation 10.8 finally gives us something tangible—activity. Counting atoms is possible, but activity is usually easier to measure. This equation allows easy experimental determinations of flux, cross section, or the amount of a particular nuclide in a sample. The last is known as **Neutron Activation Analysis**—a powerful analytical tool, especially if the target nuclide has a large neutron cross section. It is capable of detecting as little as 10^{-14} grams of target per gram of sample.

Example 10.5: A 2.50 g sample of ^{18}O water was placed in a cyclotron beam of protons (flux = 3.01×10^3 p/cm^2·s) for 5.00 h, resulting in an activity of 153 dps. Assuming that a $^{18}O(p,n)^{18}F$ is the only reaction that takes place, calculate the cross section for this reaction.

First, calculate the number of ^{18}O atoms in the target, remembering it is $H_2{}^{18}O$.

$$2.50 \text{ g H}_2{}^{18}O \times \frac{\text{mol H}_2{}^{18}O}{20.0 \text{ g H}_2{}^{18}O} \times \frac{1 \text{ mol } {}^{18}O}{1 \text{ mol H}_2{}^{18}O} \times \frac{6.022 \times 10^{23} \text{ atoms}}{\text{mol}} = 7.53 \times 10^{22} \text{atoms}$$

The half-life of ^{18}F is 1.8293 h, therefore:

$$A_0 = \sigma \Phi N_A \left(1 - e^{-\lambda t}\right)$$

$$153 \text{ dps} = \sigma \times \left(3.01 \times 10^3 \frac{\text{p}}{\text{s} \cdot \text{cm}^2}\right) \times \left(7.53 \times 10^{22} \text{ atoms}\right) \left(1 - e^{-\frac{\ln 2}{1.8293 \text{ h}} \times 5.00 \text{ h}}\right)$$

$$\sigma = 7.95 \times 10^{-25} \text{ cm}^2 = 0.795 \text{ b}$$

Note that this reaction had not yet reached saturation. We can see this by dividing the half-life into the irradiation time

$$\frac{5.00 \text{ h}}{1.8293 \text{ h}} = 2.73 \text{ product half-lives}$$

At least seven product half-lives need to pass before saturation is reached. The other way to see that saturation has not been reached is to calculate the saturation term separately. If it is very close to one, then saturation has been reached.

$$\left(1 - e^{-\lambda t}\right) = \left(1 - e^{-\frac{\ln 2}{1.8293 \text{ h}} \times 5.00 \text{ h}}\right) = 0.850$$

Not close enough.

When the radioactive product has a really long half-life the math also simplifies. When $t_{1/2} \gg t$, then $(1 - e^{-\lambda t}) \approx \lambda t$. Therefore, our equation for the activity at the time the sample is removed from the beam simplifies to:

$$A_0 = \sigma \Phi N_A \lambda t \qquad (10.10)$$

Nuclide production is linear with time! Because of the long half-life, more product is made as more time passes. As a result, the issue of saturation is a bit moot if the half-life of the product is too long. Also, irradiating for seven (or more) product half-lives may not be practical. Finally, the reader should notice that Equation 10.10 is the same as Equation 10.5. In other words, production of a nuclide with a sufficiently long half-life is the same as producing a stable nuclide.

Example 10.6: 9.507 g of thallium is irradiated in a neutron howitzer with a flux of 3.67×10^6 n/cm²·s. How long will it take to obtain a ^{204}Tl activity of 2894 dpm?

Thallium is made up of ^{203}Tl (29.52%) and ^{205}Tl (70.48%). We're only interested in the reaction of ^{203}Tl with neutrons, assuming it is an n,gamma reaction.

$$^{203}_{81}\text{Tl} + {}^1\text{n} \rightarrow {}^{204}_{81}\text{Tl} + \gamma$$

First let's figure out how many target atoms we have to shoot at.

$$9.507 \text{ g} \times \frac{\text{mol}}{204.4 \text{ g}} \times 0.2952 \times \frac{6.022 \times 10^{23} \text{ atoms}}{\text{mol}} = 8.268 \times 10^{21} \text{ atoms}$$

The cross section for this reaction is 11.4 b and the half-life of ^{204}Tl is 3.78 years. Assuming that this will be long relative to the irradiation time, we can use Equation 10.10.

$$\frac{2894 \text{ decays}}{\text{min}} \times \frac{\text{min}}{60 \text{ s}} = \left(11.4 \times 10^{-24} \text{ cm}^2\right)\left(3.67 \times 10^6 \frac{\text{n}}{\text{cm}^2 \cdot \text{s}}\right)\left(8.268 \times 10^{21} \text{ atoms}\right)\left(\frac{\ln 2}{3.78 \text{ a}}\right) t$$

$$t = 7.60 \times 10^{-4} \text{ a}$$

Our assumption that irradiation time is relatively short is correct! Let's convert the time value to something more tangible.

$$t = 7.60 \times 10^{-4} \text{ a} \times \frac{5.259 \times 10^5 \text{ min}}{\text{a}} \times \frac{\text{h}}{60 \text{ min}} = 6.66 \text{ h}$$

If a mixture is being irradiated and there are short-lived *and* long-lived nuclides being formed, we can take advantage of their difference in time to saturation. If we want the short-lived nuclide(s), we irradiate for a short time—because very little of the long-lived product will be formed. If we want the long-lived nuclide(s), we irradiate for a long time, take the sample out, and wait for the short-lived stuff to decay.

10.4 ACCELERATORS

How do you get energetic projectiles? Probably not from your favorite online super store. Aside from radioactive decay, bombarding a gas with energetic electrons can produce charged particles. The positive gas ions produced are separated by attraction to an electrode with negative voltage. They are then accelerated through electrical potentials. Passing a particle through a 1000 V potential would give it another keV per unit charge. There are a couple of ways to get this done.

A schematic for a traditional linear accelerator is shown in Figure 10.7a. Linear accelerators are often called linacs, and they accelerate charged particles in a straight line through single

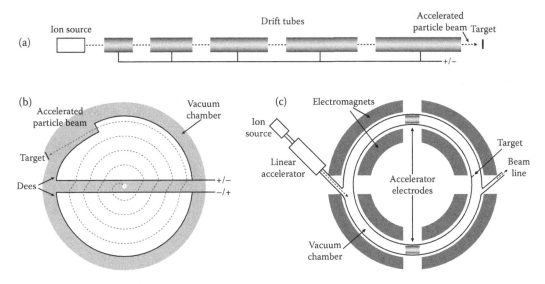

FIGURE 10.7 Simple schematic representations of (a) linear accelerator, (b) cyclotron and (c) synchrotron.

or multiple stages. Each stage is a cylindrical electrode (also known as drift tube) that has the opposite charge of the particle as the particle moves toward it, then the same charge of the particle as it is moving away. Since opposite charges attract and the same charge repels, the particle accelerates as it approaches *and* as it moves away from two adjacent drift tubes. The tubes have no charge while the particle is moving through them, thus it is coasting (or drifting!) as it passes through the tube. The final energy of the particle beam depends on the voltage applied at each stage, and the total number of stages.

While they can accelerate a variety of charged particles, linacs are most commonly used to accelerate electrons. As the electrons are accelerated, the stages get longer; this is done to ensure the electron spends the same amount of time traveling through each tube. This simplifies the electronic switching between positive and negative charges on the tubes. In order to accelerate particles to very high velocities (energies), this type of linac needs to be very large—some are even a couple miles in length. While this is fine for large research centers, more compact designs are necessary in more common settings such as a hospital.

Most **modern-day linear accelerators** use microwaves to accelerate electrons. These microwaves are often produced in much the same way they are for an ordinary microwave oven. Microwaves, like all forms of electromagnetic radiation, can only travel at the speed of light, therefore the radiation itself cannot be used directly for acceleration of the electrons—the photons are not going to slow down and push the electrons to higher and higher speeds. Instead it is the phase of these waves that is manipulated to accelerate electrons.

Remember that electromagnetic radiation behaves like waves in many respects. We can think of phase in relation to the peaks and troughs of a wave. A peak represents one phase, and the trough represents the opposite phase. The electron is pushed and pulled by these phases, so if we arrange them just right, they will accelerate the electron. It's a bit like a surfer riding a wave as it approaches the beach. The surfer is constantly moving downhill, and is being pushed forward by the wave. In our case the wave is the phase and it's accelerating.

There's another way to look at the acceleration of charged particles in this kind of linear accelerator. As the microwave moves through a metal tube, it will induce temporary positive and negative electrical charges in the walls of the tube. If these temporary charges move just right, they can be used to accelerate charged particles.

For clinical applications, such as radiation therapy, these accelerators only need to be a couple of feet long, and can accelerate electrons up to 20 MeV. The Stanford linear accelerator (SLAC) uses microwaves to accelerate electrons up to 50 GeV and is nearly two miles long!

Cyclotrons are the most commonly used accelerators for radionuclide production in hospitals and other, more ordinary places. They take advantage of the fact that charged particles float in a circle when placed in a constant magnetic field, and that the radius of the circle is directly related to the velocity (energy) of the particle. In other words, the faster the particle is moving, the bigger the circle it'll make. The size of the circle also depends on the strength of the magnetic field—the stronger the field, the smaller the circle.

A schematic for a simple cyclotron is illustrated in Figure 10.7b, although the magnets have been removed for clarity. Only the accelerator portion of the cyclotron is shown. Imagine two round magnets sitting above and below the page in Figure 10.7b—just like an Oreo™[3] cookie, where the magnets are the crunchy chocolate cookie, and the particles are accelerated in the delicious creamy filling.

Charged particles are injected into the center of the cyclotron (the small white circle), and immediately begin to move in a circle parallel to the two magnets. As it circles around, it moves in and out of two hollow, semicircular electrodes. These electrodes are called **dees** because they look like the letter D. Imagine making them by cutting a tuna can in half along the can's diameter and taking all the tuna out (somewhere, a cat is purring …). As is true for all particle accelerators, the cyclotron is kept under a vacuum so that the accelerating particles don't run into any matter along the way.

Just like the traditional linear accelerator described above, acceleration of the charged particle takes place as it travels between the dees. As it leaves one dee, that dee is given the same charge as the particle, pushing it away. At the same time, the dee the particle is approaching is given the opposite charge, attracting the particle to it. As it's accelerated, the particle moves in a larger circle, giving it a kind of spiral path as it is accelerated in the cyclotron. It is not a smooth spiral, as the particle is not accelerated as it travels through the dees and therefore moves along a semi-circular path with a fixed radius. Eventually, the circular path of the particle is large enough (energy is high enough) to allow it to exit the cyclotron.

Cyclotrons are not very good at accelerating electrons. Because of their low mass, it doesn't take a whole lot of energy (~100 keV) to get them traveling at velocities approaching the speed of light. When particles approach the speed of light increasing amounts of energy are required to accelerate the particle by increasingly small amounts (Section 7.2). In a cyclotron, this means that once the particle gets to a certain velocity, it'll be difficult to continue to accelerate it, and it will pretty much stay in a circle (not continuing to spiral). Additionally, there are practical limits to the strengths of the electromagnetic fields within the cyclotron. As a result, cyclotrons tend to have energy limits for various particles. For most particles accelerated in cyclotrons (^{1}H, ^{2}H, ^{4}He, etc.) this limit is 25–50 MeV.

A common nuclide produced by cyclotrons for clinical use is ^{18}F. As we saw in Section 10.1, the following reaction requires a minimum energy of ~2.6 MeV. Protons can easily be accelerated in cyclotrons to this energy.

$$^{18}_{8}O + ^{1}_{1}p \rightarrow ^{18}_{9}F + ^{1}_{0}n$$

Like ^{18}F, most products of cyclotron reactions tend to be proton-rich. Therefore, they tend to decay via positron emission or electron capture. Atomic number (Z) almost always changes in the reactions that occur in cyclotrons, making it easier to isolate carrier-free product because the target and the product will have differing chemical properties. Cyclotron reactions are even less efficient

---

3 Oreo is a registered trademark of Kraft Foods Inc., Northfield, Illinois.

TABLE 10.2
Some Cyclotron-Produced Radionuclides Used in Nuclear Medicine

Product Nuclide	Decay Mode(s)	Common Production Reaction(s)	Natural Abundance of Target Nuclide (%)
^{11}C	β^+, EC	^{14}N(p,α)^{11}C	99.63
		^{10}B(d,n)^{11}C	19.9
		^{11}B(p,n)^{11}C	80.1
^{13}N	β^+	^{16}O(p,α)^{13}N	99.76
		^{12}C(d,n)^{13}N	98.93
^{15}O	β^+	^{14}N(d,n)^{15}O	99.63
		^{15}N(p,n)^{15}O	0.368
^{18}F	β^+, EC	^{18}O(p,n)^{18}F	0.205
		^{20}Ne(d,α)^{18}F	90.48
^{22}Na	β^+, EC	^{23}Na(p,2n)^{22}Na	100
^{43}K	β^-	^{40}Ar(α,p)^{43}K	99.60
^{67}Ga	EC	^{68}Zn(p,2n)^{67}Ga	18.75
^{111}In	EC	^{109}Ag(α,2n)^{111}In	48.16
		^{111}Cd(p,n)^{111}In,	12.49
^{123}I	EC	^{122}Te(d,n)^{123}I	2.55
		^{124}Te(p,2n)^{123}I	4.74
^{201}Tl	EC	^{201}Hg(d,2n)^{201}Tl	13.18

than neutron activation. This makes sense because we are now trying to get two positively charged particles to hit each other. Coulomb repulsion forces must be overcome for this to happen. As a result, cyclotrons produce nuclides in pretty low yields, and they tend to be a little pricey.

Table 10.2 shows some of the more common nuclides produced by cyclotrons for medical applications. The first three nuclides listed, ^{11}C, ^{13}N, and ^{15}O, all have short half-lives (≤ 20 minutes), so they must be produced on-site, and quickly incorporated into a drug then delivered to the patient. Carbon, nitrogen, and oxygen are important elements in biological molecules—if we could place one of these nuclides in a biomolecule, then it would have all the same chemical and biological properties, but the radioactive label would allow us to see where that molecule goes in the body. This is an exciting area of current research in nuclear medicine, with the potential to develop a number of highly effective radiopharmaceuticals.

Another type of circular accelerator that can accelerate particles up to incredibly high energies (TeV = 10^{12} eV) is the **synchrotron**. A simple schematic drawing of a synchrotron is shown in Figure 10.7c. Synchrotrons use oscillating magnetic and electrical fields to accelerate particles on a fixed radius. As the particle energy increases, so does the strength of the magnetic field, keeping the particles moving through the same circle. Synchrotrons are enormous. The Large Hadron Collider (LHC) near Geneva, Switzerland, is over 27 km in circumference, and accelerates protons to 7 TeV. One of the cool aspects of the LHC is that protons can be accelerated to 7 TeV in opposite directions in the same circle before being forced to collide with each other.

10.5 COSMOGENIC NUCLIDES

Just about all of the background radiation detected on the surface of the planet originates from cosmic radiation. As depicted in Figure 10.8, cosmic radiation originates from our sun, other stars in our galaxy, and other galaxies. In terms of ionizing radiation, these stars emit mostly protons, but also alpha particles, other light nuclei, electrons (collectively, these particle emissions make up what is called the solar wind), high-energy photons, and neutrinos. While most particles are about

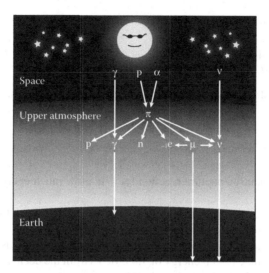

FIGURE 10.8   Cosmic radiation.

1 GeV, some are *very* high-energy (up to 10^{10} GeV). When these screaming nuclear projectiles slam into the atmosphere, they annihilate themselves, each one producing dozens of pions (π). Pions are subatomic particles that are commonly produced in high-energy (> 400 MeV) nuclear reactions. In the average nucleus, pions are exchanged between nucleons and thereby hold the nucleus together—somewhat like the way electrons hold atoms together in molecules.

Pions don't last long on their own. They decay to neutrinos and muons (μ) with a lifetime (τ)[4] of only 10^{-6} s. Muons are yet another subatomic particle and are similar to electrons. Somewhere between 50 and 80 percent of the muons formed actually make it to the Earth's surface. Muons don't last long either ($\tau = 2.2 \times 10^{-6}$ s), decaying to electrons and neutrinos. Even so, much of the cosmic radiation that reaches the surface of our planet is in the form of muons. High-energy photons (γ) and neutrinos (ν) make up most of the rest of the cosmic radiation flux at the planet's surface.

Thanks mostly to the Earth's magnetic field, and to all the matter in our atmosphere, we are shielded from all but about 5% of the cosmic radiation that bombards our planet. The significantly higher radiation flux in space is an important concern for astronauts spending considerable amounts of time there and is a major hurdle to long space missions such as travel to other planets both within and outside of our solar system.

Pions also collide with matter in the atmosphere producing electrons, neutrons, protons, and photons. Some of these particles can initiate nuclear reactions forming ^{3}H, ^{10}B, and ^{14}C, providing a constant source of these (and other!) radionuclides. As mentioned in Chapter 2, ^{14}C has a half-life of 5715 years, is a soft β^{-} emitter (E_{max} = 157 keV), and is the nuclide of interest when performing carbon dating. ^{14}C is formed by the ^{14}N(n,p)^{14}C reaction when a slow neutron is involved. If the neutron is more energetic (fast), formation of ^{12}C is preferred via ^{14}N(n,t)^{12}C.

---

4   Lifetime is the average time needed for one particle (or nuclide) to decay and is calculated by inverting the decay constant (1/λ) and is ~70% of the half-life.

QUESTIONS

10.1 Complete the following and write balanced equations (A + x →B + y).

$^{18}O(n,\beta^-)$ $^{14}N(\alpha,p)$ $^{141}Pr(\gamma,2n)$
$^{238}U(\alpha,n)$ $^{59}Fe(\alpha,\beta^-)$ $^{10}B(d,n)$

10.2 Balance the following nuclear reactions and provide the shorthand for each.

$^{20}Ne + \_\_\_ \rightarrow {}^{18}F + {}^4He$ $\_\_\_ + n \rightarrow {}^{32}P + {}^1H$
$^{14}N + {}^1H \rightarrow \_\_\_ + {}^4He$ $^{124}Te + {}^1H \rightarrow {}^{123}I + \_\_\_$

10.3 Write out the neutron capture reaction for ^{191}Ir. How much energy would be produced by this reaction?

10.4 Identify the target, projectile, and product in the reaction from Question 10.3.

10.5 Define the following: compound nucleus, endoergic, cross section, and resonance integral.

10.6 Write out the α,p reaction for ^{12}C. What is the compound nucleus formed in this reaction? If the alpha particle has an energy of 7.69 MeV (from the decay of ^{214}Po), what is the excitation energy of the compound nucleus?

10.7 What is the minimum amount of energy required for the reaction in Question 10.6 to proceed in good yield? What is the excitation plus kinetic energy of the product and emitron (B + y)?

10.8 Calculate the velocity (m/s) and wavelength (nm) of a thermal neutron.

10.9 The reaction below occurs in a small cyclotron. Calculate the effective Coulomb barrier for this reaction, the amount of energy produced or required by this reaction (Q), and the kinetic energy of the tritium ion necessary to make this reaction proceed in good yield.

$$^3_1H + {}^{124}_{50}Sn \rightarrow {}^{126}_{51}Sb + {}^1n$$

10.10 ^{111}In is a nuclide used in clinical settings for diagnostic scans. It can be made using an α,2n reaction. How much energy is required for this reaction to proceed in good yield? Could this reaction be performed using a cyclotron? Briefly explain.

10.11 A neutron treatment for cancer produces its neutrons by the bombardment of beryllium with 66 MeV protons. Assuming this is a p,n reaction, write out a balanced nuclear reaction for this process. What is the minimum energy required for this reaction to proceed in good yield? If the product of the nuclear reaction is unstable, write out a balanced decay equation.

10.12 What do threshold energy and effective Coulomb barrier have in common? What distinguishes them?

10.13 Suggest a reasonable, low-energy reaction for the production of ^{18}F from $^6Li_2CO_3$ using reactor neutrons (2 MeV). What is the excitation energy of the compound nucleus formed in this reaction?

10.14 Thermal neutron cross sections for n,gamma reactions of some lead isotopes are given in Table 10.3. Briefly explain their relative values.

TABLE 10.3
Thermal Neutron Cross Sections of Lead

^{204}Pb	^{205}Pb	^{206}Pb	^{207}Pb	^{208}Pb
0.70 b	4.5 b	0.027 b	0.70 b	0.23 mb

10.15 Define the following: flux, saturation, dee, and synchrotron.

10.16 A thin sample of Mg (0.100 g) is irradiated for 24.0 h in a nuclear reactor with a neutron flux of 5.82×10^6 n/cm²·s. Calculate the number of ^{25}Mg atoms formed.

10.17 A thin sample of Mn (549 mg) was placed in a neutron howitzer (a laboratory neutron source) with a flux of 1.50×10^4 n/cm²·s for 5.00 h. Calculate the activity of the sample when it was removed from the howitzer. If a product is formed, what is it, and how much (mg) is present when the sample was removed from the howitzer?

10.18 A 0.486 g sample of gold was placed in a cyclotron beam of helium ions (flux = 1.03×10^4 ions/cm²·s) for 67 hours, resulting in an activity of 97 dps. Assuming that the only particle emitted from the compound nucleus is ^{3}He, calculate the cross section (b) for the reaction and the total number of atoms produced (throughout irradiation). Is there a simpler way to perform this reaction? Explain.

10.19 Give at least three nuclear reactions that could produce ^{15}O (used for PET). On the basis of threshold energy and effective Coulomb barrier, which would be the best one to use?

10.20 A neutron howitzer employs 25 gallons of water for shielding and has a flux of 4.9×10^4 n/cm²·s at the neutron source. If the source is a cylinder that is 4.0 inches tall and 1.0 inch in diameter, what percentage (maximum) of the water will be DOH (D = deuterium) after one year? Would you expect a significant amount of TOH (T = tritium) or D_2O to form in this time? Briefly explain.

10.21 A 1.0 g sample of ^{14}N is placed in a neutron howitzer with a flux of 1.50×10^4 n/cm²·s for 5.0 days. Write out the most probable nuclear reaction. Assuming this is the only reaction, what percentage (atom!) of the sample was transmuted?

10.22 5.43 g of lead is irradiated in a nuclear reactor with a flux of 4.74×10^{13} n/cm²·s. How long (days) will it take to obtain a ^{205}Pb activity of 120 dpm?

10.23 A thin 17.3 mg sample of nickel is pummeled by 8.43 MeV deuterons in a cyclotron. If the emitron is a neutron and the desired product is ^{61}Cu, what is the reaction? After 95 minutes in the cyclotron beam, the sample has a count rate of 1653 cpm under the annihilation peak on a high-purity Ge detector. Assume that decay of ^{61}Cu is the only contributor to this peak. If the detector has an efficiency of 10.8% at 511 keV, and the cross section for this reaction is 78 mb, what is the deuteron flux in this cyclotron?

10.24 Briefly describe how a cyclotron works.

10.25 Cold fusion enthusiasts recently reported that the ^{133}Cs(d,γ)^{135}Ba reaction takes place close to room temperature. As the science editor for a prestigious journal, you are asked to review their work. How would you respond?

10.26 A student places a sample containing an unknown element in a neutron howitzer for 1.5 hours. Upon removal, the student observes that a nuclide was formed with a half-life of about 3.8 h, but the count rates were low. How long should the student keep the sample in the howitzer to achieve maximum activity?

10.27 It has been reported that the reaction of ^{48}Ca with ^{186}W produces ^{40}Mg. If only one other nuclide is formed by this reaction, write out a balanced nuclear reaction. What might be the driving force for the formation of these particular products?

10.28 How is tritium formed from cosmic radiation? Calculate the minimum projectile energy necessary for this reaction to proceed in good yield.

11 Fission and Fusion

Nuclear fission and fusion are special types of nuclear reactions. Because of their prominence in modern science and society, they will both be discussed in some detail in this chapter. Before we do, we need to cover one more form of decay, spontaneous fission.

11.1 SPONTANEOUS FISSION

Spontaneous fission is like alpha and cluster decay in that a nuclide splits in two. The difference is that with spontaneous fission, the nuclide splits roughly in half. It is only observed for the heavier nuclides, and even then, is only a primary decay mode for a handful of nuclides. Spontaneous fission can be generically represented by:

$$_{Z}^{A}X \rightarrow \, _{Z_1}^{A_1}L + \, _{Z_2}^{A_2}M + \nu n$$

where L and M are **fission products** (or fragments), and νn indicates the release of a small whole number (typically 0–4) of neutrons. A wide variety of fission products are formed by the decay of a particular parent, but products with magic numbers (like 50 and 82) of neutrons and/or protons have increased odds of formation.

The only naturally occurring nuclide that undergoes spontaneous fission is ^{238}U, however, its branch ratio for this form of decay is quite low (0.00005 %). ^{238}U has 92 protons and 146 neutrons and is therefore an even–even (*ee*) nuclide. Even–even nuclides are somewhat more likely to undergo fission than nuclides with odd numbers of protons and/or neutrons.

The neutrons released in fission reactions usually escape their immediate surroundings and can interact with material that is a significant distance away. The fission products formed often do not travel far. These products have a great deal of kinetic energy, which is mostly dissipated as heat. The kinetic energies of the fission products are high because of the strong Coulomb repulsion of the two fragments right after they separate. The total kinetic energy (K_T) in MeV can be estimated using Equation 11.1.

$$K_T = 0.80 \times \frac{Z_1 Z_2}{\sqrt[3]{A_1} + \sqrt[3]{A_2}} \tag{11.1}$$

where Z_1 and Z_2 are the atomic numbers of the two fission products, and A_1 and A_2 are their mass numbers. Notice that this formula is similar to the one we used to calculate the Coulomb barrier (Equation 10.2). The only difference is the constant value—it was 1.11 for the Coulomb barrier and is 0.80 here. The reason for the difference lies in the distance between the centers of the nuclei

DOI: 10.1201/9781003303787-11

(a)

(b)

FIGURE 11.1 The differences in nuclear separation for (a) collision, and (b) fission.

as they are separating or joining. When calculating the Coulomb barrier, we assume the nuclei are spheres that are just barely touching each other (Figure 11.1a). This is when the Coulomb repulsion forces are strongest, and the strong force (between the two nuclei) is just beginning to kick in. At this point, the Coulomb barrier is at a maximum. When determining kinetic energy following fission, it is better to assume some separation between the spheres, as this gives more realistic results (Figure 11.1b). With greater separation, energy (K_T) should be lower, so we use 0.80 to modify the Z/A term (Equation 11.1) rather than 1.11 (Equation 10.2).

The kinetic energies of the individual fragments can then be estimated using the same conservation of momentum formulas used to approximate recoil and alpha particle energies in alpha decay (Section 5.1).

$$K_1 = K_T \left(\frac{A_2}{A_1 + A_2} \right)$$

(11.2)

$$K_2 = K_T \left(\frac{A_1}{A_1 + A_2} \right)$$

(11.3)

Some of the variables have changed, but these are the same as Equations 5.1 and 5.2.

Example 11.1: Calculate the total energy released and estimate the kinetic energies of the fission products in the spontaneous fission reaction shown here.

$$^{238}_{92}\text{U} \rightarrow ^{134}_{52}\text{Te} + ^{102}_{40}\text{Zr} + 2\,^1_0\text{n}$$

Note that these fission products are just two of a great many possible spontaneous fission products for ^{238}U. Notice also that ^{134}Te has a magic number (82) of neutrons and therefore can be expected to be formed in somewhat higher yield. To find the total energy released by this decay, we need to determine how much mass is lost.

Mass of parent	^{238}U	238.050788 u
Mass of products	^{134}Te	−133.91154 u
	^{102}Zr	−101.92298 u
2n = 2(1.008665 u) =		−2.017330 u
Mass lost		0.19894 u

Total energy of decay:

$$E = \frac{931.5\,\text{MeV}}{\text{u}} \times 0.19894\,\text{u} = 185.3\,\text{MeV}$$

Now, let's estimate the kinetic energy of the two fragments.

$$K_T = 0.80 \times \frac{Z_{Te} Z_{Zr}}{\sqrt[3]{A_{Te}} + \sqrt[3]{A_{Zr}}} = 0.80 \times \frac{52 \times 40}{\sqrt[3]{134} + \sqrt[3]{102}} = 170 \text{ MeV}$$

$$K_{Te} = K_T \left(\frac{A_{Zr}}{A_{Te} + A_{Zr}} \right) = 170 \text{ MeV} \times \frac{102}{134 + 102} = 73 \text{ MeV}$$

$$K_{Zr} = K_T \left(\frac{A_{Te}}{A_{Te} + A_{Zr}} \right) = 170 \text{ MeV} \times \frac{134}{134 + 102} = 97 \text{ MeV}$$

These formulas provide an estimate only, so it's a good idea to keep the number of significant figures low in the answers. The heavier fragment has less energy than the light one—just like alpha decay, only now the difference in energy between the two fragments is a lot less dramatic. Note that the sum of the two fragment kinetic energies equals the total kinetic energy.

$$K_1 + K_2 = K_T$$

73 MeV + 97 MeV = 170 MeV

Notice also that the total energy of decay is significantly larger than the kinetic energy of the fragments. The remaining 15 MeV (185 MeV –170 MeV) is divided among the neutrons and gamma rays emitted as part of the decay, and the excitation energy of the products. The gamma rays emitted as part of the decay are referred to as **prompt** gamma photons and are not typically shown when writing out the decay equation.

Because of the low branch ratio, relatively few ^{238}U atoms undergo spontaneous fission. Generally speaking, the odds of spontaneous fission increase with mass number (A). For example, ^{256}Fm decays via spontaneous fission in 92% of all decays. This makes intuitive sense; the bigger the nucleus, the more likely it is to split into two fragments.

11.2 NEUTRON-INDUCED FISSION

If heavy *ee* nuclides are more likely to undergo spontaneous fission, can we get a heavy even–odd (*eo*) nuclide to split if we fire neutrons at it? You bet! In fact, this is a pretty handy way to get fission to occur, rather than sitting around waiting for a nuclide to undergo spontaneous fission. The advantage of adding a neutron to an *eo* nuclide is that we obtain an *ee* nuclide with some excitation energy. Neutron-induced fission reactions (typically referred to simply as "fission reactions") can be generically represented as:

$$A + n \rightarrow [C]^* \rightarrow B + D + vn$$

where A is the heavy nuclide, also known as the **fissionable material** (*vide infra*). The compound nucleus is [C]*, and B and D are the fission products (fragments). Neutrons are typically released by fission reactions, which is pretty handy if you've got a lot of A and want more of them to undergo fission. Such a process is known as a **fission chain reaction** and was introduced back in Chapter 4 and will be discussed some more in Chapter 12. Just as in spontaneous fission, the symbol v represents a small whole number, typically 0–4.

A fissionable material is any nuclide that will undergo fission when you throw a neutron (thermal *or* fast) at it. As you no doubt remember from Chapter 4, fissile material undergoes fission after being hit by a thermal neutron (our main examples are ^{233}U, ^{235}U, and ^{239}Pu). As such, both ^{235}U and ^{238}U are fissionable (^{238}U can undergo fission with fast neutrons, see Section 4.2), but, of the two, only ^{235}U is fissile. Further complicating fission semantics is the term fertile. As defined in Section 4.1.3, a fertile nuclide is one that is not itself fissile but can be converted to a fissile nuclide in a reactor. Examples are ^{232}Th and ^{238}U, which make ^{233}U and ^{239}Pu, respectively.

Fission reactions tend to be pretty messy—hundreds of different nuclides have been identified as fission products for ^{235}U. These products tend to be rather neutron-rich since they originate from a nuclide (^{235}U) with an *N/Z* ratio that is relatively high when compared to the stable isobars of the fission products. One example of a fission reaction for ^{235}U forms ^{94}Rb, ^{140}Cs, and two neutrons.

$$^{235}_{92}\text{U}_{143} + {}^{1}_{0}\text{n} \rightarrow {}^{94}_{37}\text{Rb}_{57} + {}^{140}_{55}\text{Cs}_{85} + 2\,{}^{1}_{0}\text{n}$$

$$N/Z \rightarrow \quad 1.55 \qquad\qquad 1.54 \qquad 1.54$$

The *N/Z* ratio for each nuclide is given directly below it. Despite the emission of two neutrons in this particular fission reaction, the *N/Z* ratios of the products are pretty much the same as the target nuclide. However, we know (Section 1.4) that nuclides with lower mass numbers will only be stable with lower *N/Z* ratios. For example, the only stable nuclides for $A = 94$ are $^{94}_{42}\text{Mo}_{52}$ (*N/Z* = 1.24) and $^{94}_{40}\text{Zr}_{54}$ (*N/Z* = 1.35). Since fission products tend to be neutron-rich, they will generally undergo β^{-} decay. In fact, each fission product can form a short (isobaric) decay series on the way to stability. The series formed by ^{94}Rb consists of ^{94}Sr and ^{94}Y before forming stable ^{94}Zr.

$$^{94}_{37}\text{Rb} \xrightarrow{\;\beta^{-}\;} {}^{94}_{38}\text{Sr} \xrightarrow{\;\beta^{-}\;} {}^{94}_{39}\text{Y} \xrightarrow{\;\beta^{-}\;} {}^{94}_{40}\text{Zr}$$

The relative amount that each nuclide is formed is known as the **fission yield** and is expressed as a percentage. It is sometimes difficult to determine the yield of a particular fission product (**independent yield**) because it may have been formed through beta decay of an isobar. Some of these products formed through beta decay series have very short half-lives, making it difficult to run a fission reaction for an amount of time, then analyze the products and get reliable independent yields. Since the products generally retain the same mass number, yields can be determined for all isobars (**cumulative yield**).

The cumulative yields for thermal neutron-induced fission of ^{235}U are collectively graphed in Figure 11.2. Fission products for this reaction distribute themselves into two large peaks. Products with mass numbers of 90–100 u and 130–145 u are formed in the highest yields and are therefore preferred. Generally speaking, the mass numbers from these two peaks add up to the mass number of ^{235}U, meaning that nuclides with differing mass numbers are usually formed. Thermal neutron fission of ^{235}U is therefore generally **asymmetric**.

Symmetric fission would be the division of the ^{236}U compound nucleus (formed by the interaction of a neutron with ^{235}U) into identical fragments. This clearly does take place, but at a much lower yield than asymmetric fission. This is indicated in the middle of Figure 11.2 by the formation of nuclides with mass numbers of approximately 118.

Why is asymmetric fission so strongly preferred? Look closely at the small spike in fission yield at $A = 132$. Something is definitely special about the formation of nuclides with this mass number. In particular, $^{132}_{50}\text{Sn}_{82}$ is formed which has magic numbers of both neutrons and protons, and therefore has greater stability. Fission reactions prefer to form at least one product with a magic number of protons and/or neutrons because of the stability of those nuclides. Many of the other nuclides

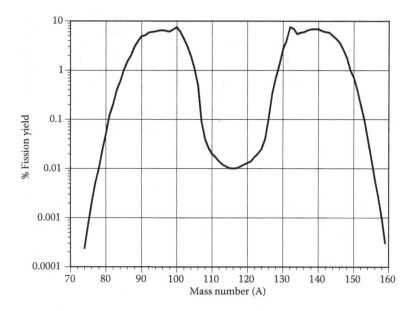

FIGURE 11.2 Fission product yields for thermal neutron fission of ^{235}U.

formed under the broad peak on the right of Figure 11.2 also have either 50 protons (tin) or 82 neutrons. Some of the nuclides under the left peak are preferentially formed because they have 50 neutrons. The rest of the graph is a mirror to these preferences. For example, the small spike at $A = 100$ reflects the spike at $A = 132$ and corresponds to a reaction like the one shown here.

$$^{235}_{92}\text{U} + ^1_0\text{n} \rightarrow ^{132}_{50}\text{Sn} + ^{100}_{42}\text{Mo} + 4\,^1_0\text{n}$$

^{100}Mo is simply what's left over after the very favorable ^{132}Sn is formed and four neutrons are emitted.

The energy of the incident neutron (the projectile) will affect the distribution of the fission products. As shown in Figure 11.3, higher energy (more than thermal) neutrons will increase the probability of symmetric fission relative to the probability of asymmetric fission for ^{235}U. The valley between the two peaks fills in as neutron energy increases. This can be understood in terms of the excitation energy of the compound nucleus ($[^{236}\text{U}]^*$). The more energy the neutron has, the greater the excitation energy of the compound nucleus. The greater the excitation energy, the less fussy the compound nucleus will be about forming products with magic numbers of nucleons. Formation of nuclides with magic numbers generates more energy for the reaction; however, if the reaction already has some extra energy, it isn't as important to generate more.

The fission product distribution also depends on the nuclide undergoing fission (the target). Figure 11.4 shows how the distribution varies between ^{233}U and ^{239}Pu. Notice that the distribution for the lighter of these two nuclides (^{233}U—solid line) is shifted to lower masses, while it is shifted to higher masses for ^{239}Pu (dashed line), when compared to Figure 11.2. Notice also that the right peak shifts much less than the left peak. This reflects the fact that nuclides formed under the right peak have two choices for magic numbers (50 protons *or* 82 neutrons) while the left peak only has one (50 neutrons).

Not all heavy *eo* nuclides undergo thermal neutron fission, and not all heavy *ee* nuclides decay via spontaneous fission. Why not? There is an energy barrier to fission, just like we saw for alpha decay in Section 6.4.2, and we can think of as the same as the Coulomb barrier discussed in Section 10.1. Just as there is an energetic barrier to bring two positively charged nuclei together, there is also a barrier to splitting them up. It is called the **fission barrier** and can be thought of as disruption of

FIGURE 11.3 Fission product distributions for ^{235}U with thermal neutrons (solid line), ~2 MeV neutrons (larger dashed line), and 14 MeV neutrons (smaller dashes). Data from Flynn, K. F. and Glendenium, L. E., Rep. ANL-7749 Argonne National Laboratory, Argonne, IL 1970.

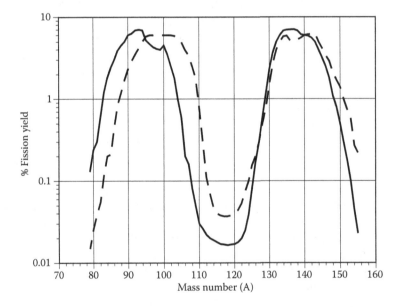

FIGURE 11.4 Thermal neutron fission product distributions for ^{233}U (solid) and ^{239}Pu (dashed).

the strong force, which holds all the nucleons together in the nucleus. In light of this barrier, it is surprising that spontaneous fission takes place at all. The only way it can take place is by avoiding the barrier. It is generally believed that spontaneous fission takes place via tunneling—the nuclide finds a way through the barrier rather than over it. As you will recall, tunneling is a path less taken, therefore spontaneous fission is a relatively rare form of decay, but can sometimes occur for some of the heaviest known nuclides.

TABLE 11.1

Relative Tendencies to Undergo Neutron-Induced Fission for Selected Heavy Nuclides

Nuclide	σ_f (b)	$\sigma\gamma$(b)	Binding E of a Thermal ^{1}n (MeV)	Fission Barrier (MeV)
^{232}Th	0.000003	7.34	4.8	7.5
^{233}U	531	46	6.8	6.0
^{235}U	585	98	6.5	5.7
^{238}U	0.000005	2.7	4.8	5.8
^{239}Pu	750	271	6.5	5.0

The binding of an additional neutron to a target nucleus can provide the energy to overcome the fission barrier. Let's look at some relevant information for common nuclides of the actinide series in Table 11.1. As discussed in Section 10.2, σ_γ is the cross section (probability) for an n,γ reaction, and σ_f is the cross section for a neutron-induced fission reaction. Both cross-sectional values are given in barns to allow easy comparison. For example, ^{232}Th is *much* more likely to form ^{233}Th when hit by a neutron, while ^{233}U is more likely to undergo fission.

For each nuclide in Table 11.1, the energy produced by binding a thermal neutron is calculated. This is the same as the excitation energy of the compound nucleus formed by neutron capture. Let's use ^{235}U as an example.

$$^{235}\text{U} + {}^1\text{n} \rightarrow [^{236}\text{U}]^*$$

$$Q = \frac{931.5 \text{ MeV}}{\text{u}} \times (235.043930 \text{ u} + 1.008665 \text{ u} - 236.045568 \text{ u})$$
$$= 6.546 \text{ MeV}$$

You may recall that calculation of the excitation energy of a compound nucleus also needs to consider the kinetic energy of the compound nucleus (E_{KC}), which is conveniently calculated with Equation 10.4.

$$E_{KC} \approx E_{KX} \times \left(\frac{A_X}{A_C}\right)$$

The kinetic energy of a thermal neutron is rather low ($E_{KX} = 0.0253$ eV), therefore, the kinetic energy of the compound nucleus formed after thermal neutron capture will be very small relative to the excitation energy. The bottom line is that excitation energy of the compound nucleus is essentially equal to Q for thermal neutron capture.

Notice that the excitation energy is lower (~4.8 MeV) for $^{232}_{90}\text{Th}_{142}$ and $^{238}_{92}\text{U}_{146,}$ while it is higher (6.5-6.8 MeV) for $^{233}_{92}\text{U}_{141}$, $^{235}_{92}\text{U}_{143}$, and $^{239}_{94}\text{Pu}_{145}$. This difference arises from the number of neutrons in each nuclide. ^{232}Th and ^{238}U have an even number of neutrons, so addition of another neutron would result in the formation of a compound nucleus with an odd number of neutrons. ^{233}U, ^{235}U, and ^{239}Pu all start with an odd number of neutrons, and adding another neutron forms a compound nucleus with an even N. Remember that even numbers of nucleons mean greater stability, and odd numbers mean less stability (Section 1.4). An appropriate analogy is a rock rolling down a hill into a ravine. Generally speaking, the taller the hill and the deeper the ravine, the more energy the rock will have when it hits bottom. Going from even N to odd N (^{232}Th and ^{238}U) means the hill is lower, and the ravine shallower, therefore less excitation energy results.

The fission barrier for each nuclide is given in the final column of Table 11.1. Those nuclides that generate enough excitation energy from thermal neutron capture to overcome the fission barrier are those that tend to undergo fission rather than an n,γ reaction. Notice that those nuclides with enough energy to overcome the fission barrier also have relatively high σ_f values. Our example nuclide, ^{235}U, looks pretty good for fission, since its excitation energy is 6.5 MeV, but its fission barrier is only 5.7 MeV. As we might then expect, its cross section for a thermal neutron-induced fission reaction is high (585 b!) relative to its cross section toward an n,γ reaction (98 b). For ^{238}U, the opposite is true. Its excitation energy is too low (4.8 MeV) to overcome its fission barrier (5.8 MeV), and it has a very low σ_f (0.000005 b!) relative to its σ_γ (2.7 b). We would therefore expect ^{238}U to predominately form ^{239}U when exposed to thermal neutrons (although not in great yield).

As we learned in Section 10.2, cross sections change with the energy of the neutron. Fast neutrons will dramatically increase σ_f while decreasing σ_γ, causing more ^{238}U to fission rather than just form ^{239}U. Again, this can be understood in terms of excitation energy of the compound nucleus. If the neutron has more than about 1 MeV of energy, the excitation energy of the compound nucleus will be greater than the fission barrier, allowing more ^{238}U fission reactions to take place. Therefore, fission is not restricted to excited *ee* nuclides, but to any large nucleus with sufficient excitation energy to overcome its fission barrier.

The energetics of a neutron-induced fission reaction can be handled the same way as for spontaneous fission. Don't forget there's now a neutron on the left side of the arrow. You can save a step in the math by canceling it with one of the product neutrons as shown in the example below.

Example 11.2: Calculate the total energy released by the following fission reaction. Estimate the kinetic energy of the fission products.

$$\,^{235}_{92}\text{U} + \,^{1}_{0}\text{n} \rightarrow \,^{141}_{56}\text{Ba} + \,^{92}_{36}\text{Kr} + 3\,^{1}_{0}\text{n}$$

Remember, this is just one of hundreds of possible reactions for ^{235}U fission. Following Example 11.1:

Mass of target	^{235}U	235.043930 u
Mass of products	^{141}Ba	−140.914406 u
	^{92}Kr	−91.92615 u
	net 2n = 2(1.008665 u) =	−2.017330 u
Mass lost		0.18604 u

Energy produced by the reaction:

$$Q = \frac{931.5\,\text{MeV}}{\text{u}} \times 0.18604\text{u} = 173.3\,\text{MeV}$$

Total kinetic energy and kinetic energies of the fission products:

$$K_T = 0.80 \times \frac{Z_{\text{Ba}} Z_{\text{Kr}}}{\sqrt[3]{A_{\text{Ba}}} + \sqrt[3]{A_{\text{Kr}}}} = 0.80 \times \frac{56 \times 36}{\sqrt[3]{141} + \sqrt[3]{92}} = 166\,\text{MeV}$$

$$K_{\text{Ba}} = K_T \left(\frac{A_{\text{Kr}}}{A_{\text{Ba}} + A_{\text{Kr}}} \right) = 166\,\text{MeV} \times \frac{92}{141 + 92} = 66\,\text{MeV}$$

$$K_{\text{Kr}} = K_T \left(\frac{A_{\text{Ba}}}{A_{\text{Ba}} + A_{\text{Kr}}} \right) = 166\,\text{MeV} \times \frac{141}{141 + 92} = 100\,\text{MeV}$$

TABLE 11.2
Approximate Average Energy Values (MeV) for the Thermal Neutron Fission of ^{235}U

Prompt energy	Fission products	167
	Neutrons	5
	Prompt gamma rays	6
Total prompt energy		178
Delayed energy	Beta particles	8
	Antineutrino particles	12
	Gamma rays	6
Total delayed energy		26
Total fission energy		204

Just as we saw with spontaneous fission, the total energy is carved up into kinetic and excitation energy of the fragments, neutron energy (velocity), and prompt gamma photons. In this example, only about 7 MeV separate the total energy produced (Q) and total kinetic energy (K_T). What more can we say about the neutrons, prompt gamma rays, and excitation energies of the fission products? Keep in mind: (1) The total kinetic energy is estimated—it could be a little bit off, and (2) the energy values for the emitted neutrons, prompt gamma rays, and excitation vary somewhat from one fission reaction to another. The math doesn't always work perfectly, but it does give us an idea of where all the energy goes.

The fission products in Example 11.2 will both undergo decay. The energy released as part of these decays is called **delayed energy**, as opposed to the **prompt energy** associated with the act of splitting the nucleus. The average values for both are detailed in Table 11.2.

The total kinetic energy for ^{235}U fission products averages about 167 MeV. ^{235}U emits an average of 2.4 neutrons per fission, each carrying about 2 MeV of energy, giving approximately 5 MeV of energy associated with the prompt neutrons. Finally, an average of six gamma rays are emitted per fission of ^{235}U, each packing about 1 MeV of energy.

Since delayed energy is all about the beta decay of the fission products, it is composed of the energy of the beta and antineutrino particles and the gamma photons emitted during decay(s) to a stable nuclide. The 6 MeV of energy listed for the delayed gamma rays consist of gamma emissions that are part of the beta decays and those that are emitted during de-excitation of the fission fragments. On average, the thermal neutron fission products of ^{235}U emit about 26 MeV as delayed energy. This brings the total average energy per fission to about 200 MeV. Let's see how our example compares.

Example 11.3: Estimate the delayed energy emitted by the fission reaction given in Example 11.2. All we need to do is look at the two fission products and add up the total energy of their isobaric decays to stable nuclides. The data below are rounded to the nearest tenth MeV because of the qualitative nature of this calculation.

$$^{141}_{56}\text{Ba} \xrightarrow[3.2\ \text{MeV}]{\beta^-} {}^{141}_{57}\text{La} \xrightarrow[2.5\ \text{MeV}]{\beta^-} {}^{141}_{58}\text{Ce} \xrightarrow[0.6\ \text{MeV}]{\beta^-} {}^{141}_{59}\text{Pr}$$

$$^{92}_{36}\text{Kr} \xrightarrow[6.0\ \text{MeV}]{\beta^-} {}^{92}_{37}\text{Rb} \xrightarrow[8.1\ \text{MeV}]{\beta^-} {}^{92}_{38}\text{Sr} \xrightarrow[2.0\ \text{MeV}]{\beta^-} {}^{92}_{39}\text{Y} \xrightarrow[3.6\ \text{MeV}]{\beta^-} {}^{92}_{39}\text{Zr}$$

| ^{141}Ba: | 3.2 MeV + 2.5 MeV + 0.6 MeV = 6.3 MeV |
| ^{92}Kr: | 6.0 MeV + 8.1 MeV + 2.0 MeV + 3.6 MeV = 19.7 MeV |

Total delayed energy: 6.3 MeV + 19.7 MeV = 26.0 MeV

Spot on!

Some of the fission products are so neutron-rich they sometimes emit delayed neutrons. Don't confuse delayed neutrons with delayed energy. As mentioned in Section 5.8.4, particles are sometimes emitted from an excited daughter, and are called delayed because they are emitted *after* the original decay has taken place. For example, 0.03% of all ^{92}Kr beta decays are followed by the emission of a delayed neutron from the excited ^{92}Rb daughter. Because a short time interval (seconds to minutes) occurs between the beta particle emission and the neutron emission, it is called a delayed neutron. These delayed neutrons are useful in controlling nuclear reactions in power plants, as discussed further in the next chapter.

Other projectiles can induce fission reactions. This makes some sense, since the reasonable occurrence of fission requires only a big, excited nucleus. How that compound nucleus is formed is somewhat irrelevant. Protons, alpha particles, deuterons, and other small nuclei have caused fission in a variety of heavy targets. Since all of these projectiles carry a positive charge, they need to be accelerated before they can strike the heavy target (to overcome the Coulomb barrier). High-energy photons can also induce fission (**photo fission**) simply by transferring their energy to the nucleus. Neutron-induced fission is the most heavily studied because of its applications in nuclear reactors and weapons (Chapters 4 and 12), and because neutrons do not need to overcome the Coulomb barrier to react with a nucleus.

11.3   FUSION

Fission reactions release energy because a large nuclide splits into two that have greater stability in terms of binding energy per nucleon (or lower average mass per nucleon, see Section 3.1). Likewise, the combination of two light nuclides forming a product with greater stability will also release energy. The combination of two light nuclides to form a heavier one is called **fusion**. Stars, like our Sun, are huge fusion reactors, and the energy they emit comes from fusion reactions. Some of the important fusion reactions taking place inside stars and the energy produced by each are given here.

$$^1_1H + ^1_1H \rightarrow ^2_1H + _{+1}e + \nu \qquad\qquad 0.42\ MeV$$
$$^1_1H + ^2_1H \rightarrow ^3_2He + \gamma \qquad\qquad 5.49\ MeV$$
$$^3_2He + ^3_2He \rightarrow ^4_2He + 2^1_1H \qquad\qquad 12.86\ MeV$$

It is so scorching hot inside the Sun that matter does not exist as atoms—nuclei and electrons are separated and therefore it is plasma. Hence, it is more appropriate to think of these reactions as taking place as nuclei, rather than atoms. The first reaction fuses two ^{1}H nuclei (protons) into a deuterium. Once formed, deuterium can combine with another proton to form ^{3}He (second reaction). If the first two reactions run twice, we'll have two ^{3}He nuclei to fuse together to form ^{4}He and two protons (third reaction). Because nuclei, rather than atoms, are reacting, it is sometimes best to use nuclear masses when performing energy calculations for fusion reactions. As an example, let's calculate the energy released by the first reaction, the fusion of two protons.

| Mass of reactants: | 2(proton mass) = 2(1.007276 u) = | 2.014552 u |
| Mass of products: | ^{2}H nucleus = 2.014102 u – 0.000549 u = | –2.013553 u |

$$\text{Positron mass} = \qquad\qquad\qquad \underline{-0.000549 \text{ u}}$$

Mass lost: 0.000450 u

Energy produced: $0.000450 \text{ u} \times \dfrac{931.5 \text{ MeV}}{\text{u}} = 0.419 \text{ MeV}$

Note that the neutrino mass is not necessary in the calculation because it is so ridiculously small. Note also that if this calculation were done with atomic masses, an incorrect value would be obtained. Using atomic masses for fusion reaction energy calculations only works when the "atoms" on both sides of the reaction arrow have the same number of electrons. Do you understand why?

The positrons produced in the proton–proton reaction will annihilate (there are plenty of electrons in the Sun!), producing even more energy.

$$_{+1}e + {}_{-1}e \rightarrow 2\gamma \qquad\qquad\qquad 1.022 \text{ MeV}$$

Taken together, the four reactions shown so far in this section make up what is called the **proton cycle**. If we add them together, we will see the *net* reaction and energy for the proton cycle. Remember that the first two reactions need to happen twice, and therefore the positron annihilation also happens twice per cycle.

$$
\begin{aligned}
&2({}_{1}^{1}\text{H} + {}_{1}^{1}\text{H} \rightarrow {}_{1}^{2}\text{H} + {}_{+1}e + \nu) && 2(0.42 \text{ MeV}) \\
&2({}_{1}^{1}\text{H} + {}_{1}^{2}\text{H} \rightarrow {}_{2}^{3}\text{He} + \gamma) && 2(5.49 \text{ MeV}) \\
&{}_{2}^{3}\text{He} + {}_{2}^{3}\text{He} \rightarrow {}_{2}^{4}\text{He} + 2{}_{1}^{1}\text{H} && 12.86 \text{ MeV} \\
&\underline{2({}_{+1}e + {}_{-1}e \rightarrow 2\gamma)} && \underline{2(1.022 \text{ MeV})} \\
&2\,{}_{-1}^{0}e + 4{}_{1}^{1}\text{H} \rightarrow {}_{2}^{4}\text{He} + 2\nu + 6\gamma && 26.7 \text{ MeV}
\end{aligned}
$$

The proton cycle accounts for roughly 90% of the Sun's energy output. Fortunately, our Sun is about 70% hydrogen (by mass), so it looks like there's still plenty of fuel up there. Notice that the Sun is also emitting a lot of gamma photons and neutrinos (see Figure 10.8). Are you scared?

We should also be thankful that the proton–proton reaction has a low cross section (probability of success). This slows down the entire proton cycle. If the cross section were higher, the Sun would burn hotter, and run out of hydrogen fuel faster—not an attractive prospect for life on Earth.

The fusion reactions shown in this section fit the general characteristics of the nuclear reactions studied in Chapter 10, but which is the projectile, and which is the target? These reactions happen inside the Sun because the reacting nuclei are all moving very quickly, i.e., they have a lot of energy. Their energy comes from the extraordinarily high temperatures inside the Sun, which allow these nuclei to overcome the Coulomb barrier and react. Because of this, fusion reactions are often called **thermonuclear reactions**. As such, the projectile/target labels seem inadequate here; it's more like two projectiles colliding with each other.

The proton cycle is generally thought to be impractical for fusion power reactors here on Earth, primarily due to its relatively high **ignition** temperature—the temperature needed to start a fusion reaction. To get an idea of how hot it needs to be to get the proton–proton reaction going, let's calculate its effective Coulomb barrier.

$$_{1}^{1}\text{H} + {}_{1}^{1}\text{H} \rightarrow {}_{1}^{2}\text{H} + {}_{+1}e$$

$$E_{ecb} \approx 1.11 \times \left(\frac{1+1}{1}\right) \times \left(\frac{1 \times 1}{\sqrt[3]{1} + \sqrt[3]{1}}\right) = 1.11 \text{ MeV}$$

This value can be converted to temperature using Boltzmann's constant.

$$1.11 \text{ MeV} \times \frac{10^6 \text{ eV}}{\text{MeV}} \times \frac{K}{8.63 \times 10^{-5} \text{eV}} = 1.29 \times 10^{10} \text{ K}$$

According to these calculations, the Sun would have to be nearly 13 billion Kelvin to get the proton cycle to proceed in good yield. The surface of the Sun is about 6000 K, and the core is believed to only reach ~15 million K. Fortunately, plasma, like other forms of matter, contains individual particles that are moving both faster and slower than they should be at a particular temperature. Temperature reflects the average kinetic energy of all particles, and there is always a distribution of energies about the mean value.[1] Also remember that both particles on the left side of the arrow are moving, so, if they hit head-on, they only need half as much energy each. Therefore, the Sun will have some particles moving fast enough to overcome the Coulomb barrier and fuse. Finally, remember that nuclear reactions can take place even if the particles lack the minimum energy— through tunneling. This has positive implications for human-built fusion reactors here on Earth. These reactors do not need to get quite as hot as calculated using Boltzmann's constant, because some reactions will take place at lower temperatures due to tunneling.

Tunneling may have led two chemists to conclude they had observed fusion in 1989. They observed heat being generated during an electrochemical experiment involving a palladium electrode in deuterated water (D_2O). They dubbed the phenomenon "cold fusion," then, circumventing the normal scientific peer review process, announced their results at a press conference. A great deal of popular media attention was given to this result, which seemed to promise an easy fix for our increasing energy needs. Several other scientists tried to duplicate these remarkable results. At best, these efforts could be described as inconsistent, and at worst they were a clear repudiation of the original results. The scientists were discredited, and the entire scientific community suffered a loss of public confidence. Some research has continued in this area, but proof of low-energy nuclear fusion remains elusive. It is unfortunate that the term "cold fusion" was applied to these poorly interpreted experiments, as it is a *bona fide* area of nuclear science (see Section 11.6).

11.4   NUCLEAR MODELS APPLIED TO FISSION AND FUSION

The liquid drop model of the nucleus (Section 6.2) works quite well when applied to nuclear fusion and fission. Remember that the liquid drop model imagines the nucleus as if it were a drop of water with each nucleon feeling an attractive force to its nearest neighbors, just like water molecules in a droplet. The liquid drop model also accounts for the Coulomb repulsion of the protons, and the instability caused when numbers of one type of nucleon start to vastly outnumber the other. Putting all of the attractive and repulsive forces together allows the calculation of nuclear binding energy, and produces Figure 3.1, the nuclear binding energy curve. This curve provides an understanding of the energy released during fission and fusion. Movement from either extreme toward A ~50–100 results in an increase in binding energy per nucleon and is the energy released by these processes.

---

1   It is an *asymmetric* distribution. It tails off more gradually at the high-energy end, insuring at least some will have sufficient energy, even at much lower temperatures.

Alternatively, the semi-empirical mass equation (Equation 6.2) can be used to determine the average mass of a nucleon, generating Figure 3.2. Instability toward fission and fusion can then be understood in terms of individual nucleons losing mass during these processes, releasing it as energy. As smaller nuclei fuse to a larger one, a larger proportion of nucleons are below the surface and the surface term becomes less significant (Figure 6.3), increasing binding energy per nucleon and decreasing average nucleon mass. When a heavier nuclide splits into smaller ones, Coulomb repulsion is dramatically decreased (Figure 6.3), again resulting in higher binding energy per nucleon and decreased nucleon masses.

The liquid drop model is also helpful in visualizing the process of nuclear fission. If a spherical nucleus starts to elongate, what happens to the various terms in the semi-empirical mass equation (Equation 6.2)? Surface area will increase, so overall the total amount of strong force holding the nucleus together will decrease. Coulomb repulsion will also decrease because the protons will, on average, be further away from each other. Whether the nucleus splits depends on the extent of elongation (how much additional energy did the neutron bring with it?), and the size of the nucleus (how many protons are there *and* how many nucleons do we still have completely surrounded by others?). It turns out that a simple combination of the Coulomb and surface terms from Equation 6.2 can give the parameter Z^2/A, which is an indicator of the nucleus' relative probability of undergoing spontaneous fission. Let's calculate Z^2/A for ^{239}Pu and ^{209}Bi.

$$^{239}_{94}\text{Pu} \quad \frac{94^2}{239} = 37 \qquad ^{209}_{83}\text{Bi} \quad \frac{83^2}{209} = 33$$

^{209}Bi is much less likely to undergo spontaneous fission than ^{239}Pu. Sure enough, spontaneous fission is a non-existent decay mode for ^{209}Bi, but occurs in $3 \times 10^{-10}\%$ of all ^{239}Pu decays.

The shell model also facilitates understanding of fission and fusion, particularly in terms of the products of these reactions. The shell model demonstrates how there's extra stability (greater binding energy or less mass per nucleon) at shell closures. This is seen in fission product distributions as a distinct preference in forming those with one or more magic numbers of nucleons. The same is true with fusion, as a doubly magic product like ^{4}He is formed in more than its share of fusion reactions.

11.5 STELLAR NUCLEOSYNTHESIS

How were the elements we find here on Earth made? In the stars! Fusion reactions can make nuclides up to ^{56}Fe. Remember, this is the pinnacle of nuclear stability (binding energy per nucleon—Figure 3.1, or mass per nucleon—Figure 3.2), so fusion up to the peak is exoergic, and is endoergic after.

After ^{56}Fe, stars rely on neutron capture to make most of the heavier nuclides. The neutrons come from nuclear reactions taking place in the Sun like this one.

$$^{21}\text{Ne} + {}^4\text{He} \rightarrow {}^{24}\text{Mg} + {}^1\text{n}$$

^{56}Fe will capture neutrons until it forms ^{59}Fe, which decays to ^{59}Co. ^{59}Co will capture a neutron and form ^{60}Co, which decays to ^{60}Ni. Through this combination of n,γ reactions and beta decay, almost all the naturally occurring nuclides from $A = 56$ through $A = 210$ can be prepared. This series of reactions is called the **s-process**, where "s" stands for slow. It's slow because the neutron flux inside of stars is not all that great, and many of the cross sections are small. It takes a long time for significant amounts of nuclides beyond ^{56}Fe to build up.

The s-process also fails to explain the existence of a few stable, but somewhat proton-rich, nuclides such as ^{124}Xe. ^{124}Xe cannot be formed through n,γ reactions followed by beta decay because it is "shielded" by stable ^{124}Te. Any neutron-rich, $A = 124$ nuclide will undergo beta decay until it forms ^{124}Te, which cannot then form ^{124}Xe through decay. Nuclides such as ^{124}Xe are formed through the **p-process**. Two types of reactions are postulated to be part of the p-process, γ,n or p,x. Either one can lead to the formation of more proton-rich naturally occurring nuclides. Some of the other nuclides in the same boat are ^{78}Kr, ^{112}Sn, ^{120}Te, ^{144}Sm, and ^{184}Os. These nuclides are all found in lower percent abundances than the more neutron-rich, stable isobars because the p-process is rather unlikely. This is generally true for pairs of naturally occurring isobars between $A = 70$ and $A = 204$—the isobar with more neutrons will often have a higher percent abundance because it can be formed through the more likely s-process.

The s-process cannot make nuclides beyond $A = 210$. It terminates in the following cycle of reactions:

$$^{209}\text{Bi} + {}^{1}\text{n} \rightarrow {}^{210}\text{Bi} + \gamma$$

$$^{210}\text{Bi} \rightarrow {}^{210}\text{Po} + {}_{-1}\text{e}$$

$$^{210}\text{Po} \rightarrow {}^{206}\text{Pb} + \alpha$$

$$^{206}\text{Pb} + {}^{1}\text{n} \rightarrow {}^{207}\text{Pb} + \gamma$$

$$^{207}\text{Pb} + {}^{1}\text{n} \rightarrow {}^{208}\text{Pb} + \gamma$$

$$^{208}\text{Pb} + {}^{1}\text{n} \rightarrow {}^{209}\text{Pb} + \gamma$$

$$^{209}\text{Pb} \rightarrow {}^{209}\text{Bi} + {}_{-1}\text{e}$$

So, how are elements like thorium and uranium made? The rapid process (**r-process**) makes them. As its name implies, it is a quick series of neutron absorptions, leading to an extremely neutron-rich nuclide, which then undergoes several beta decays until the N/Z ratio optimizes. An exceptionally intense neutron flux is needed to accomplish this, as the neutrons need to be continually added, faster than the newly formed nuclides can decay. Such a flux is not possible under normal conditions in a star. It is believed that these fluxes are possible in supernovae. Our solar system was formed from the remnants of a supernova, which explains why we have these heavy elements in some abundance. Therefore, we live in a second- (or third-) generation solar system. Makes you wonder if there are alien civilizations out there in older solar systems …

11.6 SYNTHESIS OF UNKNOWN ELEMENTS

Humans have done a pretty good job preparing isotopes of elements not found on Earth. These elements had been unknown because all of their isotopes have half-lives that are short when compared to the age of the solar system (~4.6 billion years) and they are not part of one of the naturally occurring decay series (Section 1.7). Therefore, if they were formed in the sun that preceded ours, they have since decayed away. The first "artificial" element in the periodic table is technetium (Tc, $Z = 43$), which was first prepared in 1945 by accelerating a deuteron in a cyclotron and slamming it into a molybdenum target. Its name comes from the Greek word *technetos*, which means artificial. The preparation of technetium was a validation of the configuration of the periodic table as it filled a void in the middle of the transition metals. Technetium has since been made in large quantities as a fission product in nuclear reactors. A significant amount of its chemistry has been investigated, and it generally parallels the chemistry of its heavier congener rhenium. It is an obscure irony that technetium is now more plentiful on Earth than rhenium.

Promethium (Pm, $Z = 61$) is the only other element before lead ($Z < 82$) that was "missing" from our current periodic table because it did not exist on Earth. It was first observed among the fission products of ^{235}U in 1945. It is named for the mythological Greek titan who stole fire from the gods and gave it to humans.

Both Tc and Pm have an odd number of protons, which means that any of their isotopes with an odd number of neutrons (oo) will be unstable; but why are there no stable oe isotopes of these elements? A simple way to look at Tc is to state that $^{98}_{43}Tc_{55}$ ($N/Z = 1.28$) has the best N/Z ratio and the longest half-life of all the known Tc isotopes, but it is oo, and therefore cannot be stable. What about $^{97}_{43}Tc_{54}$ ($N/Z = 1.26$) or $^{99}_{43}Tc_{56}$ ($N/Z = 1.30$)? They have even numbers of neutrons and pretty good N/Z ratios. Examination of the $A = 97$ and $A = 99$ isobars shows that $^{97}_{42}Mo_{55}$ and $^{99}_{44}Ru_{55}$ are the only stable nuclides for those sets of isobars. As we saw in Section 5.6 (The Valley of Beta Stability), when $A = $ odd there can only be one stable isobar. Additionally, a stable ^{97}Tc or ^{99}Tc would violate Mattauch's rule of no stable, neighboring isobars.

What about the anthropogenic syntheses of the elements beyond uranium? A partial answer was given in Example 10.2 where a plutonium target was hit with an alpha particle to produce a curium isotope.

$$^{240}_{94}Pu + {}^{4}_{2}He \rightarrow {}^{243}_{96}Cm + {}^{1}n$$

So long as the target nuclide can be isolated and has a reasonably long half-life, we can perform lots of reactions like this one to make nuclides of the elements beyond uranium. We can even mimic the s-process, making new nuclides through n,γ reactions and beta decay, such as the preparation of ^{239}Pu from ^{238}U below.

$$^{238}_{92}U + {}^{1}n \rightarrow {}^{239}_{92}U + \gamma$$

$$^{239}_{92}U \rightarrow {}^{239}_{93}Np + {}_{-1}e$$

$$^{239}_{93}Np \rightarrow {}^{239}_{94}Pu + {}_{-1}e$$

The r-process is mimicked in the explosions of thermonuclear weapons. Tremendous neutron fluxes are briefly created in these explosions that can add several neutrons to ^{238}U, which is often used as a casing material for these weapons. The first detonation of a thermonuclear device occurred in 1952 on Eniwetok Island in the South Pacific. Two elements that were first observed in the debris following this explosion were einsteinium (Es) and fermium (Fm). As many as 17 neutrons were absorbed by ^{238}U atoms in the bomb's casing during the explosion, followed by a series of beta decays to form the observed nuclides.

We can only get so far with the preceding reactions. Generally speaking, the half-lives of nuclides get shorter as atomic number increases beyond uranium. This makes intuitive sense—the nuclides are already too big and making them bigger should push them further from stability.[2] Larger nuclides mean the strong force is having an increasingly difficult time keeping the nucleus together under the burgeoning Coulomb repulsion of increasing numbers of protons. At a certain point, the half-lives of potential targets will get too short, so the projectile will have to be bigger if we want to make super heavy nuclides. For example, ^{254}No can be prepared by hitting ^{246}Cm with a ^{12}C projectile.

$$^{246}_{96}Cm + {}^{12}_{6}C \rightarrow {}^{254}_{102}No + 4{}^{1}n$$

Increasing the atomic number (Z) of the projectile increases the Coulomb barrier and, therefore, the minimum energy for the reaction to proceed in good yield. This also means that the excitation energy of the compound nucleus will also increase. We'll soon reach a point where the compound nucleus will have too much excitation energy and will fall apart in ways that make it difficult to

---

2   This isn't always true; an ee that is heavier than a nearby oo nuclide might have a longer half-life. We could also make new nuclides (heavier than those known today) with magic numbers of nucleons that have anomalously long half-lives.

determine if it was actually made. A more subtle approach is required. Reflecting these challenges, no new elements were discovered between 2010 and 2022.

If the projectile has less than the minimum energy required for the product to be formed in good yield, the product can still be formed, there just won't be very much of it. Thankfully, nuclear decay can be detected, even if only a handful of decays occur—so we really only need to make a few product nuclides. This approach has traditionally been referred to as "cold fusion," while the reaction of ^{246}Cm with ^{12}C would be called hot fusion. As mentioned at the end of Section 11.3, cold fusion has unfortunately become more widely applied to alleged, low-energy nuclear reactions involving low Z nuclides. The first atoms of element 111 (roentgenium, Rg) were made with the cold fusion method, bombarding a ^{209}Bi target with ^{64}Ni.

$$^{209}_{83}\text{Bi} + ^{64}_{28}\text{Ni} \rightarrow ^{272}_{111}\text{Rg} + ^{1}\text{n}$$

QUESTIONS

11.1   Calculate the partial half-lives for the two modes of decay (alpha emission and spontaneous fission) for ^{238}U. How many alpha and spontaneous fission decays take place every minute inside a 100 g block of ^{238}U?

11.2   Which is more likely to undergo spontaneous fission, ^{256}Fm or ^{254}No? Briefly explain.

11.3   Which is more likely to undergo spontaneous fission, ^{256}Fm or ^{258}Md? Briefly explain.

11.4 $^{256}_{100}$Fm can decay by spontaneous fission yielding two atoms of $^{128}_{50}$Sn. Calculate the total energy released by this decay and estimate the kinetic energy of the tin fragments. Why is symmetric fission of ^{256}Fm a reasonable possibility?

11.5 $^{240}_{94}$Pu can decay by spontaneous fission yielding ^{103}Mo and ^{134}Te. Write a balanced equation and estimate the kinetic energy of the fragments. Would you expect the fission yield for these products to be high? Why?

11.6   Why are nuclides with N or $Z = 50$ or 82 more likely to be formed in fission reactions?

11.7   Is ^{238}U fissile, fissionable, and/or fertile? Briefly explain.

11.8   Just like fission, there's a barrier to alpha decay. Explain the nature of this barrier and why alpha decay spontaneously occurs despite this barrier.

11.9   Write out the decay series that is generated by the formation of ^{140}Cs as a fission product. Calculate the N/Z ratio for each isobar formed.

11.10 Briefly explain why not all heavy, eo nuclides undergo thermal neutron fission.

11.11 Calculate the excitation energy of the compound nucleus formed by the capture of a thermal neutron by ^{239}Pu and ^{240}Pu.

11.12 If the fission barrier for ^{240}Pu is 5.5 MeV, decide which (if either) nuclide, in the previous question, will fission. If either nuclide will not fission, how much energy would need to be brought in by the neutron to overcome the fission barrier?

11.13 Calculate the minimum neutron energy needed for fission of ^{238}U to proceed in good yield. Don't forget to account for the kinetic energy of the compound nucleus.

11.14 ^{233}U has a critical mass that is much lower than ^{235}U. What can you conclude about the number of neutrons produced per fission of ^{233}U?

11.15 One of the possible reactions for thermal neutron fission of ^{239}Pu produces ^{134}Te and ^{103}Mo. Write out a balanced equation, calculate the total (prompt and delayed) energy released, and estimate the kinetic energies of the products.

11.16 Using the reaction in the previous question, calculate the nuclear binding energy for ^{239}Pu and the products. Are the differences between reactants and products consistent with the results above? Explain.

11.17 Ten years after some pure ^{235}U undergoes fission, what are the major radioactive nuclides present?

11.18 The three fission reactions given in Section 9.4.1 are reproduced here. Would you expect these reactions to have high fission yields? Briefly explain.

$$^{235}U + {}^1n \rightarrow {}^{135}Sn + {}^{99}Mo + 2{}^1n$$

$$^{235}U + {}^1n \rightarrow {}^{131}I + {}^{102}Y + 3{}^1n$$

$$^{235}U + {}^1n \rightarrow {}^{133}Xe + {}^{100}Sr + 3{}^1n$$

11.19 Define the following: fission products, fission yield, thermonuclear reaction, and cold fusion.

11.20 Why are fission of heavy nuclides and fusion of light nuclides generally exoergic?

11.21 Calculate the amount of energy produced in the following fusion (net) reaction based on the mass differences. Also, determine how many MeV/u are produced by this reaction. How does this value compare to a typical fission reaction?

$$4{}_1^1H \rightarrow {}_2^4He + 2{}_{+1}e + 2v$$

11.22 Home fusion enthusiasts claim they can perform the reaction below in their garage by accelerating deuterons through a 45 kV electric field. Estimate the effective Coulomb barrier for this reaction and comment on the validity of their claims. If this reaction is taking place in someone's garage, what safety concerns might result? Also calculate the energy produced by this reaction.

$$2{}_1^2H \rightarrow {}_2^3He + {}^1n$$

11.23 Why can't atomic masses be used in energy calculations for certain fusion reactions?

11.24 In 2002, a scientist at Oak Ridge National Laboratory claimed he was observing the fusion reaction below by exposing deuterated acetone $((CD_3)_2CO)$ to 14 MeV neutrons and ultrasound. Suggest another reaction that could've produced the tritium.

$$2{}_1^2H \rightarrow {}_1^3H + {}_1^1H$$

11.25 Calculate Z^2/A for ^{232}Th, ^{236}U, ^{246}Cm, ^{252}Cf, and ^{256}Fm. What trend is observed?

11.26 Starting with ^{131}Xe, give the likely nuclear reactions that might produce ^{132}Xe and ^{132}Ba in stars. Look up the percent abundances of ^{132}Xe and ^{132}Ba. Can you explain the difference between them?

11.27 Rationalize why no stable isotopes of Pm exist.

11.28 Compared to other elements near it on the periodic table, why is gold relatively rare in our solar system? Why are iron and lead so common?

11.29 ^{244}Pu was detected for the first time following the thermonuclear detonation on Eniwetok Island. Show the probable nuclear reactions and decays that likely led to the formation of this nuclide from ^{238}U. Also show how ^{235}U might be formed in a supernova from ^{209}Bi— this time, simplify the process by writing it as a single nuclear reaction.

11.30 ^{266}Mt can be prepared by the reaction of ^{209}Bi with ^{58}Fe. Write out the complete nuclear reaction for this synthesis. What projectile energy would be required to prepare ^{266}Mt in good yield using this reaction? What would the excitation energy of the compound nucleus be in this case? Do you think this reaction is an example of hot or cold fusion? Briefly explain. If the excitation energy of the compound nucleus is limited to 13 MeV, what would the projectile energy be?

12 Applications of Nuclear Science III

More About Nuclear Reactors

Chapter 4 provides an introduction to nuclear power and this chapter seeks to explore it a little deeper. This chapter follows those on nuclear reactions (Chapter 10) and nuclear fission (Chapter 11) because it applies some of that stuff to nuclear power reactors. We'll also take a more in-depth look at the different types of nuclear reactors, their safety systems, and the causes of some of the more significant nuke plant accidents. We'll also look at what makes modern nuclear power plants better than those built in the 1960s and 1970s. Finally, we'll examine ideas for future nuclear power reactors including fusion power.

12.1 REACTIONS IN REACTORS

It may seem obvious that "reactions" would take place in "reactors." Even so, a little redundancy is usually a good thing when it comes to nuclear power. The most important reaction that takes place inside a nuclear reactor is the neutron-induced fission of ^{235}U. Not only does it produce a lot of energy by splitting the ^{235}U nucleus into two fission fragments, it also frees up a couple (on average) of neutrons. These newly freed neutrons can then go on to cause more fission reactions to take place, creating a nuclear chain reaction as illustrated in Figure 12.1 (yes, it's the same as Figure 4.1, but it *is* quite pretty!).

Whether a neutron captured by ^{235}U leads to fission or to an n,gamma reaction is governed by their respective cross sections. Table 12.1 (partly a reproduction of Table 11.1) gives the fission and n,gamma cross sections for various fissile and fertile nuclides. Notice that both cross sections are large for ^{235}U (the most common nuclear fuel today), but fission is clearly the more probable outcome of a thermal neutron running into a ^{235}U nucleus, which is why it is fissile. Table 12.1 also lists the ratio of fission cross section to n,gamma cross section (σ_f/σ_γ) to provide an easy comparison of the relative probabilities of these two reactions. Notice that, of the fissile nuclides, ^{233}U has the highest ratio, making it an attractive nuclear fuel. More on this in Section 12.2.5.

Remember also (Section 4.1.3) that nuclear fuel nowadays is only ~4–5% ^{235}U, the rest is ^{238}U. What will ^{238}U do with all the neutrons flying around inside of a nuclear reactor? A quick look at Table 12.1 tells us ^{238}U will most likely undergo an n,gamma reaction, eventually forming ^{239}Pu (Section 4.1.3). Even so, the cross section for this reaction is not very high, so only a relatively small percentage of the ^{238}U reacts.

Successful management of a nuclear power reactor is all about what happens to the neutrons, i.e., how many induce a fission reaction versus how many end up doing something else. To properly maintain the chain reaction, at least one of the neutrons produced by a fission reaction should cause another fission reaction to occur. If too many neutrons are lost to other reactions, the chain reaction will die out. If too many cause new fission reactions, the chain reaction can get out of control,

DOI: 10.1201/9781003303787-12

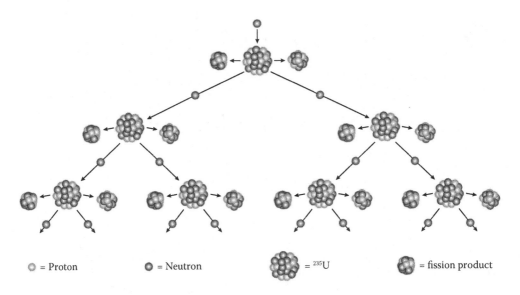

○ = Proton   ● = Neutron   ◉ = ^{235}U   ◉ = fission product

FIGURE 12.1   Nuclear chain reaction.

TABLE 12.1
Select Data for Some Fissile and Fertile Nuclides

Nuclide	σ_f (b)	σ_γ (b)	σ_f/σ_γ	$\overline{v}$
^{232}Th	0.000003	7.34	4×10^{-7}	
^{233}U	531	46		3.1
^{235}U	585	98		2.4
^{238}U	0.000005	2.7	2×10^{-6}	
^{239}Pu	750	271		2.9

and the reactor will overheat. Therefore, it's probably a good idea to know the average number of neutrons produced per fission reaction. These values ($\overline{v}$) are listed in the final column of Table 12.1, but only for the fissile nuclides.

We now have another way to think of **critical mass**. Remember (Section 4.1.1) that it is the minimum amount of material necessary to obtain a self-sustaining fission chain reaction. So long as we're talking about the same overall shape of pure material, this value will depend on the average number of neutrons produced per fission, and the probability of fission (σ_f). For example, ^{239}Pu produces more neutrons per fission than ^{235}U, and has a higher fission cross section (Table 12.1), and therefore has a smaller critical mass than ^{235}U.

In order for a reactor to run for any reasonable amount of time, much more than a critical mass of fuel needs to be placed into the reactor. Placing a supercritical mass all in one place would result in a rapidly accelerating fission chain reaction, a lot of neutrons and heat, and would make a bit of a mess. To prevent this, the fuel rods are spread out in the reactor, and, initially, are separated by control rods. As the reactor starts up, the rods are partially removed, just enough to obtain criticality, and they are gradually withdrawn over the following year or so as the ^{235}U gets used up. Since control rods need to absorb neutrons, they are typically made of elements that have a high cross section for various n,x reactions, such as cadmium ($\sigma_a = 2520$ b), boron ($\sigma_a = 760$ b), and gadolinium ($\sigma_a = 49,000$ b). These elements (B and Gd in particular) can also be dissolved in the moderator and/or coolant and their concentration adjusted to maintain the proper neutron flux within the reactor.

Remember that the purpose of the moderator in a nuclear reactor is to slow the neutrons down (Section 4.1.2). This is because slower neutrons are generally more likely to initiate a reaction (Section 10.2). Unfortunately, the average energy of a neutron produced in the fission of ^{235}U is 2 MeV (Section 11.2). If more fission reactions are desirable, it'd be best to slow them down. This is another good reason to separate the fuel rods. Aside from being able to place the control rods between the fuel rods, it would also be a good idea to put some stuff in there that will slow down the neutrons. An ideal moderator has a relatively low neutron absorption cross section (σ_a) yet has a high neutron scattering cross section (σ_s), and is relatively inexpensive. Water (H_2O) is commonly used as a moderating material as its σ_a is 0.66 b, its σ_s is 49 b, and it is very cost-effective. Other moderator materials are heavy water (D_2O, where D = ^2H, σ_a = 0.0013 b, σ_s = 10.6 b) and graphite (a form of pure carbon, σ_a = 0.0035 b, σ_s = 4.7 b). Based solely on their very low neutron absorption numbers, both D_2O and graphite would appear to be superior moderators than regular, or **"light" water**, but H_2O is much less expensive, and can also be used as a coolant for the reactor.

The coolant carries the thermal energy away from the fuel to be used in power generation (Section 4.1.2). Water (H_2O) is by far the most common coolant used because of its great thermal properties. Water is somewhat unusual in that it takes a great deal of thermal energy (heat) to raise its temperature by a degree. In other words, it has a rather high specific heat capacity. To transfer heat, water's a great choice because it soaks up heat like a sponge! Water is also easy to come by.

That's wonderful, but what about neutron reactions? Well, water has two elements, and either can react with neutrons. Naturally occurring hydrogen consists of two stable isotopes (99.99% ^1H and 0.01% ^2H) and a ridiculously small amount of tritium (^3H, $t_{1/2}$ = 12 a). Oxygen occurs as three stable isotopes (99.76% ^{16}O, 0.04% ^{17}O, and 0.20% ^{18}O). ^1H, ^{16}O, and ^{17}O can all undergo n,gamma reactions to form another stable isotope, and, given the high percent abundances of ^1H and ^{16}O and reasonable cross sections, these are the most likely neutron-induced reactions involving the coolant. The three most likely reactions yielding *radioactive* products are listed here.

$$^2_1H + {}^1n \rightarrow {}^3_1H + \gamma$$

$$^{17}_8O + {}^1n \rightarrow {}^{14}_6C + {}^4_2He$$

$$^{18}_8O + {}^1n \rightarrow {}^{19}_8O + \gamma$$

Deuterium can capture a neutron, forming tritium. ^{17}O undergoes an n,alpha reaction, forming ^{14}C ($t_{1/2}$ = 5715 a). Finally, ^{18}O reacts with neutrons to form ^{19}O ($t_{1/2}$ = 27 s). All three radioactive products decay by beta emission to form stable daughters. The first two (^3H and ^{14}C) are also naturally occurring. ^{19}O has a relatively short half-life, and therefore does not represent a long-term radioactive hazard in the coolant. Another source of radioactive contamination could be small amounts of fission products leaked into the coolant from faulty fuel rods.

Neutrons can also interact with structural components of the reactor. Most significant is the neutron activation of cobalt in steel.

$$^{59}_{27}Co + {}^1n \rightarrow {}^{60}_{27}Co + \gamma$$

^{60}Co is particularly nasty because it has a half-life of 5.3 years and emits two high-energy gamma photons with every beta decay.

Finally, we should consider the fact that neutrons need to pass through the zirconium alloy (commonly called zircaloy) that makes up the fuel rods. Zirconium typically makes up at least 95% of the alloy because Zr is nearly transparent to neutrons (σ_a = 0.18 b) and maintains its structural integrity even under high temperatures and in high-radiation environments. While Zr may be unre-active toward neutrons, it can react *chemically* with water to form zirconium oxide and hydrogen.

$$Zr(s) + 2H_2O(g) \rightarrow ZrO_2(s) + 2H_2(g)$$

While this reaction is spontaneous at room temperature, it is inhibited by the formation of a thin layer of zirconium oxide on the surface of the metal. It only becomes a problem in a nuclear reactor at relatively high temperatures (above ~1200 °C), i.e., when the fuel rods are no longer underwater, but are in contact with really hot steam. This reaction is rather exothermic, which exacerbates situations where heat removal from the reactor core is impaired. We're talking nuclear meltdown.

As mentioned in Section 11.2, not all neutrons produced in a nuclear reactor originate in the fission reaction. Some of the fission products emit delayed neutrons a period of time after the fission reaction occurs. These neutrons can also cause more fuel to undergo fission and are important in controlling the rate of fission reactions in the reactor. If fission were the only source of neutrons (**prompt neutrons**), the reactor operators would only have a fraction of a second to respond if the rate of fission reactions suddenly started to increase. Even computer-operated control rods would not have enough time to respond, and meltdowns would be common. The fact that a small fraction of all the neutrons flying around the reactor are produced some time after fission occurs allows operators and automated safety systems a minute or two to respond to a sudden escalation in the chain reaction. The delayed neutrons provide a nice buffer to unanticipated changes in the number of fission reactions per unit time.

12.2   OTHER REACTOR TYPES

The attentive reader will have noticed that pressurized water reactors (PWRs) and boiling water reactor (BWRs) only make up roughly 84% of the operable nuclear power plants in the world as of mid-2022. The remaining 16% are mostly pressurized heavy water reactors (PHWRs, ~11% of all reactors), gas-cooled reactors (GCR, ~2% of all reactors), and light water graphite reactors (RBMKs,[1] ~2% of all reactors). The basic operating principles of these reactors will be discussed in this section along with ideas for reactors that are currently being researched and developed. Along the way, we'll also learn a little bit more about PWRs and BWRs.

12.2.1   PRESSURIZED HEAVY WATER REACTORS (PHWRs)

The only difference in nomenclature from pressurized water reactors (PWRs) is insertion of the word "heavy." P*H*WRs use heavy water (D_2O, where $D = {}^2H$) rather than regular (or "light") water as a moderator, and, typically, as a coolant. The advantage to using deuterium rather than hydrogen in the moderator is that it has a much lower neutron absorption cross section (σ_a). This means that far fewer neutrons will be absorbed by the moderator/coolant and are therefore still available to cause fission. This is often referred to as an improved "neutron economy."

There are a few significant ramifications to having a good neutron economy. Most important is that these reactors can run on natural uranium, i.e., they don't require enriched uranium. With more neutrons flying around, the concentration of fissile ${}^{235}U$ doesn't need to be as high, enough will still get hit by a neutron and split. Enrichment is expensive, so PHWRs are cheaper to fuel. Greater numbers of neutrons also means more ${}^{238}U$ atoms will get hit, form ${}^{239}U$, then decay to form fissile

---

1   RBMK is the acronym commonly used to characterize reactors cooled with light water and moderated by graphite because almost all reactors of this type were designed in the old Soviet Union. RBMK stands for reaktor bolshoy moschnosty kanalny which translates to English as high-output channel reactor.

^{239}Pu. In fact, this happens roughly twice as often in a heavy water reactor than in a light water reactor. This means more power could be generated from the same amount of uranium.

Sounds great! So why aren't there more PHWRs? The heavy water doesn't come easily. Even though deuterium is a naturally occurring hydrogen isotope, separating it from the rest of hydrogen is costly and a little messy. Since several hundred tons are required for moderator and coolant, this significantly adds to the cost of constructing a PHWR. These reactors also tend to be more complex than other types, which also increases the upfront costs. These additional initial expenses are offset by saving on fuel costs, so PHWRs end up costing about the same to build and operate as other reactors. Another drawback is that with so many ^{2}H atoms being bombarded with neutrons, some tritium (^{3}H) is formed. As a result, PHWRs need to routinely remove the ^{3}H as it builds up. Tritium is also formed in light water reactors, but at a much lower rate than in heavy water reactors.

The most common commercial PHWR is the Canadian deuterium uranium (CANDU) reactor, which is represented schematically in Figure 12.2. The CANDU reactor is very similar to other pressurized (light) water reactors (PWRs) in that the coolant is kept under pressure, exchanges its heat with another water loop, which boils and drives a steam turbine to generate electricity. The differences are all in the reactor. The main difference is that, instead of enclosing all the fuel in one pressure vessel, the fuel is divided up among hundreds of pressure tubes (only three are shown in Figure 12.2). These tubes are placed horizontally (instead of vertically, as in other reactors) inside a large tank filled with D_2O. This tank is called a **calandria**. The calandria is thermally insulated from the pressure tubes by a larger tube surrounding each pressure tube filled with CO_2. The heavy water in the calandria does not heat up much, it simply acts as the moderator for neutrons traveling through it.

Each pressure tube is loaded with several **fuel bundles**. These are made up of several short fuel rods, bundled together in the shape of a log. The coolant (also D_2O, but could be H_2O or some other stuff) flows over the fuel bundles in the pressure tubes. With this design, moderator and coolant are both water (D_2O), but are physically separated.

The multiplicity of pressure tubes in a CANDU reactor has some advantages. Other reactor designs only have a single reactor pressure vessel and must shut down every year or two to refuel. This takes some time and may not always be convenient. Having hundreds of pressure tubes allows CANDU reactors to refuel while still running. Just shut down one pressure tube and swap out its fuel bundles. CANDU operators also move fuel around within the reactor to maintain more consistent power production throughout the reactor.

FIGURE 12.2   Schematic of a CANDU reactor.

With all the shuffling of fuel bundles, the total amount of fissile material in a CANDU reactor stays relatively constant. This is not the case in other reactor designs which are loaded with enriched uranium at startup, which gradually gets "burned up" until the reactor is shut down for refueling a year or two later. To maintain consistent power production in a PWR or BWR during this time, control rods are (very!) gradually removed during that time. Because the CANDU has roughly the same amount of fissile material all the time, there's not much need for control rods. They are only needed to shut the reactor down. Again, this improves the CANDU's neutron economy, i.e., fewer neutrons are absorbed by the control rods during operation of the reactor and are therefore available to cause fission or create more fissile material ($^{238}U \rightarrow \rightarrow ^{239}Pu$).

12.2.2 GAS-COOLED REACTORS (GCRs)

As their name implies, the coolant is now a gas rather than a liquid. In the reactors we've considered so far, the coolant and moderator have been the same, either H_2O or D_2O. Because of its low density, a gas is not a very efficient moderator, so something else must be used as a moderator, and graphite is the material of choice. Both CO_2 and He have been used as coolants in GCRs, but He is preferred because of its extreme lack of chemical reactivity. Helium also has a low neutron absorption cross section (σ_a = 7 mb for thermal neutrons). Other than the gas coolant and the graphite moderator, these reactors are very similar to PWRs.

The main advantage of a gas-cooled reactor is that a gas can be heated much higher than water. This makes some of the thermal transfer from fission to electricity more efficient (less heat lost elsewhere). Heating to about 700°C is common. The super-hot gas can also be used to run turbines directly to achieve even higher thermal efficiencies. Finally, the heat could be used industrially (to heat chemical reactions, e.g., splitting water into H_2 and O_2), or to simply provide heat for a group of buildings.

Like PHWRs, gas-cooled reactors have a good neutron economy. Because of this GCRs can use unusual fuel mixtures such as fissile ^{235}U and fertile ^{232}Th. Fuel pellets can be coated with ceramic materials such as pyrolytic carbon or silicon carbide. These materials are stable to much higher temperatures than the zircaloy used in water-cooled reactors. Alternatively, $^{235}U/^{232}Th$ carbide microspheres can be incorporated into a ~6 cm sphere (about the size of a tennis ball) of graphite, then coated with silica or silicon carbide. This latter form is called a "pebble," even though it is more the size of a rock. These pebbles can be slowly loaded, and, when spent, removed from a reactor on a conveyor system. Such a reactor is called a pebble bed reactor. Another advantage of combining fuel and moderator in a relatively stable ceramic form is that it can then be disposed of directly, since the fuel is already isolated in a chemically inert form.

12.2.3 LIGHT WATER GRAPHITE REACTORS (RBMKs)

Almost all reactors of this type were built in what once was the Soviet Union. These reactors are very similar to the PHWRs except that the moderator is graphite and the coolant is regular (light) water (Figure 12.3). Instead of a huge tank filled with heavy water (the calandria), RBMK reactors surround their pressure tubes with tons of graphite. RBMK reactors have hundreds of pressure tubes (~1500 in a 1000 MW reactor) containing a couple fuel bundles each. As with the PHWRs, this allows for refueling while the reactor is still running, and for use of only slightly enriched uranium. Unlike the PHWRs, the fuel bundles are vertical in RBMK reactors and the coolant water is allowed to boil in the pressure tubes, making these a boiling water type reactor. It's not shown in Figure 4.3, but all boiling water reactors have a steam separator, or dryer. Its function is to separate the gaseous water (steam) from the liquid, so only steam goes on to the turbine. The BWRs such as those pictured in Figure 4.3 have their steam dryer in the top of the reactor vessel.

FIGURE 12.3   Schematic diagram for a RBMK reactor.

Most nuclear reactors have what is known as a negative **void coefficient**. This means that the rate of fission reactions decreases as the density of matter (usually water) around the fuel decreases. This makes some intuitive sense, as the amount of matter (per unit volume) decreases, there's less for neutrons to interact with, lowering the probability of slowing them down. Faster neutrons means fewer fission reactions taking place (lower σ_f). This is most significant for boiling water reactors. As water boils, bubbles of steam form, lowering the density of water in the volume that contains the bubble. As such, BWRs tend to be a little sluggish in responding to an increase in power. As the water heats, and more boiling takes place, more bubbles form, lowering density and lowering neutron moderation—power then drops.

RBMK reactors are unusual in that they have a large *positive* void coefficient when operating at low (less than 20%) power levels. This stems from the fact that graphite moderates the neutrons. After they are slowed down by the graphite, these neutrons need to travel through the coolant water to get to the fuel, and some are absorbed. When the water is in the gas phase (steam), not as many neutrons get absorbed, and more fission reactions take place. This allowed the Chernobyl 4 reactor to run its power up from below 20% to hundreds times normal in a matter of seconds, leading to two explosions, and a core meltdown. Direct exposure of the red-hot core launched 1.4×10^{19} Bq of radioactive nuclides into the atmosphere. It has been widely assumed that the spread of radioactive material was facilitated by burning graphite, but the nuclear power industry disputes this saying that it is nearly impossible to burn reactor-grade graphite. Regardless, reactors with graphite moderators have become a lot less popular since the Chernobyl accident.

12.2.4   SMALL MODULAR REACTORS (SMRs)

This isn't really a new idea; hundreds of small reactors have been built for the propulsion of naval vessels and as research reactors located at many universities. What's new is thinking of small nuclear reactors for power and/or steam generation in isolated areas, or for a smaller community or an industrial facility. It also represents a rethinking of the economics of nuclear power.

Nuclear power plants are large, complex, and highly regulated. As a result, they are rather expensive to build. Compared with other major power-generating plants (coal and natural gas), nuclear plants are far more costly. Where they tend to make this up is in their lower operating and fuel costs. Coal and natural gas plants require a lot more fuel to operate, and therefore cost more than nuclear to run. Fossil fuel plants also save money by dumping most of their wastes directly into the environment. Because of the high upfront, or capital, costs for nuclear, larger and larger plants were built thorough the 1970s, trying to realize some economy of scale. This was also a time of high inflation, which significantly added to the costs (when you borrow tons of money at a high interest rate, it takes a lot more to pay it back).

Building a large, complex plant is also prone to delays, and, therefore, cost overruns. Let's say that you're building a nuke plant and you put the wrong kind of rebar into the concrete footings for the plant. You did it because it met the current construction standards for large buildings, but unfortunately doesn't match the specifications for the nuke plant that were submitted to the regulator ten years ago. It doesn't matter if the new standards are better, you still need to tear it up, then rebuild. Everything else gets delayed. Really big parts have to be stored, construction crews deferred. Now let's say you've also got some changes to the plant design because the folks building an identical plant in China figured out that the changes would make it safer. You can't make the changes until approved by the regulator, which again idles crews and postpones shipments. It's a huge puzzle with a billion pieces, and the time frames keep changing with every misstep or tweak. Quite a headache, eh?

This is where small nukes suddenly look attractive. They can be built in a factory and shipped by rail. They can be modular so the whole reactor can be replaced instead of refueled (refueling would take place back at the factory), and multiple units can be located at sites with greater demand. Because of their simplicity, several can be operated through a single control room. Just dig a hole in the ground, drop in the reactor, and hook it up to the local grid. Economy of scale is provided by the quantities produced. While a great deal of research and development is taking place, a few of these sorts of reactors are already operating or are under construction in Russia, India, and China. The Russian reactors have been providing steam and electricity to a remote corner of Siberia since 1976.

12.2.5 THORIUM IN REACTORS

As we saw back in Section 4.1.3, thorium exists entirely as ^{232}Th ($t_{1/2} = 1.4 \times 10^{10}$ a) in Nature and is fertile in nuclear reactors undergoing conversion to ^{233}U.

$$^{232}_{90}\text{Th} + {}^{1}\text{n} \rightarrow {}^{233}_{90}\text{Th} + \gamma \qquad \sigma_\gamma = 7.34 \text{ b}$$

$$^{233}_{90}\text{Th} \rightarrow {}^{233}_{91}\text{Pa} + {}_{-1}\text{e} \qquad t_{1/2} = 22 \text{ min}$$

$$^{233}_{91}\text{Pa} \rightarrow {}^{233}_{92}\text{U} + {}_{-1}\text{e} \qquad t_{1/2} = 27 \text{ d}$$

^{232}Th is much more likely to undergo an n,gamma reaction than split when it absorbs a neutron. The ^{233}Th product can then decay via beta to ^{233}Pa, which can also spit out a beta, forming fissile ^{233}U. ^{232}Th could therefore be used to some extent in just about any reactor. So, what's the big deal? Thorium gets its own subsection because of recent (rather zealous) advocacy for designing and building reactors that primarily use thorium as a fuel source.

^{232}Th is roughly three times more abundant on Earth than uranium, with the most significant reserves in India, Brazil, Australia, and the United States. At the time of this writing, the United States, India, Canada, and China have shown the most interest in using thorium in nuclear reactors. The United States used thorium blended with HEU or ^{233}U in four different reactors in the 1970s and 1980s. India currently includes thorium in the fuel mix of some of its heavy water reactors (PHWRs).[2] A company in Norway[3] has recently developed fuel pellets that are a mix of ^{232}Th and ^{239}Pu for use in current reactors.

---

2   India uses ^{239}Pu and ^{233}U produced in other reactors to mix in with ^{232}Th. The reactors "burning" thorium will get ~75% of their power from thorium.

3   Ironic, since there are no nuclear power plants in Norway.

Not only is there a lot of thorium lying around, but ^{232}Th is a superior fertile material than ^{238}U. The cross section for ^{232}Th(n,γ)^{233}Th with a thermal neutron is ~7 b, but increases to ~26 b (due to some resonance) for reactor neutrons. These values beat ^{238}U, as shown in Table 12.1. The ^{233}U formed from ^{232}Th is also superior to the ^{239}Pu formed from ^{238}U. ^{233}U cranks out an average of 3.1 neutrons per fission ($\bar{v}$, in Table 12.1), while ^{239}Pu generates only 2.9. ^{233}U also has a better odds of undergoing fission than an n,gamma reaction. This is represented in Table 12.1 as the ratio of the fission cross section to the n,gamma cross section (σ_f/σ_γ).

Another advantage to thorium is that it is all burned in the reactor, generating more energy per kg of fuel. This also means the waste will almost entirely be made up of fission products, which *generally* have shorter half-lives.

Sounds great! There's plenty of thorium, it produces a better fission fuel more easily than from ^{238}U, we get more energy, and waste that might be more socially acceptable. What are the drawbacks? First, uranium is pretty cheap, making the cost of fuel at a nuke plant a very small percentage of its total costs. Also, we need to remember that we're asking our neutrons to do double duty. Not only do we need to have enough neutrons around to cause fission reactions, but they also have to start thorium on its way to becoming fissile by making the ^{232}Th(n,γ)^{233}Th reaction happen. In light water reactors, there aren't enough neutrons to convert all of the thorium—too many are soaked up by the water and the control rods. The only currently operating reactors with enough neutrons to spare are the heavy water and gas reactors (PHWRs and GCRs). In light water reactors, this problem can be circumvented by blending in more ^{239}Pu. Can you explain why?

Another problem using thorium as fuel is the formation of so many fission products. Some of these have rather high neutron cross sections. If accumulated in sufficient quantities in the reactor core, they will absorb enough neutrons to significantly affect the rate of fission and conversion reactions (**poison** the reactor). To maximize thorium utilization, the fuel will need to be reprocessed to remove the fission products. Reprocessing is also desirable because of the 27-day half-life of ^{233}Pa. ^{233}Pa is formed along the way of converting ^{232}Th to ^{233}U (see equations above). Because of its long half-life and high cross section toward an n,gamma reaction (the thermal neutron cross section is 1500 b!), some of it will likely absorb a neutron and form ^{234}Pa. ^{234}Pa then decays to ^{234}U, which absorbs a neutron and forms fissile ^{235}U.

$$^{233}_{90}\text{Th} + {}^1\text{n} \rightarrow {}^{234}_{90}\text{Th} + \gamma \qquad\qquad \sigma_\gamma = 1500 \text{ b}$$

$$^{234}_{90}\text{Th} \rightarrow {}^{234}_{91}\text{Pa} + {}_{-1}\text{e} \qquad\qquad t_{1/2} = 24 \text{ d}$$

$$^{234}_{91}\text{Pa} \rightarrow {}^{234}_{92}\text{U} + {}_{-1}\text{e} \qquad\qquad t_{1/2} = 27 \text{ d}$$

$$^{234}_{90}\text{U} + {}^1\text{n} \rightarrow {}^{235}_{90}\text{U} + \gamma \qquad\qquad \sigma_\gamma = 100 \text{ b}$$

In the end we still have a fissile nuclide, but we need to use up two more neutrons in the process. We'll also get fewer neutrons from fission of ^{235}U than from fission of ^{233}U (Table 12.1).

In an attempt to address these issues, some have proposed using a liquid fluoride thorium reactor (LFTR) like the one developed at Oak Ridge National Laboratory in the 1960s. In this reactor, the ^{232}Th and ^{233}U form compounds that are soluble in a molten salt. Yes, even salts will form liquids if you give them enough heat. In this case the salt is made up of lithium, beryllium, and fluoride ions. This soup circulates through the reactor in a number of small tubes separated by the graphite moderator. It heats up, then transfers the heat, which eventually produces electricity. When returning to the reactor, some of the circulating molten salt can temporarily be removed from the loop to separate out fission products and ^{233}Pa. The ^{233}Pa can then be held until it decays to ^{233}U, then reinserted into the molten salt loop. This way waste is continuously removed, and fresh fuel can continuously be introduced. Somewhat sophisticated separations systems will be needed to accomplish all of this under relatively high radiation conditions. Alternatively, solid particles of the fuel, wrapped in graphite, can be suspended in the liquid salt forming an emulsion. However, this would not address the issue of ^{235}U formation. There are also corrosion issues when using molten salts—a leak in this system would not be pretty. Considerable shielding will also be necessary on the molten salt loop, since it will also be circulating the fuel.

An experimental LFTR started up in 2022 at the Shanghai Institute of Applied Physics (SINAP). It started with uranium enriched to 20% ^{235}U and only contains 50 kg of thorium initially, but the relative amount of thorium will gradually be increased. The reactor is cooled by air, instead of water, making it attractive for more arid locations. Separations and fuel reprocessing will be done on a batch basis initially, but with transition to a continuous process.

As discussed in Section 4.1.7, significant nuclear weapons proliferation risks exist around enrichment and reprocessing activities. An important downside to the use of thorium in power reactors is that one or both of these activities are necessarily involved. However, some ^{232}U is also formed when using ^{232}Th in nuclear fuels. ^{232}U is pretty darned toasty, making it difficult to work with the fissile ^{233}U and it also makes detection of these operations easier.

12.2.6 GENERATION IV REACTORS

The liquid fluoride thorium reactor is one of several new commercial reactors that are being proposed for development. Collectively, these new reactor types are known as fourth-generation nuclear reactors. So, what were the first three generations? Early prototypes were the first generation of commercial nuclear power reactors, built in the 1950s and early 1960s. The bulk of the ~440 operable reactors today[4]—PWRs, BWRs, CANDU, and RBMK reactors comprise the second generation and were built from the late 1960s, through the mid-1990s. Reactors that have recently been constructed, are currently under construction, or planned for construction in the next few years are third generation. These reactors are basically the same as the second generation, but have improved safety, efficiency, and reliability. Most significantly, some have passive cooling features that would operate without power or operator intervention.

Generation III reactors represent the summit of our ability to engineer nuclear reactors that use water as a coolant. Generation IV reactors aim beyond water-cooled reactors, look to use more fertile fuel, and to generate less long-lived radioactive waste. Molten salt-cooled reactors (MSRs), like the LFTR described in Section 12.2.5, are examples of fourth-generation nuclear power reactors. The following paragraphs briefly describe the other reactor types under active research and development.

The **very-high-temperature reactor** (VHTR) is essentially the same as the gas-cooled reactors (GCRs) described in Section 12.2.2, except that these are designed to run hotter (900–1000°C) for greater thermal efficiency, and to allow the possibility of splitting water into H_2 and O_2. They are helium-cooled and graphite-moderated and use enriched uranium as the fuel. Since a number of GCRs have been built and operated around the world (albeit at lower temperatures), this technology appears relatively accessible. In fact, China connected the first high-temperature gas reactor to the electrical grid at the end of 2021. It is in Shandong province.

The **gas-cooled fast reactor** (GFR) is very similar except it lacks a moderator, and the neutrons are therefore "fast." This allows fission of various actinides that comprise much of the long-lived radioactive waste generated in more conventional nuclear reactors. Various isotopes of Np, Pu, and Am are produced from neutron capture reactions initiating at ^{238}U (see Questions 12.1 and 12.14 at the end of this chapter). These nuclides make up most of the long-term radioactive components of spent fuel, so burning them with fast neutrons would help make the nuclear waste problem more tractable. Fast reactors can also burn ^{238}U and ^{232}Th. These reactors would use He as a coolant and

---

4   September 2022.

operate either at ~850°C or at 600–650°C. They could be fueled by ^{239}Pu (15–20%) mixed with depleted uranium, or reprocessed spent fuel from water-cooled nuke plants. Spent fuel from GFRs would be reprocessed, and any fissile or fertile stuff recovered will be fed back into the reactor, limiting waste to almost entirely shorter-lived fission products. While this is not a new idea, no prototype had been built as of 2022.

Metal-cooled fast reactors, like GFRs, also eschew moderators and can burn nuclear "trash" with their zippy neutrons but are cooled by sodium (SFR) or lead (LFR—actually either pure lead or a lead/bismuth 45:55 mix). In both cases the coolant heats up to about 550°C. Experimental and demonstration models of both reactor types have been built and run, so, these types of reactors are closest to commercialization. Small modular reactors of these types are being considered as well as large (1400–1500 MW) plants. Russia has two sodium-cooled reactors and has started construction on a lead-cooled reactor.

An **accelerator-driven system** (ADS) slams really fast protons into high Z material (W, Pb, etc.), knocking some of their neutrons loose in the process. The high Z stuff is surrounded by fissionable material which splits, produces heat, which is transferred to a coolant, probably lead or a lead–bismuth mix. Like GFRs and metal-cooled fast reactors, ADSs do not moderate the neutrons, so they can burn ^{232}Th, ^{238}U, other actinides in nuclear wastes, as well as the usual fissile nuclides. In fact, an accelerator-driven system (notice "reactor" isn't being used here) could be set up to run on spent fuel from generation I–III reactors. Since the accelerator is providing some of the neutrons, subcritical amounts of fuel can be used. This is a relatively new concept for nuclear energy, and we are not likely to see commercial application before the mid-2030s (France). A demonstration system is under construction in Belgium in 2022.

The **supercritical water-cooled reactor** (SCWR) is like most of the generation I–III reactors. The distinctive difference is that the water is kept under higher pressure and temperature than conventional modern reactors. The water is pushed past the critical point and is therefore supercritical. This means it is no longer a liquid, but it is not a gas either—it is something in between.[5] Cranking up the pressure and temperature in this way provides greater thermal efficiencies, roughly one-third higher than today's nuke plants. Like a boiling water reactor, the supercritical water heated in the reactor core would directly drive the turbine, generating electricity. Because of the lower density of the coolant, this reactor could be built using thermal or fast neutrons. It can also be built with a single pressure vessel, or hundreds (like the CANDU).

As a nuclear nerd, it is easy to get excited about nifty new nuke plant designs. Keep in mind that the very short descriptions above are limited to how they work and what they could do. When generation I and II nuclear power plants were being built, we were still learning how to build and operate them efficiently and safely. Generation IV plants present a whole new set of problems to solve before we can decide that they will be economical, safe, and reliable.

12.3   REACTOR SAFETY SYSTEMS

This section will focus on the safety systems in the most common reactor types: pressurized water (PWR), boiling water (BWR), and pressurized heavy water (PHWR, which are generally CANDU) reactors. The basic operating principles for PWRs and BWRs are presented in Section 4.1.2, and all other types in Section 12.2. Nuclear power plant safety systems can broadly be divided into **active** and **passive**. While the lines between these two are sometimes a little fuzzy, active systems can be defined as those that require human intervention and/or electrical power to function. Passive systems are therefore those that would function without humans or electricity—they are inherent in

---

5   This is not new in power generation. Hundreds of coal-fired power plants heat water past its critical point.

the design of the reactor. Generally speaking, the more passive systems deployed in a reactor, the safer it will be.

The most important safety systems in nuclear reactors are those that shut down (scram) the fission chain reaction. This can be accomplished by reducing the neutron flux within the reactor. As explained in Section 12.2.3, when the water in most light water reactors (LWRs) heats up, its density decreases. This decreases neutron moderation, leading to fewer fission reactions, eventually shutting down the fission chain reaction. This is not a particularly good way to shut down the chain reaction because it requires running the reactor hotter than normal and the neutron flux will likely return to normal after the water begins to cool and higher density is restored. Even if shutdown were successful, there is still plenty of residual heat from the decay of the fission products in the fuel rods. This heat is typically equal to roughly 7% of the reactor's power on shutdown (Figure 12.4). Still plenty toasty to boil off the water and melt the fuel assemblies. Figure 12.4 shows how the heat from fission product decay tapers off with time. It decreases rather rapidly after shutdown, cooling to 4% of full power after only a minute—this is still pretty toasty—it would be 60 MW for a reactor that was running at 1500 MW. After an hour, it's down to ~1.5% and it is about 0.4% after a day. Note that the rate of cooling slows considerably with time. This is why spent fuel must be stored for a few years under water.[6] Once again, water's amazing ability to absorb heat comes in handy!

A better way to scram a nuke plant is by inserting the control rods all the way into the reactor. Remember that control rods are made from elements with high neutron absorption cross section such as Cd, B, and Gd (Section 12.1). These elements could also be injected into a reactor core in some water-soluble form (e.g., sodium borate or gadolinium nitrate). Note that inserting control rods into PWRs and BWRs will meet with resistance as these reactor vessels are both pressurized. BWR control rods are also typically located at the bottom of the reactor vessel, and therefore also need to fight gravity (active!). CANDU (and other PHWR) reactors have a distinct advantage here in that their control rods drop into the non-pressurized calandria, and therefore could be rigged to drop automatically upon loss of power (passive).

Figure 12.4 tells us that cooling of the fuel rods is still necessary after the fission chain reaction is shut down. This cooling is typically accomplished by pumping water through the reactor and dumping that heat into a convenient heat sink like a nearby body of water, or into the air through

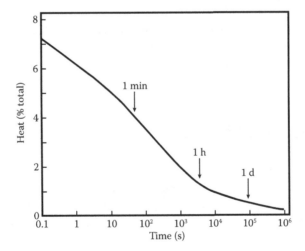

FIGURE 12.4   Residual heat in a nuclear reactor after shutdown.

---

6   The U.S. Nuclear Regulatory Commission has allowed removal in as little as 3 years, but prefers at least 5 years, and states that the industry norm is 10 years.

cooling towers. Extra water (as a liquid or a solid) is typically stored in one or more places in the reactor building so there is enough to cover modest losses due to a leak. Many reactors store these in tanks located higher than the reactor so gravity can feed it into the reactor, if necessary. CANDU reactors cleverly build a large water tank into the top of containment building. This feature has been duplicated in some of the Generation III reactors such as the Westinghouse AP1000 and the AREVA EPR. The AP1000 also uses natural air convection to passively cool the inside of the reactor containment structure, indirectly exchanging heat with air outside of the containment building. Additionally, water can be sprayed onto the outside of the reactor vessel to facilitate cooling. The steam that is generated can then flow out of the building.

In addition to water stored above the reactor vessel, boiling water reactors (BWRs) use a large body of water in the bottom of the reactor building called a suppression pool (Figure 12.5). Excess heat, in the form of steam inside of the primary containment, is vented into the suppression pool, where it condenses. This pool can serve as a passive, temporary heat sink until more active cooling (pumping water through the core) can be started. The suppression pool also serves as a trap for radioactive materials potentially being released by the core.

Nuclear reactors protect workers, those living nearby, and the environment from ionizing radiation by putting several barriers between the radioactive material and the outside. First, the uranium fuel is in a ceramic oxide form. This is similar to how uranium is found naturally and is relatively insoluble and unreactive. Second, the fuel pellets are sealed inside of zirconium tubes (fuel rods). The idea here is to contain the fission products within the fuel rods. Some still leak out, even under normal conditions. Those that get out tend to be isotopes of elements that are gases, namely Xe and Kr. Under more extreme conditions (fuel rod distortion or melting), isotopes of volatile[7] elements (I, and Cs) are also released. The fuel rods are grouped into assemblies, which are surrounded by water, and all of this is sealed inside the reactor vessel. The reactor vessel (and heat exchangers, for PWRs) are surrounded by a containment structure. This structure is usually constructed from thick concrete and steel, and is sturdy enough to withstand a plane crash, or a core meltdown (preventing

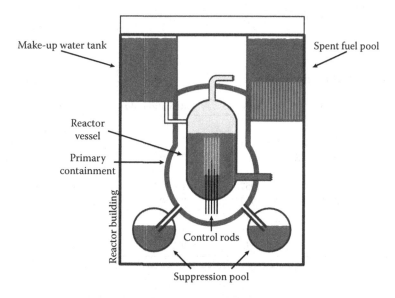

FIGURE 12.5   Some safety systems in a BWR.

---

7   In this context, volatile means simply "easily evaporated." Don't start thinking these elements are somehow explosive.

the molten core from getting into the ground). Boiling water reactors also have secondary containment (and additional shielding) from the reactor building (Figure 12.5).

12.4   NUCLEAR POWER PLANT ACCIDENTS

This section will focus on how the safety systems described in the previous section worked or failed in the accidents at Three Mile Island (TMI), Chernobyl, and Fukushima. These accidents are generally described in Section 4.1.4, so the focus here will be on reactor safety systems and operator responses to the accidents. Other nuclear reactor accidents and incidents have occurred, but these three are the most commonly known and resulted in some of the most significant radiation releases. To put the amount of radiation emitted from these accidents in context, these, and other, significant releases of radioactive material that have occurred are listed in Table 12.2.

By far, the most significant releases have been the detonation of nuclear weapons above ground, first over Hiroshima and Nagasaki, then primarily in the United States and the Soviet Union. Remember that nuclear weapons are inefficient (Section 4.2.1), so a good deal of the fissile material in the weapon is spread around the blast zone and launched into the atmosphere. There's also an intense neutron flux during detonation, which can cause rapid, multiple neutron absorptions (Section 11.6). When these neutron-rich products undergo beta decay, a variety of actinide and other nuclides are formed.

Looking through Table 12.2, it is clear that 1957 was a bad year for radiation releases. Not only were nuclear weapons being detonated above ground, but there were two other large emissions. The Windscale accident was caused by a sudden burst of energy in a graphite-moderated, air-cooled reactor in the UK. This caused the fuel to overheat and started a fire, releasing gaseous and volatile fission products into the atmosphere. Some fission products were also released in 1957 at a high-level waste storage facility near Chelyabinsk (southern Russia, near the Kazakh border) after what was believed to have been a chemical explosion caused by mixing nitric acid with organic chemical wastes. The contamination was spread over approximately 1600 km^2. A similar accident may also have taken place in 1967 at the same site.[8]

TABLE 12.2
Significant Releases of Radioactivity

Source	Country	Year	Radiation (Bq)	Important Nuclides
Hiroshima/Nagasaki	Japan	1945	4×10^{16}	Fission products and actinides
Above ground tests	USA/USSR	1945-63	2×10^{20}	Fission products and actinides
Windscale	UK	1957	1×10^{15}	Noble gases, ^{131}I
Chelyabinsk	USSR	1957	8×10^{16}	^{90}Sr, ^{137}Cs
TMI	USA	1979	1×10^{12}	Noble gases
Chernobyl	USSR	1986	1.4×10^{19}	^{131}I, ^{137}Cs
Fukushima	Japan	2011	9.4×10^{17}	^{131}I, ^{137}Cs

8   Yet another similar incident took place in the Waste Isolation Pilot Plant (WIPP—a deep geological repository in the U.S.) in 2014. Organic (instead of inorganic) kitty litter was mistakenly used to pack a drum containing a mixture of radioactive nitrate salt waste. While only trace amounts of radioactive materials were released to the surface, WIPP was shut down for 3 years at a cost of 500 million U.S. dollars.

12.4.1 THREE MILE ISLAND

The meltdown at Three Mile Island (TMI) started with the failure of some relatively minor pumps on the secondary coolant loop (the loop connecting the heat exchanger and turbine—see Figure 4.2). These pumps run the water through a loop (not shown in Figure 4.2) that removes impurities common to hot water running through pipes. One of these pumps was clogged and being repaired when all the others went down. Normally a valve that would bypass this cleaning loop would open at this point but didn't. This prevented circulation through the secondary coolant loop and caused the turbines to automatically shut down. This also prevented heat from being properly transferred in the heat exchanger, causing temperature and pressure to rise in the reactor. This caused the reactor to scram by automatically inserting the control rods, shutting down the fission chain reaction. However, the reactor still needed to deal with decay heat from the fission products.

The critical point in this crisis occurred when a pressure relief valve automatically opened to relieve some of the pressure in the reactor coolant loop but failed to close properly when the pressure was sufficiently low. As a result, coolant drained out of the reactor for a couple of hours. A similar incident happened two years earlier at a reactor near Toledo, Ohio. This reactor was the same design as the one at TMI but did not melt down because one of the operators figured out what had happened before too much coolant had been lost. Unfortunately, the Toledo incident was not well communicated to other plants of the same design. The confusion at both plants is understandable because the control room indicator for this valve showed it was closed. Additionally, they had no gauge to tell them water levels in the reactor vessel were lowering, it had to be inferred from other readings, which gave apparently contradictory information.

As water drained from the reactor, alarm systems began to go off, but none directly told the operators that the problem was a stuck valve. The operators became focused on the fact that water levels were too high in one part of the loop (near the stuck valve!) and began shutting off pumps circulating water through the reactor. At about this time, one of the operators determined that the pressure relief valve was stuck open, and they closed another valve below it to stop the leaking. Unfortunately, they did not then consider that a great deal of water may have drained out of the reactor (about one-third of the total). Shortly thereafter, the reactor got hot enough to vaporize enough of the remaining water to expose the fuel rods and begin a meltdown. The zirconium cladding around the fuel pellets reacted with the steam to form hydrogen gas and the fuel rod assemblies began to slump, break apart, and collapse into the bottom of the reactor.

Once the cladding around the fuel is breached, the more volatile fission products are released into the coolant. Radiation alarms started going off all over the place, which generally added to the confusion. It wasn't until ~21 hours after the initial pump failures that one of the reactor coolant pumps was successfully restarted and progress toward stability began. Unfortunately, a fair amount of pressure had built up inside the buildings from the release of fission products from the meltdown. This pressure was vented, resulting in the release of radioactive isotopes of Xe and Kr, as well as a very small amount of ^{131}I. As can be seen in Table 12.2, the amount of radioactive material released was low (compared to others on the table), and, because it was almost entirely noble gases, it was of negligible health concern. Because they are chemically unreactive, they do not accumulate in living things. Unfortunately, the releases were not always well communicated, resulting in significant public concern and sometimes gave the appearance that those managing the crisis were incompetent or even malevolent.

12.4.2 CHERNOBYL

Like TMI, the Chernobyl disaster was the result of inadequacies in the reactor design and the inability of its operators to fully understand the consequences of their actions. The reactor was scheduled to shut down for routine maintenance, and the operators thought it would offer an opportunity to test how long the turbine would provide power to the pumps after a sudden shutdown. In other words, when the steam is cut off from the turbine, it continues to spin for a while, generating electricity, until it coasts to a stop. The Chernobyl operators wanted to know if it would give them enough time to start up the emergency backup (diesel) generators and thereby have no interruption of power to the pumps. They had done this experiment before, but the turbine spun down too quickly. They had since changed some of the electrical equipment, thinking this would give them more time.

The shutdown of the reactor ended up being delayed a few hours because the power it was generating was still needed by the grid. When approval for shutdown came, a different shift was operating the reactor instead of the one that had planned to perform the experiment. The first step of the experiment was to slowly lower power to about 25–30% of maximum. Due to an operator error, it instead dropped to about 1%. Because it was a light water graphite moderated reactor (RBMK), whose main flaw is a large positive void coefficient at low power levels (< 20%; see Section 12.2.3), the operators should have terminated the experiment at this point and completed reactor shutdown.

The operators instead removed control rods to try to bring the power back up. This type of reactor responds a bit sluggishly under these conditions, so the operators bypassed safety protocols and pulled out more control rods than they were normally allowed. Even so, they were only able to bring the reactor power up to ~6% of capacity. Again, safety protocols dictated they should've shut down at this point as the reactor was in a dangerous state (low power, large positive void coefficient—if this were a movie, you'd be hearing the ominous music playing in the background!). Instability in the amount and pressure of the steam developed, and should've triggered an automatic shutdown, but didn't because the operators had also bypassed this safety feature so it wouldn't mess up their experiment.

They started their experiment by shutting off the steam to the turbine. This reactor was a boiling water reactor, so there's only one coolant loop. Shutting off the steam to the turbine lowered the pump output to the core, resulting in a rise in temperature, creating steam bubbles (voids!), increasing the rate of fission reactions. Reactor power and temperature increased rapidly,[9] and the operators tried to insert the control rods, but they were too slow. It appears that most only made it in about halfway before the reactor started to blow itself apart, and a few jammed due to a faulty design. The core overheated, vaporizing the coolant which exploded with considerable force. A few seconds later a second explosion occurred. It's not entirely clear what caused the second explosion, but it is widely assumed it involved hydrogen produced from the reaction of the super-heated steam with the zirconium fuel cladding. However, the second blast could've been another steam explosion.

The explosions destroyed the reactor core and the upper part of the reactor building. The Soviets did not build containment buildings around their RBMK reactors, so this exposed the red hot reactor materials to the atmosphere, and the second explosion launched some of the core onto nearby buildings, causing a number of fires. The fires were contained within a few hours, but only after many of the firefighters received very large doses. Over the next several days, helicopters dropped boron carbide (absorbs neutrons), dolomite (to absorb heat), lead (radiation shielding), and sand and clay (to inhibit release of radioactive materials to the atmosphere). Additionally, cold nitrogen was pumped in to cool and blanket the core materials (to inhibit combustion), and a tunnel was dug under the reactor to install a massive concrete slab with a cooling system. Nine days after the accident, all of these measures finally significantly reduced the temperature of the core materials and sharply curtailed the release of radioactive materials.

A massive concrete structure with a cooling system (called the "sarcophagus"), was built over the trashed reactor. It took seven months to complete and was designed to shield and contain. Unfortunately, it was not designed to last. It started allowing rainwater into the destroyed reactor, and was weakening to the point that it is threatening to collapse. A larger, more permanent cover (The New Safe Confinement, or NSC) was built next to the reactor and slid into place in 2016. This structure also has remote control equipment to dismantle the sarcophagus, the old reactor, and remove the radioactive material from the destroyed reactor.

---

9   Reactor power was estimated to be some hundreds times normal at the time of the explosions.

Like TMI, the Chernobyl disaster could've been averted had the operators taken a safer path at any number of points leading up to the explosions. There were two prior, similar incidents in the Soviet Union with RBMK reactors. They occurred on a much smaller scale, with very little radiation released, but these incidents were kept secret instead of shared with other RBMK reactor operators. Unlike TMI and the previous Soviet incidents, Chernobyl released a considerable amount of radioactive material. It is estimated that all the xenon, half of the iodine and cesium, and roughly 5% of the rest of the radioactive material in the core at the time was released into the environment. Much of it was deposited locally, but measurable quantities were spread over much of the Northern Hemisphere. As described in Section 4.1.4, Chernobyl is the only nuclear power plant accident known to have caused fatalities due to exposure to ionizing radiation. Explosively exposing the reactor core makes a big difference—TMI was pretty mild in comparison to Chernobyl.

About 200,000 people were eventually relocated away from contaminated areas around Chernobyl. Many of the evacuated areas have since reopened for settlement. Roughly 600,000 "liquidators" were brought in during the years after the accident to decontaminate the area near the reactor, and build the various structures, including the sarcophagus. Epidemiological studies of these groups of people in the 36 years since the accident have revealed very few radiation-related deaths.

12.4.3 FUKUSHIMA

Unlike TMI and Chernobyl, it is difficult to point to specific deficiencies in reactor design as major contributors to the radiation releases. As we'll see, the main factors that led to radioactive material being released at Fukushima came largely from poor site design, and (like TMI and Chernobyl) inadequate operator training for such an incident. The reactor design is implicated in some reports, usually simply by stating that they were old with little or no further explanation. It is true that they were built in the late 1960s and early 1970s but they did not contain substantially different safety systems than most other BWRs. The only real argument to be made is that they lacked sufficient passive safety systems that could've brought them under control without electrical power. However, such a statement is true of almost all of the hundreds of Generation II reactors still in operation today.

The main reason the Fukushima reactors ended up releasing radioactive materials was the site design (Figure 12.6). There were 15 BWRs along the Japanese coast that were affected by the tsunami, but only three reactors, all at the Fukushima Daiichi[10] site, that suffered meltdowns and released radioactive material. In fact, there were six reactors at the Fukushima Daiichi site—the

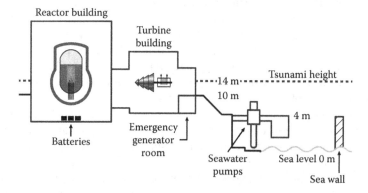

FIGURE 12.6 Schematic view of part of the Fukushima Daiichi site. Not drawn to scale.

---

10 There are two nuclear power plant sites in the Fukushima prefecture: Daiichi and Daini. Daiichi means "first" and Daini "second." To state the obvious, the Daiichi site was developed first.

other three did not experience meltdowns. Four of the six reactors (reactors 1–4) at Fukushima Daiichi were located near the ocean. When the tsunami waves hit, they flooded the lower levels of the reactor and turbine buildings disabling the emergency diesel generators and the backup batteries (Figure 12.6) because they were located in the basements of these buildings. The designers of this site had not thought to place these critical backup systems on higher ground, or to waterproof their locations. Only one of the 13 diesel generators at the Fukushima Daiichi site survived the tsunami, but, because electrical switching equipment was also knocked out, it could only supply power to reactors 5 and 6. Because they had power, these two reactors were able to safely achieve cold shut-down nine days after the tsunami.

This was generally true of the other Japanese reactors that were able to survive the catastrophic events. They all had some backup power which enabled them to shut down their reactors. South of Fukushima, the Tokai Daini plant had just finished waterproofing one of its diesel generators, and it was the only one to survive—but it was enough to safely shut down the reactor. Further north, the Onagawa site was originally planned to be located on lower ground but shifted to higher ground after a local engineer pointed out the tsunami hazard.

It should also be noted that the sea wall at Fukushima was only four meters tall, while the highest tsunami wave to hit the site in 2011 was about 14 m high (Figure 12.6). There was historical evidence of tsunami waves of this height hitting this part of Japan. Japanese nuclear authorities were also twice warned by the International Atomic Energy Agency about the potential dangers of flooding. The first came after some flooding at a nuke plant in France in 1999 illustrated vulnerabilities to floods, and the second after a tsunami caused an Indian plant to shut down in 2004. One of the most significant outcomes from the Fukushima disaster is the importance of a strong, independent nuclear regulatory agency. A clear flaw prior to the 2011 earthquake and tsunami was the cozy relationships between and insular confidence of the Japanese government, the nuclear regulatory agency, and the nuclear industry.

The earthquake and tsunami knocked out all power to reactors 1–4 at Fukushima Daiichi (all BWRs). Reactor 4 had no fuel in its reactor vessel at the time as it was undergoing routine mainten-ance. As described in Section 12.3 and illustrated in Figure 12.5, these reactors could transfer excess heat from the reactor to a suppression pool without the need for electrical power. Reactors 2–4 had an additional safety system which used the steam produced from the decay heat to inject water into the core. Both of these safety systems operated in reactors 2 and 3 for about 1.5 and 3 days, respectively—which is significantly longer than they are designed to run.

The valves connecting the containment structure with the suppression pool in reactor 1 were closed when the tsunami hit, which meant the decay heat had no place to go. Heat and pressure built up within the reactor vessel and within a few hours the reactor vessel had lost enough coolant that the fuel began to melt, and the zirconium cladding started reacting with steam to form hydrogen. The pressure (from the heat and the release of gases) soon overwhelmed the primary containment struc-ture, sending volatile fission products and hydrogen into the reactor building. Twenty-five hours after the tsunami, a hydrogen explosion blew the top and much of the sides of the reactor building off, releasing a fair amount of radioactive Xe, I, and Cs.

Somewhat similar sequences of events occurred in reactors 2 and 3 when the suppression pools and steam-driven water injection failed. These systems can only operate for a limited amount of time because there's only so much water in the make-up tank and the suppression pool eventually gets too hot to absorb more reactor heat. Normally the excess heat would've been transferred to the ocean, but the seawater pumps were knocked out by the tsunami. Like reactor 1, the top of part of the reactor 3 building blew out with a hydrogen explosion (three days after the tsunami). You'd think that any additional hydrogen gas produced by reactor 3 would've vented directly to the atmosphere at this point. It was instead vented into the reactor 4 building through some shared ductwork. The top of the reactor 4 building exploded (Figure 4.5) a day after the reactor 3 explosion. Reactor 2 managed to avoid a hydrogen explosion because a large panel in the exterior wall near the top of the building was removed, allowing the hydrogen (and volatile fission products) to escape to the atmosphere. It is believed that most of the radioactive releases came from building 2. Ironic, since it was the only one that did not explode.

The Fukushima reactors had other emergency core cooling systems designed to remove decay heat and add water as necessary, but these systems required electrical power, which was not available for several days following the earthquake/tsunami. The operators also had an option to pump water into the reactors from firetrucks, which were on the scene shortly after the tsunami. This required depressurizing of the reactor vessel, which the operators were unable to do because the valves required electricity or compressed air to operate. In some instances, there were valves that could be opened manually, but the workers were prevented by high levels of radiation or debris from the tsunami. By the time they could start depressurizing, it was too late, containment was already lost. Had they been able to act more quickly to depressurize the reactor vessels, and pump water in from the trucks, the reactors may have been saved with little to no release of radioactive material.

It wasn't until 5–6 months after the tsunami that adequate cooling was established at all four damaged reactors. It was a full nine months before radioactive emissions from the plant were minimized and "cold shutdown" was declared. The top floors of each of the damaged reactors (also known as the service floor) have since been rebuilt to provide containment, protection from the elements, and allow the removal of the stored spent fuel (in the spent fuel pools—see Figure 12.5). All of the spent fuel that was in the reactor 3 and 4 pools has been removed, and the reactor 1 spent fuel removal was started in 2021. Removal of the melted fuel also started in 2021.

12.5   FUSION REACTORS

The proton cycle discussed in Section 11.3 is generally thought to be impractical for fusion power reactors here on Earth, primarily due to its relatively high ignition temperature—the temperature needed to start a fusion reaction. Back in Section 11.3, we calculated the temperature needed for ignition of the proton–proton fusion reaction as nearly 13 billion Kelvin! In reality, fusion reactors do not need to get quite this hot, because some reactions will take place at lower temperatures due to the distribution of particle energies at a particular temperature, or because tunneling will occur. At a certain point (several million Kelvin), enough reactions are running that they generate enough heat for the reactions to be self-sustaining. Even so, reaction ignition is a major roadblock in the development of practical fusion power reactors. All of the test reactors built so far require more energy to be put in for ignition than is obtained from the subsequent fusion reactions.

Currently, the best hope for a practical fusion reactor lies in the so-called "D + T reaction." It is the reaction of deuterium (^{2}H) and tritium (^{3}H).

$$^2_1H + ^3_1H \rightarrow ^4_2He + ^1n$$

This reaction produces 17.6 MeV of energy, and its Coulomb barrier is only 0.68 MeV, thereby minimizing the ignition temperature. The D + T reaction has been successfully performed in a number of research reactors. A problem here is sustaining the fusion reaction. For this to happen, fuel needs to be fed into the reactor while products are removed. One way around this problem is to build a very large reactor, one that can hold enough fuel to provide more energy than was necessary to initiate fusion.

Some fusion enthusiasts like to state that it does not produce radioactive waste. This is not true. Short half-life nuclides with low mass numbers will be formed in any fusion reactor. The fact that they are short-lived means that fusion plant operators will just need to wait a little while for them to decay to stable nuclides. The neutrons produced in the D + T reaction are of greater concern. As with fission reactors these neutrons could interact with some of the metals used to build the reactor, making them radioactive through n,gamma reactions. For example, cobalt (often used in high-performance metal alloys) would form ^{60}Co, a beta- and high-energy gamma-emitting nuclide with a half-life of 5.3 years.

$$^{59}\text{Co} + {}^{1}\text{n} \rightarrow {}^{60}\text{Co} + \gamma$$

$$^{60}\text{Co} \rightarrow {}^{60}\text{Ni} + {}_{-1}\text{e} + 2\gamma$$

Another big technical challenge for a practical fusion reactor was how to confine the very hot plasma that would be generated. The confinement issue has been solved by using very strong magnetic fields. At these temperatures the fuel exists as plasma, bare nuclei, and free electrons in a gaseous soup. Since everything is charged, it can be suspended in a magnetic field and will never touch the walls of its container. The optimal shape for this plasma is like a donut (toroid).[11]

Because of the enormous challenges in researching and developing a practical fusion reactor, an international consortium is working together to build the International Thermonuclear Experimental Reactor (ITER), a prototype designed to demonstrate the feasibility of fusion as a power source. The ITER is currently under construction near Cadarache, in the south of France, and is planned to be fully operational by 2035.

A major fusion research facility in the U.S. is the National Ignition Facility (NIF) at Lawrence Livermore National Laboratory (LLNL). It uses 192 powerful laser beams and inertial confinement to attempt the D + T reaction. During the summer of 2021 a shot generated a net 1.3 MJ. This marked the first time a fusion reactor produced more energy than was needed to initiate a "burning plasma."

Finally, you probably noticed that fusion produces a lot less energy per reaction than fission. The D + T reaction is one of the best, but only generates 17.6 MeV per reaction, while a typical fission reaction produces roughly 200 MeV. This may just be a matter of perspective. If we look at how much energy is released in terms of MeV per mass unit of "target" nuclide, fusion rules! Fusion cranks out ~6 MeV/u while fission only produces ~0.8 MeV/u. This argument is only important for weapons, where delivering a payload (mass) costs money and energy—placed on the same rocket, more bang can be generated from a fusion weapon than a fission warhead with the same mass.

QUESTIONS

12.1   Starting with ^{239}Pu(n,γ), show the most likely pathway forming ^{241}Am and ^{244}Cm in a nuclear reactor. Include appropriate cross sections and/or half-lives.

12.2   Explain why ^{235}U is a better fuel for a nuclear reactor than ^{238}U.

12.3   Of the nuclides listed in Table 12.1, which has the smallest critical mass?

12.4   Beryllium has a low neutron absorption cross section ($\sigma_a = 0.0076$ b) and a high neutron scattering cross section ($\sigma_s = 5.9$ b). Would it be a good material for a fuel rod, control rod, moderator, and/or coolant in a nuclear reactor? Briefly explain your answer.

12.5   Beryllium is used as a neutron "mirror" around the core of reactors that need a higher neutron flux. Using the information from Question 12.4 briefly explain how this works. Is "mirror" a perfect term? Briefly explain.

12.6   In a nuclear reactor, what are delayed neutrons? Why are they helpful?

12.7   Why can't light water be used as a moderator/coolant in CANDU reactors?

12.8   Explain why moving fuel bundles around in a CANDU reactor is necessary to maintain more consistent power production throughout the reactor.

12.9   What is the main advantage in using light water in a reactor? Heavy water?

---

11 German scientists are working on a variation of the toroid, called a stellarator. It's a toroid with twists and curves—a cruller? A U.K. company is exploring a small reactor that uses spherical confinement—a donut hole?

12.10 Why is tritium produced at lower rates in light water reactors than in heavy water reactors?

12.11 Define the following: calandria, fuel bundle, void coefficient, and reactor poison.

12.12 Why can't RBMK reactors use natural uranium as a fuel?

12.13 Based on their design, would you expect CANDU reactors to have a positive or negative void coefficient? Briefly explain.

12.14 Assuming the initial reaction is n,2n, give the likely series of reactions and decays that results in the formation of ^{232}U in thorium reactors.

12.15 Why does ^{239}Pu make a good "mix-in" with ^{232}Th fuel?

12.16 Why are elements like He, Pb, and Bi selected as coolants in generation IV reactors?

12.17 Why don't ADSs need control rods?

12.18 Define "scram" in the context of nuclear power plants. After a scram, is the reactor then safely shut down? Explain.

12.19 Classify the scram methods described in Section 12.3 as active or passive.

12.20 Assuming all the radioactive material released from TMI was ^{133}Xe, calculate its volume under standard temperature and pressure. If a typical nuclear medicine lung scan uses 12.5 mCi of ^{133}Xe, how many lung scans would the TMI releases represent?

12.21 Assuming all the radioactive material released from Fukushima was ^{131}I, calculate the total mass emitted. Calculate its activity after three months.

12.22 The reaction in Question 12.23 has been proposed for use in reactors here on Earth. Its proponents say the only drawback is the lack of ^{3}He. What is the percent abundance of ^{3}He on Earth? Would you expect natural abundance of this nuclide to be low? Explain. The moon is known to have ample supplies of ^{3}He at or near its surface. Why?

12.23 Based on the energy produced by and the effective Coulomb barrier of this reaction, does it look like a viable reaction for fusion power production?

$$^2_1H + {}^3_2He \ \rightarrow \ {}^4_2He + {}^1_1H$$

12.24 The most common alpha particles emitted by ^{235}U have an energy of ~4.4 MeV. Is this enough energy for ^{9}Be to be an effective neutron initiator in a fission weapon?

12.25 Why is a ^{239}Pu fission bomb so much more efficient than one using ^{235}U?

12.26 Calculate the energy produced by the reaction below, which is part of how fusion weapons detonate. Is a similar reaction also likely for the other naturally occurring isotope of lithium? Briefly explain.

$$^6_3Li + {}^1n \rightarrow {}^4_2He + {}^3_1H$$

13 Radiation Protection

Ionizing radiation is frightening, in part, because the risks of exposure are expressed as probabilities. If a person dies of cancer it is impossible to point to an exposure to ionizing radiation 20 years in that person's past and state with absolute certainty that the cancer was a result of that exposure. Despite this uncertainty, radiation hazards and risks can be understood and controlled. This chapter will help you understand how ionizing radiation interacts with humans, when to get excited about it, and what to do about it. This chapter builds on what you've learned about the detection of ionizing radiation (Chapter 8) and its interactions with matter (Chapter 7).

13.1 TERMS

It is important to distinguish between exposure and dose. **Exposure** is the amount of ionizing radiation striking an object. **Dose** is the amount of energy absorbed by an object as a result of radiation exposure. As humans, we are exposed to a lot of radiation in the form of high-energy photons, which are part of cosmic radiation. Because they are in the X-/gamma-ray part of the electromagnetic spectrum, their odds of interacting with us (we are made up of mostly low Z atoms!) are relatively low. Therefore, our *dose* from cosmic radiation is relatively low, even though our *exposure* is high.

As noted back in Section 1.6, the biggest dose to humans living in the U.S. (and most other places in the world) comes from radon (^{222}Rn). Radon is in the air we breathe and in the water we drink (and it has always been there!), so we ingest a fair bit. ^{222}Rn is an alpha-emitting nuclide, with a relatively short (3.8 d) half-life—so there's probably some inside of us right now, decaying away. Those alpha particles will deposit all of their energy inside our bodies, since alpha particles have high specific ionization (SI) and linear energy transfer (LET) values and very short ranges (Section 7.2). Even though it is the source of our largest dose, it is not because we are exposed to relatively large quantities of radiation from ^{222}Rn, it is because the energy of the decay is rather efficiently transferred to our bodies. Remember that exposure and dose are different—exposure is all ionizing radiation around something, but dose will depend on the type of radiation and the nature of the matter it interacts with.

Exposure can be more strictly defined as the amount of charge (ion pairs!) delivered to a mass (or volume) of air. Its International System of Units (unfortunately, also uses the acronym "SI") are Coulombs per kilogram (C/kg). It is sometimes expressed in units of röntgen (R). One röntgen is equal to 2.58×10^{-4} Coulombs per kilogram and produces about 2×10^9 ion pairs in 1 cm^3 of air.

$$1R = \frac{2.58 \times 10^{-4}\, C}{kg}$$

DOI: 10.1201/9781003303787-13

Likewise, dose can be defined as the amount of energy delivered per unit mass. The SI unit for dose is the gray (Gy), which is equal to a joule of energy dumped into one kilogram of matter:

$$1\frac{J}{kg} = 1\,Gy$$

In the United States, units of *r*adiation *a*bsorbed *d*ose (rad) are commonly used. It is defined as 100 erg/g, which happens to be equal to one centigray.

$$1\,rad = \frac{100\ erg}{g} = \frac{10^{-2}\,J}{kg} = 1\,cGy$$

Example 13.1: How much energy (MeV) is deposited in 91 kg by a 1.00 Gy dose?

$$1.00\ Gy = \frac{1.00\ J}{kg}$$

$$\frac{1.00\ J}{kg} \times \frac{MeV}{1.60 \times 10^{-13}\ J} \times 91\ kg = 5.7 \times 10^{14}\ MeV$$

What is this dose in rad?

$$1.00\ Gy \times \frac{rad}{10^{-2}\,Gy} = 100\ rad$$

We know that ionizing radiation interacts differently with different materials. For example, relatively low-energy X-rays are much more likely to interact via the photoelectric effect with high Z materials, and via Compton scattering with low Z materials. We also know that alpha particles will dump all of their energy in a very short distance, even in gases, while gamma photons *could* pass right through without interacting at all. Dose is nice, but how can we account for the different interactions we see with different types of ionizing radiation with different materials? Specifically, can we know what kind of dose humans get?

Dose equivalent is the answer. We really just need to modify dose so that it depends on the type of radiation. This is done by the **quality factor**, Q, also known as the weighting factor or relative biological effectiveness (RBE). It's a bit like linear energy transfer (LET). Remember that, in terms of LET, alpha particles deposit the most energy per centimeter, followed by beta, while X-rays and gamma rays deposit the least. The quality factor (Q) weights the dose for damage to biological systems. For γ- and X-rays, and most (> 30 keV) beta particles, the value of Q is 1. Beta particles with less than 30 keV have a Q of 1.7, because they are more likely to interact with matter. For thermal neutrons (relatively low-energy, only 0.025 eV!), it is 5. The Q value varies significantly with the energy of the neutron. The Q value for alpha particles is 20.

We can now define dose equivalent as the amount of biological damage caused by ionizing radiation. Mathematically, it is simply dose (sometimes called "absorbed dose") times Q.

$$H = dose \times Q \tag{13.1}$$

The SI units for dose equivalent is the sievert (Sv):

$$Sv = Gy \times Q$$

In the United States, units of rem (*röntgen equivalent man*) are sometimes used. One rem is the same as one centisievert or 10 millisieverts.

$$1 \text{ rem} = 10^{-2} \text{ Sv} = 1 \text{ cSv} = 10 \text{ mSv}$$

Since Q is 1 for γ-rays, X-rays, and most beta particles, dose is numerically equal to dose equivalent for these types of radiation. For this reason, the simpler term "dose" is a common colloquialism for "dose equivalent."

$$1 \text{ Sv} = 1 \text{ Gy and } 1 \text{ rad} = 1 \text{ rem}$$

We've defined a lot, and it should be enough … but there's another way to look at dose and dose equivalent: **kerma**. This stands for *k*inetic *e*nergy *r*eleased in *m*edi*a*. As its name implies, kerma is another way of looking at dose, but instead of energy absorbed by the material, we're looking at energy released by the radiation—a subtle distinction at best. It even has the same units as absorbed dose (Gy). Differences between kerma and absorbed dose are only seen when significant amounts of secondary radiation (bremsstrahlung, secondary electrons, …) escape the material, thereby getting some of the energy out of the material that was originally transferred to it. Under most circumstances, kerma ends up numerically equal to dose and exposure.

To relate different media, we can use a simple ratio to adjust dose, or kerma, to a medium. That ratio is called the *f* **factor** (also known as f_{medium} factor), and is similar to the quality factor, Q.

$$f = \frac{\text{dose in medium}}{\text{dose in air}} \tag{13.2}$$

Figure 13.1 shows how the *f* factor varies with photon energy for water, muscle, and bone. Dose depends mostly on the mass attenuation coefficient (μ_m—Section 7.4), which will depend on the atoms (Z) and the density of the matter. If we are only concerned with humans, we know that water

FIGURE 13.1 Variation in the *f* factor for water, muscle, and bone with photon energy. Data from Hubbell, J. H. and Seltzer, S. M., *Tables of X-Ray Mass Attenuation Coefficients and Mass Energy-Absorption Coefficients* (version 1.4). National Institute of Standards and Technology, Gaithersburg, MD, 2004. www.nist.gov/pml/x-ray-mass-attenuation-coefficients.

and muscle are made up of atoms with similar atomic numbers (Z). Therefore, we would expect them to have similar f factors. Using the y-scale on the left side of Figure 13.1 and plotting the ratios of the mass attenuation coefficients vs. photon energy, the f factor is very close to 1 for most biological materials, regardless of photon energy. This should not be much of a surprise; the atomic numbers of atoms in air molecules are pretty much the same as those in water and our muscles.

The f factors for bones are also plotted on Figure 13.1 using the right y-scale. Notice that it is also close to 1 for all but the lowest energy photons. Therefore, the f factor is close to 1 for most biological systems under most conditions, just as Q is close to 1 for common forms of ionizing radiation.

As is evident from Figure 13.1, each of the body's organs will respond a little differently to ionizing radiation. While the differences are subtle, they are sometimes distinguished. **Effective dose equivalent** (sometimes just called **effective dose**) is a dose equivalent weighted according to susceptibility of a particular organ to ionizing radiation. Going through detailed weighting schemes for different organs is beyond the scope of this text. The reader should simply know that dose can be further refined to a specific organ.

13.2   REGULATIONS AND RECOMMENDATIONS

The International Commission on Radiological Protection (ICRP) develops recommendations and standards for radiological protection, which can then inform national regulatory and legislative bodies. For example, in the United States, the National Council on Radiation Protection and Measurement (NCRP) uses ICRP standards and recommendations to develop its own recommendations. The primary regulatory authority in the United States is the Nuclear Regulatory Commission (NRC). If all the acronyms are making you a big dizzy, get used to it, such is the way of regulatory bodies.

The NRC *regulates* based on *recommendations* by the ICRP and the NCRP. Table 13.1 shows NCRP recommendations for dose limits. The NRC controls the use and distribution of radioactive materials in the United States. It has primary regulatory authority, and is a federal agency. Many individual states have been delegated some of the NRC's responsibilities—Wisconsin is one. In Wisconsin, the state Department of Natural Resources (DNR) handles much of the regulation of radioactive materials.

Actual regulations will vary from state to state and will depend on the nature of the facility. For example, the regulations will be much more extensive for a nuclear power plant with tons of toasty fuel than for an academic laboratory that handles only very small amounts of radioactive materials.

All doses in Table 13.1 are over and above those from natural background sources. The first set of limits is for radiological workers—notice that the dose for extremities (500 mSv) is much higher than the annual (whole body) dose (50 Sv). Our hands, feet, and skin can take a much greater dose because the cells in these parts of our body can better repair or replace damaged ones.

The "emergency occupational exposure" is designed for life or death situations. If a life can be saved, or the spread of significant radioactive contamination be contained, then up to a 500 mSv (one-time, whole-body) dose can be allowed.

Public exposures refer to non-radiological workers who may be exposed while touring a facility where they might get a little extra ionizing radiation. It is also exposure to those that live nearby due to releases from these facilities. Again, these exposures are over and above those from natural background.

Educational dose limits are designed mostly to allow students to perform experiments with small amounts of radioactive material, or radiation-generating devices. They are also set low because, while the body is still growing it is more susceptible to radiation damage. The human fetus is an extreme example, so dose limits are set quite low for pregnant females—to protect the fetus. Note that the human fetus is surrounded by a fair amount of shielding (also known as Mom!), and is therefore naturally somewhat protected from external ionizing radiation.

TABLE 13.1
Effective Dose Limits Recommended by the NCRP

Occupational exposures	
Effective dose limits	
(a) Annual	50 mSv
(b) Cumulative	10 mSv × age
Equivalent dose limits for tissues and organs (annual)	
(a) Lens of eye	150 mSv
(b) Skin, hands, and feet	500 mSv
Guidance for emergency occupational exposure	500 mSv
Public exposures (annual)	
Effective dose limit, continuous or frequent exposure	1 mSv
Effective dose limit, infrequent exposure	5 mSv
Effective dose limits for tissues and organs	
(a) Lens of eye	15 mSv
(b) Skin, hands, and feet	50 mSv
Education and training exposures (annual)	
Effective dose limit	1 mSv
Equivalent dose limit for tissues and organs	
(a) Lens of eye	15 mSv
(b) Skin, hands, and feet	50 mSv
Embryo-fetus exposures (monthly)	
Equivalent dose limit	0.5 mSv

Source: NCRP Report #116, Limitation of Exposure to Ionizing Radiation, 1993.

How do we know what kind of dose we're getting? By measuring exposure! Exposure is measured with TLD (Section 8.3.2) or film badges, which are worn between the neck and the waist. This gives a pretty good idea of whole-body exposure (and dose if we're dealing with photons and/or beta particles). Exposure to the hands is often measured separately using a TLD ring. The ring should be rotated so that the TLD crystals are closest to the source of ionizing radiation. Pocket dosimeters can also be used, especially if the potential for significant short-term exposure is high. These are typically ionization chambers (Section 8.1.1) for analog measurement, or compact Geiger-Müller (G-M) counters (Section 8.1.3) for digital measurement.

In areas where radioactive materials are used, regular swipe tests are performed to test surfaces for radioactive contamination. Typically, a round piece of filter paper is wiped over a random spot on a surface then counted using an organic scintillation counter (Section 8.2.3).

Personnel leaving an area where radioactive materials are handled are required to frisk (monitor) their hands and feet with an appropriate detector (usually a G-M tube). If the potential for contamination of other parts of the body exists, those parts must also be monitored. In these situations, a monitoring portal is typically used. Monitoring portals surround all or most of the body with detectors. It was monitoring portals at Swedish and Finnish nuclear power plants that first alerted the world to the Chernobyl disaster. Workers at these plants were *arriving* at work with enough contamination to set them off.

In areas where radiation-generating devices are used, exposure is determined by performing a radiation survey around the equipment while energized. In some cases, TLD badges may be placed in certain locations to measure exposure over longer periods of time.

Records are meticulously maintained for all measurements of exposure. As is typically required by law, radiological workers must receive regular (usually annual) reports of their exposure. You should always know what you are getting, and it will likely be well under the limits.

At what dose levels do bad things happen? Data are available from the survivors of the Hiroshima and Nagasaki bombings, the Chernobyl disaster, occupational exposures, as well as a number of rather questionable experiments performed by the United States government in the 1950s and early 1960s.[1]

The following lists human health effects at various acute dose equivalent levels. Keep in mind that these are dose equivalents for a single exposure event with the entire dose received more or less instantaneously. Exposure over time allows the body to repair damage, so each of the thresholds would need to be higher for chronic exposures.

- **250–500 mSv** (25–50 rem): This is when the first measurable biological effects are observed in humans. A decrease in the white blood cell count is seen in a few hours. Later, a measurable decrease in the red blood cell count and in the number of platelets will be observed. Temporary sterility (both genders) is also commonly observed. Complete recovery can be expected in a matter of weeks or a couple months. No one dies from this dose.
- **2 Sv** (200 rem): Reversible bone marrow damage occurs. Nausea, vomiting, fatigue, and hair loss also happen. Some will die.
- **4–6 Sv** (400–600 rem): A complete shutdown of the bone marrow function (production of blood cells!) happens, but is only temporary. Severe gastrointestinal distress is experienced. There's only a 50:50 chance of survival.
- **>7 Sv** (> 700 rem): The bone marrow function is irreversibly damaged. Death is almost certain.
- **10 Sv** (1000 rem): You will get really, really sick and die. Death will occur within 1–2 weeks.
- **20 Sv** (2000 rem): The nervous system is damaged. Loss of consciousness within minutes, death typically follows in a matter of hours (a day or two at the most).

Ew, sounds bad! So, what are we exposed to in daily life? Figure 1.8 gives the average dose equivalent to humans living in the United States from a variety of sources. Note that the units on the y-axis are mSv per year. Radon alone gives the average American a 2.3 mSv dose every year. The total average dose in the U.S. is about 6.2 mSv per year. Roughly 50% of that dose comes from natural sources, and about 48% from medical procedures. Almost all of the remaining 2% is due to consumer products containing radioactive materials.[2] Only about 0.05% comes from humans spreading radioactive materials around the planet, such as nuclear power plant accidents, weapons testing, etc.

Not all medical procedures zap you to the same extent. Computed tomography (CT) scans generally result in the highest doses (30–100 mSv, or 3–10 rem)—depending on how much of you gets scanned. A chest X-ray gives you only ~0.1 mSv (~10 mrem). Nuclear medicine procedures result in doses ranging from 1 mSv (0.1 rem) up to 10 mSv (1 rem).

We can compare these procedures to other doses you might experience in daily life. The new airport security scanners that peek under your clothes using X-rays[3] deliver a dose less than 0.25 μSv (weak!). During an airline flight humans receive ~5–10 μSv/h (~0.5–1 mrem/h) because planes fly up where there's not much atmosphere to shield cosmic rays. Fallout from aboveground nuclear weapons testing is responsible for less than 3 μSv/year of our annual dose. The worst doses received as a result of the Three Mile Island accident in 1979 were about 1 mSv (~100 mrem)—well below the threshold to cause immediate health effects.

---

1   Some argue that the U.S. government is continuing to experiment. Veterans of the Persian Gulf wars are currently being monitored for long-term health effects due to exposure to depleted uranium used in armor-piercing weapons (Section 4.1.3). So far, there is no epidemiological evidence to suggest these veterans are suffering any adverse health effects due to ionizing radiation.

2   Mostly from building materials, air travel, cigarettes, mining and agriculture, fossil fuels, glass, and ceramics.

3   The other type uses microwaves—see Question 1.7. For more on this, see the paper by Hoppe and Schmidt in the bibliography.

Individual doses will vary depending on which medical procedures are performed on them, whether they fly a lot, or where they live. In the United States, those that live in parts of the mountain west see the highest levels of background radiation. This is due to a combination of high uranium concentrations in the ground (higher ^{222}Rn), and higher altitude (higher cosmic radiation levels from less atmospheric shielding). For example, folks living in Denver experience ~2–4× as much background radiation than people living at sea level in the United States. There are parts of the world where background radiation levels are higher: Finnish citizens average roughly 8 mSv per year, while parts of France, China, Iran, and India see natural doses upwards of 50 mSv each year. It should be noted that people living in these areas of the world do not have higher cancer rates, nor is any adverse health effect observed.

13.3 RISK

Taken collectively, the numbers at the end of the preceding section should provide us with some reassurance concerning the doses received in medical procedures, or even living through a nuclear power plant accident of the same magnitude as TMI or Fukushima. Even with this safety in numbers, some apprehensions likely remain. Risk is something that can be quantified and understood, but for most of us, it is our *perception of risk* that affects our behavior. We'll eventually come back to a more quantitative understanding of the risks associated with ionizing radiation, but first let's examine how it is perceived.

Table 13.2 gives the relative risk of various activities and technologies. Starting from a list of 30, a group of 234 Wisconsin college students in five radiation physics classes were asked to quickly rank them from 1 to 30, with 1 being the riskiest, and 30 being the safest. Their collective results appear under "Class." Only the top ten, and selected others are shown. These results can be considered "perceived" risk.

Table 13.2 also gives rankings of some of the same activities and technologies by professionals that study risk ("Experts"). These rankings can be considered "actual" risk. By and large, the collective intelligence of the class correctly identified most of the top ten riskiest things. There are,

TABLE 13.2
Rankings of Risky Activities and Technologies

Class	Experts
1. Smoking	1. Motor vehicles
2. Motorcycles	2. Smoking
3. Handguns	3. Alcoholic beverages
4. Firefighting	4. Handguns
5. Motor vehicles	5. Surgery
6. Police work	6. Motorcycles
7. Surgery	7. X-rays
8. Mountain climbing	8. Pesticides
9. Alcoholic beverages	9. Electric power (non-nuclear)
10. Large construction	10. Swimming
11. Pesticides	13. Large construction
14. Electric power (non-nuclear)	17. Police work
16. Nuclear power	18. Firefighting
20. X-rays	20. Nuclear power
26. Swimming	29. Mountain climbing

Source: "Expert" rankings are from Slovic, P., *Radiat. Prot. Dosimetry*, 68, 165, 1996. With permission.

however, some fascinating anomalies in the overall rankings by the two groups. These anomalies are discussed in the following paragraphs.

Familiarity breeds contempt. The class ranked swimming as much safer than the experts, probably because it is something most have done safely, and likely with some pleasure. The stats say otherwise. Lots of people drown every year while swimming, making it an inherently dangerous activity. The students also ranked X-rays quite a bit lower than the pros. Again, this is because many have experienced X-rays, and suffered no ill effects. However, X-rays are a form of ionizing radiation, and are potentially dangerous when very large doses are delivered to biological systems like us humans. Similar arguments could be made for the students ranking motor vehicles and alcoholic beverages lower than the experts.

Lack of control creates fear. Most people would rank nuclear power as much riskier (higher on the list) than the experts. In fact, in Slovic's work (cited in Table 13.2) the populations he surveyed ranked nuclear power at, or very near the top, of the list. The Wisconsin radiation physics students more closely matched the experts' assessment. With the exception of X-rays, anything involving ionizing radiation creates an unusual amount of apprehension in the general population. People tend to view nuclear power as more dangerous than is borne out by experience. Ionizing radiation is feared, in part, because we cannot directly sense it, and we know that it *can* cause adverse health effects, sometimes decades in the future. Our lack of control over exposure to and effects of ionizing radiation causes it to be seen as a higher risk than is warranted. Additionally, nuclear power is feared because of its potential for catastrophe—a nuke plant burning is potentially more dangerous to the surrounding public than a coal plant catching fire. Nuclear power has proven itself much safer than other major forms of electricity generation—even taking into account the deaths caused by Chernobyl.

Training and experience make certain activities safer than we think. The students ranked police work, firefighting, large construction, motorcycles, and mountain climbing as more dangerous than they really are. While there is real risk associated with these activities, they are less dangerous than we think because of the extensive training, detailed procedures, and considerable regulation associated with each. The same is true of radiation therapy and nuclear medicine. Because both professions involve work with ionizing radiation, and because we (as a society) fear radiation, they are highly regulated. Only well-educated professionals are allowed to perform these procedures, and they do so with a remarkable safety record. These professions are among the safest of all careers! The same can be said of nuclear power. Because it is more extensively regulated than many other forms of power generation, it is inherently safer.

We've discussed risk in general terms and have tried to show that there is often a difference in the perception of risk and the actual risk. We've also looked at the health effects of receiving a single large dose. Those of us working with radioactive materials and radiation-generating devices would much rather know: what are the health risks at low doses, especially those received over a long period of time?

As discussed in Section 1.6, there is still some debate to the answer of this question. The uncertainty is illustrated nicely in Figure 13.2 (a reproduction of Figure 1.6). The solid line represents acute doses greater than 250 mSv, where measurable biological effects begin. As we learned in the previous section, the severity of the effects increases with dose, in a (very!) roughly linear fashion. It is below the 250 mSv threshold where there is still controversy. Regulatory agencies assume a linear extrapolation of this trend (straight dashed line in Figure 13.2), which is called the linear non-threshold (LNT) hypothesis. While there is no scientific evidence to support this hypothesis, it is used by regulators as a conservative approach to working with ionizing radiation. Unfortunately, it leads to the assumption that any exposure to ionizing radiation can have deleterious effects, and therefore should be avoided. This generates unnecessary fear and anxiety, even among well-educated professionals.

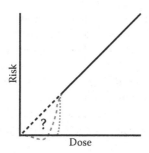

FIGURE 13.2 Risk vs. dose.

Scientific evidence tends to suggest that either there's a threshold, below which no harmful effects are observed (dotted gray line in Figure 13.2), or that there's a health benefit (radiation hormesis) to low doses of ionizing radiation (dashed gray line). Evidence comes from the relatively large number of people that live in parts of the world with higher background radiation. Even though ionizing radiation levels are 10–30 times higher than where most other people live, the folks getting zapped more do not show any less longevity or higher cancer rates. In fact, the opposite is sometimes observed. The same is true with other groups that received higher doses than normal, yet lower than what is known to cause adverse health effects. This includes many of the atomic bomb survivors in Nagasaki and Hiroshima who received low-level doses.

The Chernobyl accident provides some of the best evidence against the LNT model. Hundreds of thousands of people received low-level doses because they lived in an area that became contaminated or they worked to help clean up after the disaster. Based on the LNT model thousands should've died in the decades since the accident, yet only about a dozen have. These victims generally died from thyroid cancer (or complications resulting from it) brought about from drinking milk heavily contaminated with [131]I. This only happened because the Soviet government was in deep denial in the first couple of days following the accident and failed to warn its citizens about potential contamination in local milk.

Studies on people who work around ionizing radiation show no difference with standard populations, or experience an apparent health benefit. This positive result may be due to higher education and/or greater access to health care in the group studied. What's important is that no additional risk appears to be associated with the sorts of low-level doses experienced by those working with ionizing radiation. This makes some sense. Our bodies have mechanisms to deal with the billion or so radioactive decays that naturally take place inside of us every day, as well as damage done from external (natural) sources of ionizing radiation. Like all living things on Earth, we've adapted to being radioactive and living in a radioactive environment.

Our cell repair/replacement mechanisms are fairly quick, therefore a significant dose (more than 250 mSv) received over a period of time is less likely to cause biological damage than if it was received all at once. In fact, the same dose delivered over hours or days, instead of instantaneously, reduces the chance of cancer ~4 times. That same dose, delivered over longer periods of time results in a ~10 times lower cancer risk.[4] This is important for those living in, or returning to, areas with some contamination from Chernobyl or Fukushima. We already know that people do fine in parts of the world where external sources of background radiation are 10–40 times normal.[5] It is possible that humans can live in areas with external sources 50 or even a few hundred times normal without

---

4 These factors apply to gamma, X-ray, and beta, but not alpha or other "heavy" energetic particles (protons and other nuclei).
5 "Normal" is about 0.9 mSv/a worldwide for all external sources. Total average dose from all natural sources (internal and external) is 2.4 mSv/a.

ill effect. The Chernobyl contaminated areas open for resettlement in the Ukraine and Belarus are well within this limit (< 1 mSv/a additional dose). The Japanese government is allowing people to return to areas with less than 20 mSv/a additional dose. In light of the above, these levels seem rather conservative.[6]

Likewise, an additional occupational dose limit of 50 mSv/a also seems conservative, but this is done to provide an ample margin between what workers receive and the somewhat uncertain threshold below which ill effects might be observed. Because radiation workers have a low annual dose limit, they must be conscious in their work of ways to minimize dose, even if it appears that a higher dose poses no greater risk.

All radiological workers face exposure to ionizing radiation—how do we manage it? As mentioned earlier, we monitor it closely through badges, rings, swipe tests, and radiation surveys. When working with ionizing radiation, we practice **ALARA**. As mentioned in Section 1.6, ALARA stands for "*as low as reasonably achievable*". We want to apply this concept to ourselves, those we work with, and to the environment in terms of contamination. Just because safety limits are set conservatively doesn't mean that we have to run our doses up to them. For example, if there is a procedure than can be done more than one way with equal effectiveness, cost, and benefit, we should pick the procedure that results in the lowest dose and contamination.

Exposure can be lowered by minimizing **time** near the radioactive source, maximizing **distance** to the source, and increasing **shielding** around the source. Exposure and dose will change in a linear fashion with time. If time near a source is cut in half, dose will also be halved.

Example 13.2: If the dose rate near a source is 6.6 mSv/h, what dose will be received over 30 minutes?

$$\frac{6.6 \text{ mSv}}{h} \times 0.50 \text{ h} = 3.3 \text{ mSv}$$

Exposure or dose varies inversely with the square of the distance. If distance is doubled, exposure is decreased by a factor of four. If distance is tripled, exposure decreases by a factor of 9. This effect is known as the **inverse square law**.

Example 13.3: If the dose rate 1.2 m from a source is 6.6 mSv/h, what will the dose rate be 4.5 m from the source?

$$\frac{6.6 \text{ mSv}}{h} \times \frac{1}{\left(\dfrac{d_2}{d_1}\right)^2} = \frac{6.6 \text{ mSv}}{h} \times \frac{1}{\left(\dfrac{4.5 \text{ m}}{1.2 \text{ m}}\right)^2} = \frac{0.47 \text{ mSv}}{h}$$

We can combine time and distance into a single equation:

$$X = \frac{\Gamma \cdot A \cdot t}{d^2} \tag{13.3}$$

where X is exposure in units of Coulombs per kilogram (C/kg), Γ is the exposure rate constant which *can* have units of C·m²/kg·MBq·s. With these units of the exposure rate constant, A is activity in

---

6   Some have suggested that we return to the threshold set in the 1920s of 680 mSv/a, i.e., anything below that is safe. The safe threshold dose is likely higher, probably between 1 and 10 Sv/a. Applying a 1 Sv/a standard to Fukushima would mean *all* contaminated areas were safe for resettlement within a year of the meltdowns.

TABLE 13.3
Exposure and Dose Rate Constants

Nuclide	Exposure rate constant (C·m²/kg·MBq·s)	Dose rate constant (mSv·m²/MBq·h)
^{13}N	1.13×10^{-12}	1.53×10^{-4}
^{18}F	1.10×10^{-12}	1.49×10^{-4}
^{60}Co	2.50×10^{-12}	3.38×10^{-4}
^{99m}Tc	1.54×10^{-13}	2.06×10^{-5}
^{125}I	3.38×10^{-13}	4.34×10^{-5}
^{131}I	4.26×10^{-13}	5.65×10^{-5}
^{133}Xe	1.10×10^{-13}	2.78×10^{-5}
^{137}Cs	6.64×10^{-13}	8.90×10^{-5}
^{192}Ir	8.91×10^{-13}	1.19×10^{-4}
^{201}Tl	8.72×10^{-14}	1.16×10^{-5}

Source: Smith, D. S., and M. G. Stabin. 2012. Exposure rate constants and lead shielding values for over 1,100 radionuclides. *Health Phys.* 102(3):271-291. With permission.

megabecquerels (MBq), t is time in seconds (s), and d is distance in meters (m). Any of these units can be changed (e.g., mCi instead of MBq for activity), but should be consistent with the units on the exposure rate constant. Equation 13.3 can be used for dose or dose equivalent, again, so long as the units are changed appropriately. Exposure and dose rate constants for selected nuclides are listed in Table 13.3.

Example 13.4: A human sits 1.5 m from 925 MBq of ^{137}Cs 40 hours each week. What is her exposure each week?

The exposure rate constant is in Table 13.3. Using Equation 13.3:

$$X = \frac{6.64 \times 10^{-13} \dfrac{C \cdot cm^2}{kg \cdot MBq \cdot s} \times 925\ MBq \times 40\ h \times \dfrac{3600\ s}{h}}{(1.50\ m)^2} = 3.9 \times 10^{-5} \frac{C}{kg}$$

Shielding is a little less straightforward. Appropriate shielding should be added when necessary. Low Z materials (like plastics) should be used to shield pure beta-emitting nuclides. If there's a fair amount of the beta-emitting nuclide, significant bremsstrahlung photons may be produced from the interactions of beta particles with higher Z materials that may be nearby. The use of a low Z material (like plastics) for shielding will therefore minimize the amount of bremsstrahlung. If bremsstrahlung is unavoidable, or the nuclide emits gamma rays as part of its beta decay, a high Z, high-density material like lead, should surround the plastic shielding.

High Z, high-density materials are the best choice for shielding energetic photons such as X-rays and gamma rays. Remember from Section 7.4 that the probability of photon interaction increases with the atomic number (Z), and the amount of material it passes through (density). Also recall that half value layer (HVL) is the thickness of material that will attenuate half of the photon beam passing through it; and that TVL is the tenth value layer—the thickness of material that will cut a photon beam to one tenth of its original intensity. If two TVLs are used as shielding, the photon beam will be cut to 1/100th of its original intensity.

Example 13.5: How much lead is needed to cut the exposure in Example 13.4 to 5.2×10^{-7} C/kg per week? The linear attenuation coefficient is 0.964 cm^{-1} for Pb exposed to gamma rays from ^{137}Cs. ^{137}Cs is a beta- and gamma-emitting nuclide. Let's assume the beta particles are all absorbed and don't create any bremsstrahlung. This allows us to deal with ^{137}Cs as a pure gamma source, and to use Equation 7.16.

$$\ln \frac{I_0}{I} = \mu x$$

$$\ln \left(\frac{3.9 \times 10^{-5} \ \frac{C}{kg}}{5.2 \times 10^{-7} \ \frac{C}{kg}} \right) = 0.964 \ \text{cm}^{-1} x$$

$$x = 4.5 \ \text{cm}$$

How many HVLs is this?

$$\text{HVL} = \frac{\ln 2}{\mu} = \frac{\ln 2}{0.964 \ \text{cm}^{-1}} = 0.719 \ \text{cm}$$

$$\frac{4.5 \ \text{cm}}{0.719 \ \text{cm}} = 6.2 \ \text{HVLs}$$

How many TVLs is this?

$$\ln \frac{10}{1} = \mu \times \text{TVL}$$

$$\text{TVL} = \frac{\ln 10}{\mu} = \frac{\ln 10}{0.964 \ \text{cm}^{-1}} = 2.39 \ \text{cm}$$

$$\frac{4.5 \ \text{cm}}{2.39 \ \text{cm}} = 1.9 \ \text{TVLs}$$

We can incorporate shielding into the exposure equation (Equation 13.3):

$$X = \frac{\Gamma \cdot A \cdot t}{d^2} B \tag{13.4}$$

where B is the **transmission factor**—the fraction of radiation making it through the shielding. Transmission factor (B) has no units and is equal to the intensity of photons making it through divided by the initial photon intensity (I/I_0).

Example 13.6: A human works 40.0 hours per week 1.7 m from a 0.12 mg ^{60}Co source. What thickness of lead would lower his weekly exposure to 2.58×10^{-5} C/kg? The half-value layer for lead is 15.9 mm when exposed to radiation from ^{60}Co.

First, we need to calculate the activity of the source in MBq:

$$A = \lambda N$$

$$= \frac{\ln 2}{5.271 \text{ a} \times \left(5.259 \times 10^5 \dfrac{\text{min}}{\text{a}}\right)} \times \left(0.12 \text{ mg} \times \frac{\text{g}}{1000 \text{ mg}} \times \frac{\text{mol}}{59.9 \text{ g}} \times \frac{6.022 \times 10^{23} \text{ atoms}}{\text{mol}}\right)$$

$$= 3.0 \times 10^{11} \text{ dpm}$$

$$\frac{3.0 \times 10^{11} \text{ decays}}{\text{m}} \times \frac{\text{m}}{60 \text{ s}} \times \frac{1 \text{ Bq}}{1 \text{ dps}} \times \frac{1 \text{ MBq}}{10^6 \text{ Bq}} = 5.0 \times 10^3 \text{ MBq}$$

Now we can solve Equation 13.4 for the transmission factor (B):

$$B = \frac{X \cdot d^2}{\Gamma \cdot A \cdot t} = \frac{\dfrac{2.58 \times 10^{-5} \text{ C}}{\text{kg}} \times (1.7 \text{ m})^2}{\left(2.50 \times 10^{-12} \dfrac{\text{C} \cdot \text{m}^2}{\text{kg} \cdot \text{MBq} \cdot \text{s}}\right) \times (5.0 \times 10^3 \text{ MBq}) \times 40 \text{ h} \times \dfrac{3600 \text{ s}}{\text{h}}} = 0.041$$

Before we proceed, we'll need to calculate the linear attenuation coefficient (μ):

$$\mu = \frac{\ln 2}{\text{HVL}} = \frac{\ln 2}{15.9 \text{ mm}} = 0.436 \text{ mm}^{-1}$$

Since $B = I/I_0$:

$$\ln \frac{I}{I_0} = -\mu x \quad \ln 0.041 = -0.436 \text{ mm}^{-1} \times x \quad x = 73 \text{ mm} = 7.3 \text{ cm}$$

Let's look at one more problem, but this time let's look at dose equivalent rate (DR) rather than exposure. Keep in mind that dose equivalent rate is dose equivalent divided by time (t). Γ is now the dose equivalent rate constant.

$$\frac{\text{dose equivalent}}{t} = DR = \frac{\Gamma \cdot A}{d^2} \tag{13.5}$$

Example 13.7: Calculate the dose equivalent rates at 10 cm and 300 cm from a syringe containing 27 mCi of ^{99m}Tc.

First, we'd better convert activity to MBq.

$$A = 27 \text{ mCi} \times \frac{\text{Ci}}{1000 \text{ mCi}} \times \frac{3.7 \times 10^{10} \text{ dps}}{\text{Ci}} \times \frac{1 \text{ Bq}}{1 \text{ dps}} \times \frac{\text{MBq}}{10^6 \text{ Bq}} = 1000 \text{ MBq}$$

At 10 cm = 0.10 m:

$$DR = \frac{2.06 \times 10^{-5} \dfrac{\text{mSv} \cdot \text{m}^2}{\text{MBq} \cdot \text{h}} \times 1000 \text{ MBq}}{(0.10 \text{ m})^2} = 2.1 \frac{\text{mSv}}{\text{h}}$$

At 300 cm = 3.0 m:

$$DR = \frac{2.06 \times 10^{-5} \frac{mSv \cdot m^2}{MBq \cdot h} \times 1000\ MBq}{(3.0\ m)^2} = 0.0023 \frac{mSv}{h} = 2.3 \frac{\mu Sv}{h}$$

QUESTIONS

13.1 Define the following and give their SI units: exposure, dose, and dose equivalent.

13.2 Briefly explain why most humans receive their greatest radiation exposure from cosmic rays yet their greatest dose comes from radon.

13.3 Calculate the number of ion pairs formed in 1.00 cm³ of air (density = 1.29 g/L) exposed to 1.00 R of X-rays.

13.4 Assuming that all of the X-ray energy in Question 13.3 is deposited in the air, and that the average ionization energy of air can also be expressed as 33.85 J/C, what dose is deposited? Calculate your answer in both Gy and rad.

13.5 The average American receives a 2.28 mSv dose equivalent from radon each year. Assuming you receive this dose, and that it all comes from the 5.49 MeV alpha particle emitted by ^{222}Rn, how much energy is deposited in your body each year from radon? Approximately how many decays does this represent?

13.6 How much energy (MeV) is deposited in a 75 kg human from a 0.055 cGy dose of alpha particles? What is the maximum number of ion pairs that could be formed in air from this same dose?

13.7 Calculate the dose equivalent in humans, in both rem and mSv, for the dose in the previous question of α radiation. Do the same for 0.055 cGy of X-radiation.

13.8 A scientific paper measured doses received from external sources by people in Ramsar, Iran. The highest external dose is 260 mSv/a in a part of Ramsar noted for its hot springs. The paper noted no positive or negative health effects among people living in this part of town. How many times higher is background in this part of Ramsar than the global average? Which dose/risk hypothesis does this work most closely support? What are the ramifications for the values in Table 13.1 or the people living in areas contaminated by Chernobyl and Fukushima?

13.9 During a routine radiation therapy procedure, a therapist is exposed to 8.0 × 10⁻⁵ C/kg·hr. If the procedure takes 4.0 minutes, what is her exposure? These data were obtained while she was standing 0.91 m from the source. If she stands 0.61 m from the source, she can complete the procedure in 1.5 minutes. Would her exposure change? If so, by how much?

13.10 A concrete and steel cask is used to store fuel assemblies after they were used in a nuclear reactor. The dose rate 0.61 m from the cask is 10 μSv/h. If a person lives exactly 0.5 km from this cask, what annual dose are they receiving from it? Assume this person spends exactly half their time in the house; the rest is spent far from the cask. Compare this to something more tangible.

13.11 How does ALARA apply to a patient about to undergo a radiological procedure?

13.12 Using Figure 1.8, estimate the average total dose for someone living in the United States. What percentage of this dose is anthropogenic? Compare this to the annual public effective dose limit established by the NCRP. Briefly explain any discrepancy.

13.13 The recommended annual occupational effective dose limit for the whole body is 50 mSv, but is 500 mSv for skin, hands, and feet. Briefly explain why these numbers are so different.

13.14 By what factor will dose change if the distance to the source is doubled and the time spent is halved?

13.15 How far would a human need to be from a 925 MBq ^{137}Cs source to keep the exposure rate below 0.31 mR/h?

13.16 A person works 7.32 m from a 1.14 mg (tiny!) sample of ^{60}Co. Assuming a 40.0 h work week with two weeks off per year, and no shielding, what would her annual dose be (mSv)? The TVL in lead for the gamma rays emitted by ^{60}Co is 45.3 mm. Calculate the thickness of lead needed for a 10 mSv annual dose. Briefly state two other ways exposure could be decreased.

13.17 A patient is apprehensive about receiving the 0.15 mSv dose you are about to administer. What should you do?

13.18 A patient receives a 570 kBq of ^{131}I as part of a medical procedure, then goes home to sleep. What maximum dose equivalent could the patient's wife receive overnight if they sleep 48 cm apart for 8 hours and 20 minutes? Assume that the gamma emissions from the ^{131}I travel an average of 15 cm before leaving the patient's body. The linear attenuation coefficient for ^{131}I in humans is 0.11 cm^{-1}.

13.19 The exposure 182 cm from a source is 5.4×10^{-6} C/kg over 5.0 minutes. What is the exposure 122 cm from the same source for 1.5 minutes?

13.20 The dose 0.91 m from a source is 2.6 mSv over 4.0 hours. What is the dose 0.33 m from the same source for 1.7 hours?

13.21 Calculate the dose equivalent rate 3.00 m from a 1900 MBq ^{192}Ir source.

14 X-ray Production

X-rays are used extensively in diagnostic and therapeutic medical applications. This chapter starts by looking at how X-rays are generated for diagnostic applications such as those performed in hospitals and dentist offices, and then moves into how high-energy X-ray beams are produced for radiation therapy. Interestingly enough, knowledge of how electrons interact with matter (Section 7.2) is key to understanding how X-rays come about.

14.1 CONVENTIONAL X-RAY BEAMS

For diagnostic medical applications, such as an X-ray of your foot to see if any bones are broken, we'd like to consistently produce high-energy photons on demand—a flip of the switch to provide the same energy and intensity every time would be great. It is also desirable for our photon generator to be safe and affordable. The only source that satisfies all these requirements is a particle accelerator. A **conventional X-ray tube** is a small particle accelerator (Section 10.4). Let's see how it works.

A schematic of a typical conventional X-ray tube is shown in Figure 14.1. The filament is a tungsten wire, much like those used in (old-school) incandescent light bulbs. If we run a little electric current through the filament at a high enough voltage, we can get electrons to boil off the wire. The filament is therefore a cathode.

After they jump off the cathode, the electrons are accelerated toward the positively charged anode (target). The maximum energy they attain will be equal to the potential difference (tube voltage) between the two electrodes. For all this to happen, both electrodes need to be under vacuum, so the whole X-ray tube is enclosed, and pumped out. This is important because the electrodes would rapidly corrode if exposed to oxygen, and because the electrons could interact with any matter (gases) on their way from the filament to the target.

Electrons won't boil off the cathode unless the filament is pretty hot. The current of electrons flowing through the wire (**filament current**) does this—just like a toaster! Since it will get hot, it's important to make the filament out of a metal with a high melting point. Tungsten melts higher than any other metal (3370°C!!), so it works quite well. The filament current is usually set at a few amps. The other current flowing in the tube is made up of the (relatively few) electrons moving from the cathode (filament) to the anode (target). This is called the **tube current** and is usually a few tenths of an amp (a few hundred milliamps).

The **focal spot** is the place on the target where the electrons hit the target. A smaller focal spot creates a smaller beam, which is sometimes desirable. For example, minimizing the dose to surrounding tissue when trying to kill a small tumor. A smaller focal spot can be accomplished by using a smaller filament (makes for a smaller electron beam). Unfortunately, this also creates a weaker (less intense) X-ray beam.

DOI: 10.1201/9781003303787-14

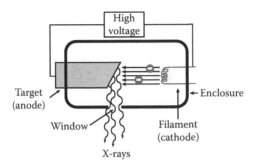

FIGURE 14.1   Schematic of a conventional X-ray tube.

So, how does an X-ray tube make X-rays? When the electrons hit the target, there's two ways they can make X-rays (see Section 7.2 for more details).

1. Bremsstrahlung—the continuous energy spectrum of X-rays produced through inelastic scattering of the electrons in the tube current by the nuclei in the target. This can always happen, regardless of the energy of the incoming electrons. The maximum possible X-ray energy would be equal to the energy of the electron (tube voltage).
2. Characteristic X-rays—if the beam electrons have enough energy to excite or ionize a K shell electron in the target material, an electron vacancy is created in that shell. When an L or M shell electron drops down to the K shell to fill the vacancy, a characteristic X-ray is emitted. Remember, a characteristic X-ray has a specific energy, because the target electron energy levels are quantized.

When we put these two types of interactions together, we see an interesting distribution of photon energies—as shown by the solid line in Figure 14.2. The rather broad part of this spectrum (almost all of it!) is bremsstrahlung radiation. Notice that it maxes out at fairly low energy then gradually tails off as energy increases until just below the tube voltage. The probability of forming lower energy photons via bremsstrahlung is higher because it is more likely that an electron will interact with an atom at a longer rather than at a shorter distance. The bremsstrahlung feature tails off sharply at low-energy because the very low-energy X-rays are more efficiently filtered by the tube enclosure. The dashed line in Figure 14.2 shows what the bremsstrahlung spectrum would look like without this filtration.

The sharp peaks on top of the bremsstrahlung spectrum are the characteristic X-rays produced from ionizations and excitations. The energies of these peaks depend on the target material. Tungsten is used as a target for medical applications because it is high melting, and because its characteristic X-rays are relatively high in energy. The energy increases because the spacing between electron orbitals generally increases with atomic number. Higher energy means more penetrating—less likely to interact with matter. If only low-energy (soft) X-rays were used, they would be almost entirely absorbed by the patient. A higher atomic number for the target also means that the intensity of the X-rays produced via bremsstrahlung will be higher. Remember that as Z increases, the more likely it is for an electron to interact (Section 7.2).

X-rays can be produced at different depths inside the target—not just at the surface. If one is produced a little way in, it could interact (scattered or absorbed) with the target material on its way out.

We've already seen that the X-ray tube enclosure filters the X-rays to some extent. Because the heat produced in the process of making X-rays needs to be dissipated, X-ray tubes are usually surrounded by oil or water that is circulated. Finally, there is usually an exit window for the X-rays to travel through before leaving the tube. The total attenuation due to the target, enclosure, coolant,

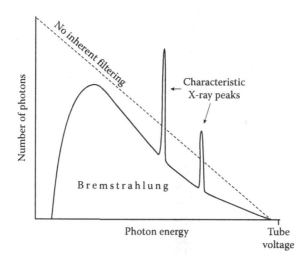

FIGURE 14.2 X-ray spectra with (solid line) and without (dashed line) inherent filtering.

and exit window is referred to as the tube's **inherent filtration**. There's nothing we can do about this; it is a result of the tube's design.

The inherent filtration is usually equivalent to placing ~1 mm of aluminum in the X-ray beam. Like all forms of photon filtration, inherent filtration will harden the X-ray beam. Remember that this means the overall intensity is lower, and the average energy is higher after filtration (Section 7.4). Additional hardening can be performed after the beam exits the tube by placing additional filters in the beam's path. This is desirable for most medical applications since the low-energy photons will tend to deposit all of their energy in the first cm or so of the matter (flesh!).

What happens to the X-ray spectrum when we mess with the current? Basically, it just changes the overall intensity. Intensity increases with current. Increasing filament current increases tube current—more electrons bombard the target, creating more X-rays.

How about voltage? When voltage is increased, the average energy of the photons produced is also increased. Sometimes the low-energy X-rays are desirable. If we wanted to treat a surface malady, such as skin melanoma, then we'd want as many low-energy photons as possible. Surface treatments are usually performed with a tube voltage of 5–15 kV. If we want to go a little deeper (~5 mm below the surface), then 50–150 kV would be appropriate. If we need to penetrate further, we need to keep cranking up the voltage. Note that people working with X-rays will typically refer to the photon beam energy using the tube potential, e.g., "treatment will be with a 150 kV X-ray beam." While conventional, it is misleading and incorrect. The photons are not 150 keV (notice corrected units); rather they are a broad continuum of energies with a maximum a little below 150 keV.

The shape of the target is also important to consider. In Figure 14.1, the electron beam hits the anode at an angle. This angle serves two purposes. First electrons with less than 1 MeV of energy (true for all conventional tubes) *tend* to produce X-rays in a direction that is 90° from the electron beam. This process is called "**reflectance**" because it *looks* like light reflecting on a mirror. You should know this is not true! First, we're starting with electrons, not photons, and second the photon direction is a tendency, in reality they are produced in all directions.

Second, a focal spot on an angle to the electron beam will produce a narrower, more intense X-ray beam. The angle allows a wider electron beam to hit the target (higher intensity X-rays produced), while producing an X-ray beam that is smaller in area—as pictured in Figure 14.3. The length A of the focal spot (approximately equal to the width of the electron beam) is larger than the length a (width of the X-ray beam). In fact, the mathematical relationship between the two lengths is:

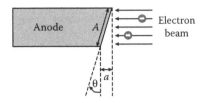

FIGURE 14.3 Geometric reduction of the beam size relative to the focal spot size.

FIGURE 14.4 X-rays produced by a transmission target.

$$a = A \times \sin\theta \qquad (14.1)$$

where the angle θ is the angle of the anode, which is typically 6° to 17° for a diagnostic unit. In Figure 14.3, this angle is drawn at 17°. If A is 0.10 cm, then:

$$a = 0.10 \text{ cm} \times \sin 17° = 0.029 \text{ cm}$$

As mentioned above, reflectance only happens with electron energies below 1 MeV. When electrons have more than 1 MeV, X-rays *tend* to be produced in the same direction as the electron beam. This process is called "**transmission**" because it appears that electrons are "transmitted" through the target, magically turning into X-rays (Figure 14.4). As you might guess, the process of "converting" an electron beam to an X-ray beam is a bit more complex (Section 7.2) than suggested by Figure 14.4!

Keep in mind that X-rays only *tend* to be traveling in the directions shown in Figures 14.3 and 14.4. In reality, X-rays will travel out of the target in lots of different directions, regardless of how they are made. We are interested in generating a *beam* of X-rays. That means we just want those photons traveling in a particular direction and to soak up all the others. For this reason, X-ray tube housings are usually pretty thick and are made up of high Z metals. To de-select the photons that are traveling close to the correct direction, they are passed through a metal collimator as shown in Figure 14.5. The length and aperture of the collimator will determine the consistency (how parallel are the photons?) and size of the beam.

Only part of the projectile electrons' kinetic energy goes into X-ray production. Remember that the maximum electron energy in keV is equal to the tube's electric potential in kV. We can then get a handle on efficiency using:

$$f = 3.5 \times 10^{-4} \, ZE \qquad (14.2)$$

where f is the fraction of electron energy that is turned into photon energy, Z is the atomic number of the target material, and E is the maximum electron energy in MeV. For example, with a tungsten anode and a tube potential of 140 kV:

$$f = 3.5 \times 10^{-4} \times 74 \times 0.140 \text{ MeV} = 0.0036 \text{ or } 0.36\%$$

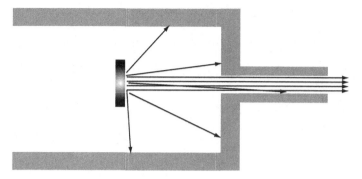

FIGURE 14.5   The use of a collimator to form an X-ray beam.

Only 0.36% of the energy being deposited by the electron beam into the anode ends up as X-ray photons! This is so inefficient it makes SUVs look pretty green. The rest of the electrons' energy ends up as heat. This is the source of all the heat that needs to be dissipated by the coolant mentioned earlier. If we repeat this calculation for a 4 MeV electron beam, we'll get 10.4% efficiency, and 25 MeV gives 65%. As the tube potential increases, so does the efficiency—in rather dramatic fashion. If more photons are desirable, then cranking up the voltage is one way to get 'em.

14.2   HIGH-ENERGY X-RAY BEAMS

As described in Section 9.2, the history of the use of X-rays for radiation therapy shows ever-increasing photon energies. Increasing photon energy means more efficient X-ray production (higher intensity) and allows for treatment deeper inside the patient.

Conventional tubes were commonly used in the first half of the 20th century. Instruments using conventional tubes have a number of names, usually related to the energy of the X-rays produced. Superficial therapy machines operate in the 50–150 kV range and were only useful in treating the surface or near surface of the patient's body. **Orthovoltage** machines were in common use in the middle of the last century but are generally limited to about 400 kV. They were also known as "deep therapy" machines, because of their ability to treat lesions below the skin. They have now been completely replaced by higher-energy linear accelerators, the so-called **megavoltage** machines (or "linacs"). Why?

Reason #1: Lower energy photons are more likely to interact with matter. Basically, they are not penetrating enough—too much energy is dumped in the first cm (or so) of tissue. If the tumor is deeper than that, we'll end up giving the healthy tissue between the skin and the tumor too much of a dose. To deliver more of a dose deeper into the patient, we need more power!

We already know this is an issue, but how can we imagine this in terms of dose delivered to humans? This is pictured in Figure 14.6. **Percent depth dose** (PDD) is the percentage of the maximum dose remaining at a particular depth in a human. If the maximum dose is delivered 1 cm below the surface, and the dose at 2 cm is 80% of the maximum, then the PDD at 2 cm is 80. Figure 14.6 shows how PDD drops off, as the high-energy photons travel through matter. This is exactly what we expect—the thicker the matter, the greater the beam attenuation (Section 7.4).

Several different curves are represented in Figure 14.6. Going from the orthovoltage curve to the 18 MV curve, the average energy of the photon beam increases, and so does the degree of penetration—greater doses are delivered to greater depth. These data were collected by placing a detector at the appropriate distance underwater while exposing the water to the X-ray beam. Water is an excellent surrogate for humans since we are composed of mostly water, and it generally interacts with high-energy photons like human tissues (similar Z and density).

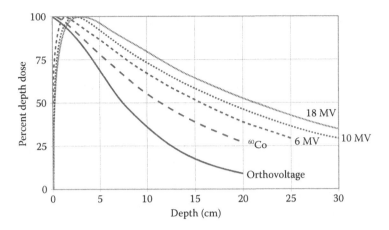

FIGURE 14.6 Variation of percent depth dose (PDD) with depth in water and energy of the X-ray beam.

The ^{60}Co curve lies just above the orthovoltage curve. ^{60}Co emits photons with energies of 1.2 and 1.3 MeV, and therefore represents a higher energy beam than an orthovoltage machine. The remaining three curves are from megavoltage machines of increasing energies. Notice that the percent depth dose starts at 100% at the surface of the water (or the patient's skin) for the orthovoltage and ^{60}Co curves. It is not until we get to the higher-energy machines that the peak moves below the surface, i.e., a greater dose is deposited at 1–3 cm depth than at the surface. Since most tumors are below the surface, this is desirable.

As you might remember from Section 9.2, frying the patient's skin was used as an indication they had been zapped enough back in the early days of radiation therapy. Delivering the maximum dose below the surface of the skin is known as **skin sparing**. This phenomenon is simply illustrated in Figure 14.7, the darker the shading, the higher the dose delivered at that depth.

If you stop and think about it, this doesn't make sense. Based on what we learned about photons interacting with matter (Section 7.4), photon intensity should decrease as it travels through matter. Therefore, the highest dose should always be delivered at the skin because that's where photon intensity is always greatest. With high-energy photons it is a little more complex than just attenuation.

Higher-energy photons interact with low Z matter almost exclusively via Compton scattering (Figure 7.9). When this happens, both the scattered photon and the Compton electron tend to move in the same general direction as the incident beam. Because of their higher probability of interacting with matter, the Compton electrons will deposit their energy fairly close to where they are generated, but it will be *deeper* into the body. Patient dose at depth is therefore due to both photon *and* Compton

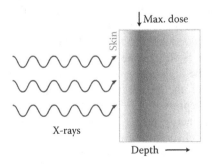

FIGURE 14.7 Skin sparing by higher-energy X-ray beams.

TABLE 14.1
Maximum Dose Depth in Water for Various Photon Energies

Photon Energy	Depth of Maximum Dose (cm)
Orthovoltage	0.0
^{137}Cs (0.7 MeV)	0.1
^{60}Co (1.2 MeV)	0.5
4 MV	1.0
10 MV	2.5
25 MV	5.0

electron interactions within their body. More photons and electrons are depositing dose at a depth of a few centimeters than just photons at the surface.

Higher-energy beams will be more penetrating and will transfer more energy to the electrons they knock loose. Higher-energy electrons will travel further into the patient's body before they deposit much of their energy (under the Bragg peak, Section 7.2). As a result, the depth of the maximum dose increases with the energy of the incident beam. Table 14.1 illustrates this trend—photon energy increases going down the first column and the depth of maximum dose increases down the second column.

It turns out that the Compton electrons are key to figuring out where the maximum dose is delivered. The maximum dose is deposited at the point inside the body where the number of electrons knocked loose is equal to the number of electrons being stopped. This is called **electron equilibrium**. Before this point more electrons are being knocked loose and after it fewer electrons are stopping. The tissue between the skin and the maximum dose is appropriately known as the **dose buildup region**.

Reason #2: Penumbra size (or beam edge unsharpness) is larger for lower energy beams. Penumbra is the fuzzy edge around the beam. This is true for any beam of photons traveling through matter (Figure 14.8a). Even laser light has a fuzzy edge to it. Penumbra is bad for radiation therapy because the tumor needs to be treated with a uniform dose, and penumbra creates non-uniformity at the beam edge.

Some of the penumbra is due to scattering of photons as they travel through matter (usually air); the rest is due to the size of the source. As can be seen in Figure 14.8b, a certain area of the skin will see the entire source and will receive the maximum dose. As we begin to move outside of that

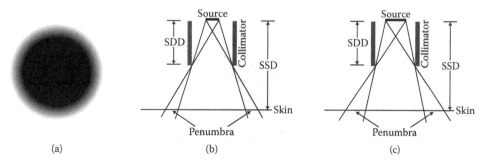

FIGURE 14.8 (a) Penumbra viewed around a cylindrical beam. (b) Geometric origin of a penumbra. (c) Effect on a penumbra of increasing source size.

area, the collimator blocks parts of the source, so that area will receive a lower dose. Eventually we move too far away, and the source is completely blocked by the collimator. This area sees no dose. Increasing the size of the source (focal spot in X-ray machines) will increase the size of the penumbra (compare Figures 14.8b & c).

Decreasing the **source to collimator distance** (SDD) also increases the penumbra. Finally, increasing the **source to surface distance** (SSD) increases the penumbra. Decreasing the collimator-patient distance decreases the penumbra, but a certain amount of distance (> 15 cm) is a good idea. Photons hitting the collimator can kick out electrons—if the collimator is too close to the patient, they can get a pretty good skin dose from these secondary electrons. Having some distance also facilitates movement of the source around the patient.

Reason #3: Can't use **isocentric techniques**. Rotating the source around the patient with the target (what you're aiming your photons at, i.e., the tumor) at the center of the circle will maximize the dose to the tumor while minimizing the dose to the healthy tissue surrounding the tumor. The point (or volume) of the beams' convergence is the **isocenter**. Figure 14.9 shows the effect of using four different source positions on the dose received—again, the darker the shading, the greater the dose. As you might imagine, using more source positions will further maximize dose to the area of convergence (the tumor) while minimizing it elsewhere (healthy tissue).

Why can't orthovoltage machines use isocentric techniques? Remember that X-rays are very inefficiently produced at orthovoltage energies. Therefore, the intensities of these beams are also low—in order to get a decent dose, orthovoltage machines need to be close to the patient (~50 cm) and cannot be safely rotated around them. Efficiency and intensity increase dramatically at the voltages used by megavoltage machines, allowing them to be far enough (~100 cm) from the patient to be rotated. An added bonus to moving the source further from the patient is that the extra air the beam now travels through will act as a filter and harden it (lowering the surface dose).

At this point you should be pretty well sold on the fact that X-ray machines operating in the MV range will be better for treating most tumors than X-rays generated by kV machines. So, how are we going to generate these higher energy photons? Unfortunately, conventional tubes cannot accelerate electrons into the MeV range. We need a more powerful particle accelerator. A quick glance back to Section 10.4 reminds us that we can use linear accelerators to accomplish this.

As we saw in Section 10.4, most modern-day linear accelerators manipulate the phase (peaks and troughs) of microwaves to accelerate electrons. The microwaves can be generated using the same technology as a microwave oven. **Magnetrons** are used in microwave ovens to heat up your leftovers, *and* in 4–10 MV accelerators. **Klystrons** are used in higher voltage machines. Both use electrons to create microwaves via resonance. It works very much the same as blowing air across a pop bottle top, or air being blown through a pipe organ. Some of the energy of the moving air gets converted into sound through resonance. The same thing happens inside a magnetron or a klystron. Instead of air, it is electrons moving near a cavity, and microwaves are created instead of sound.

FIGURE 14.9   The effect of using isocentric techniques to maximize dose in the area of convergence, while minimizing elsewhere.

FIGURE 14.10 An electromagnetic wave moving through a waveguide.

Once generated, we want to make sure our waves don't lose any energy. When a broadcast antenna emits radio waves, they spread out in all directions at once. The farther away from the antenna, the lower the intensity of the waves, and the harder it is to tune in on your radio. In a linear accelerator the microwaves are moved from one point to another, with no loss of intensity, by **waveguides**. These can be as simple as copper tubes but can be more complex (Figure 14.10). The waveguide does just what its name implies; it guides the wave along a specific path, rather than allowing it to spread out in all directions at once. Thereby, the same intensity of waves is seen at the end as is put in at the beginning.

Remember that electromagnetic radiation behaves a lot like waves in water. We can think of phase in relation to the peaks and troughs of a wave. A peak represents one phase, and the trough represents the opposite phase. The electron is pushed and pulled by these phases, so if we can arrange them just right, they will accelerate the electron. As also mentioned in Section 10.4, there's another way to look at the acceleration of charged particles inside a waveguide. As the EM wave moves through the guide, it will induce temporary positive and negative electrical charges on the inside surface of the guide. If these temporary charges move just right through the waveguide, they can be used to accelerate charged particles.

Figure 14.10 shows a wave moving from left to right through a cylindrical waveguide. This can be used simply to transfer the wave's energy from one point to another, or it can be used to accelerate a charged particle, such as an electron. When used to accelerate, it is called a **traveling wave accelerator guide**. We can also use two waves traveling in opposite directions through the same waveguide to set up an interference pattern as illustrated in Figure 14.11. This is called a **standing wave accelerator guide**. If the interference is done just right, electrons will be accelerated as they move through the guide.

Remember that *constructive interference* happens when two waves are "in phase"—e.g., the peak of one wave is at the same point as the peak of the other wave. At this point the peak height would be the sum of the two peaks. *Destructive interference* happens when the waves are "out of phase"—the peak of one wave coincides with the trough of the other. In this case, the two would cancel each other out. When two waves interfere as shown in Figure 14.11, a complex pattern results. Our concern is not with how to get this just right to accelerate an electron to specific velocity, but rather to know that this is how it is done.

Conventional X-ray tubes can only accelerate an electron to a velocity of approximately one half that of the speed of light ($0.5c$), and cyclotrons can't do much better. Linear accelerators are needed to get 'em going faster than that. Keep in mind that particles cannot travel at, or faster than, the speed of light, and as we accelerate a particle to speeds approaching that of light, increasing amounts of energy are required. Using the formulas from Section 7.2, we can calculate the relativistic kinetic

FIGURE 14.11 A standing wave accelerator guide.

TABLE 14.2
Relativistic Kinetic Energies for an Electron at
Velocities Approaching the Speed of Light

v/c	KE (MeV)
0.50	0.079
0.90	0.66
0.990	3.11
0.9990	10.92
0.9999	35.63

FIGURE 14.12   A cathode ray tube, also known as an electron gun.

energy for an electron at various velocities (Table 14.2). We can see that as the velocity approaches the speed of light ($v/c \rightarrow 1$, moving down the first column), the amount of energy (second column) goes through the roof! More and more energy is required for smaller increases in velocity. In the context of generating X-rays, we're not all that interested in the velocities of the electrons as we are their energies. It is the electron energy that determines the energy spectrum of X-rays that are produced.

Where do we get our electrons for linear accelerators? The same way they are produced for conventional tubes—boil them off a hot, high-voltage filament under vacuum. This time we drill a hole in the anode right where the focal spot would be so the accelerated electron passes right through (Figure 14.12) into the accelerator guide. This is the same technology used in cathode ray tubes (CRT—old-style televisions and computer monitors). When placed at the end of a linear accelerator, it is called an **electron gun**. Muy macho.

We can now put all of our components together. Figure 14.13 shows a schematic for a megavoltage X-ray machine that uses a magnetron microwave generator. The microwaves travel through the waveguide to the accelerator guide. There, they accelerate the electrons emerging from the electron gun, and slam them into the target. A transmission target is used for all megavoltage machines. Its thickness is a careful balance between the need to stop all the electrons (thicker is better) and the desire to avoid absorbing too many of the photons that are generated (thinner is better). Targets for higher-energy machines will be thicker than those operating at lower energies. Do you know why?

Notice how the accelerator guide fits vertically in the machine's head in Figure 14.13. For higher-energy accelerators, a longer accelerator guide is needed, so it is placed horizontally, as pictured in Figure 14.14. Both types of machines use a thin ionization chamber (Section 8.1.1) as a **monitor** to measure the radiation flux, and thereby determine the patient's dose. An ionization chamber is used precisely because it is very inefficient in detecting high-energy photons. We want to get an idea of the flux without significantly attenuating the beam. Both Figures 14.13 and 14.14 have collimators (also known as jaws) to help shape the beam. They are typically made from high Z materials (like Pb). Finally, there's a treatment platform (slab!) for the patient to lie on.

In higher-energy machines (Figure 14.14), the electron beam needs to be bent by 90° so it will hit the target at an appropriate (right?) angle. Remember from our study of cyclotrons (Section 10.4)

FIGURE 14.13 A MV X-ray machine with a magnetron and a vertical accelerator guide.

FIGURE 14.14 A MV X-ray machine with a klystron and a horizontal accelerator guide.

that magnetic fields will cause charged particles to move in a circular path. If we choose our magnet carefully (field strength) we can get the electron beam to go through a 90° *or* a 270° turn, sending them toward the target.

Unfortunately, the electrons are not all traveling at exactly the same velocity—they have slightly different energies. When they hit the 90° magnetic field, those that are traveling a little faster than the rest (*E*+) are bent through an angle slightly larger than 90°. Those that are traveling a little slower (*E*–) are bent through an angle that is somewhat less than 90°. This causes the electron beam

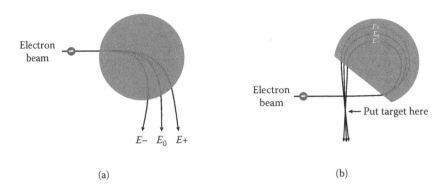

(a)   (b)

FIGURE 14.15   (a) Energy dispersal effect of a 90° bending magnet. (b) Energy convergence effect of a 270° bending magnet.

to broaden a bit with lower-energy electrons at one end, and higher-energy electrons at the other (Figure 14.15a). The effect is similar to that of a prism, which spreads out visible light according to energy. For this reason, 90° bending magnets are called **chromatic**.

The same problem is encountered in 270° bending magnets; the electrons are still spread out according to energy. However, running them through 270° causes them to converge at a certain point after exiting the magnet (Figure 14.15b). If left alone, the electrons would again diverge after this point. This point is an excellent place for the target. By doing this, we can obtain a very small focal spot which results in a narrow X-ray beam and minimal penumbra. Because of this, 270° bending magnets are called **achromatic**. As you may have already figured out, the focal spots, and therefore penumbra, on chromatic magnet machines are larger.

Achromatic magnets are more commonly used in medical linear accelerators, and since such a narrow beam is rarely needed, it needs to be broadened. This is done by the **flattener** (or flattening filter), which is sandwiched between the target and the monitor in Figures 14.13 and 14.14. Another reason to use a flattening filter is that the photon beam tends to be rather intense in the center and drops off rapidly from that point. For therapeutic applications it is better to have a more uniform beam. Flattening filters accomplish both tasks. Because they need to scatter high-energy photons (which have a low probability of interacting with matter), flatteners are chunks of metal that have different shapes depending on the energy of the photon beam—generally speaking, they are thicker in the center where the photon beam is most intense (Figure 14.16).

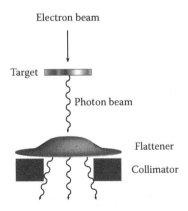

FIGURE 14.16   The dispersal effect of a flattening filter on an X-ray beam.

FIGURE 14.17   Measurement of (a) beam flatness and (b) beam symmetry.

After going through the flattener, the resulting X-ray beam should be flat and symmetric. They are both measures of the uniformity of the beam. **Flatness** is measured by looking at the variations in the intensity over the center 80% of the beam. Intensity should not vary by more than ±3% across this region (Figure 14.17a). Notice that the beam cross sections in Figure 14.17 are not perfectly flat across the top. The greater intensities near the beam edge are known as "horns."

Beam **symmetry** is determined by looking at the overall intensity between two halves of the beam—it should not vary by more than 2% (the area under each half of the beam cross section in Figure 14.17b). If the flattener is not properly aligned to the beam, an asymmetric beam will result.

We should also mention **safety interlocks**—an important component of any radiation-generating device. Safety interlocks are switches that will shut down the X-ray beam or the entire machine if a particular safety mechanism is not in place. For example, there is a switch on the door to the treatment room to ensure it is closed before exposing the patient. This way no one can wander in during therapy and be exposed. There are also switches to detect proper placement of filters and collimators, as well as functioning warning lights. Instrument software will also have safety interlocks checking flatness, symmetry, and patient dose.

Finally, you should feel a little uncomfortable with the energy characterization of the photon beams emerging from megavoltage machines. They are commonly referred to as "X MV photon beams," where X typically ranges from 4 to 25. As mentioned earlier (Section 14.1), this nomenclature comes from the tube potential settings of orthovoltage lower energy X-ray machines. With megavoltage machines, there isn't a tube potential the electrons are accelerated through; rather this energy value is referring to the energy of the electrons hitting the target and should therefore use units of MeV rather than MV. The reader should also remember that a broad energy spectrum of photons is produced, with a maximum slightly below the electron energy.

Electron beams can be generated by the same machines discussed here—just remove the target so the electron beam hits the patient instead of X-rays. Electron beams will also need to be broadened. As they emerge from the accelerator, these beams are only 1–2 mm in diameter. A thin metal foil will generally do the trick (think back to your freshman chemistry and physics classes—remember Rutherford's gold foil scattering experiment?). These are called **scattering foils**. They are thin because electrons have a relatively high probability of interacting with matter, and thicker chunks of metal would absorb all of the electrons. Additionally, an **electron applicator** or **cone** is used to help collimate the beam beyond the jaws. Electrons are more likely to be scattered by the air they travel through, so these cones will need to be fairly close to the patient.

QUESTIONS

14.1   Increasing the current on a conventional tube increases the X-ray intensity. If greater intensity is needed, why can't the current be cranked up as needed?

14.2 Define the following: focal spot, tube current, transmission target.

14.3 Why is bremsstrahlung more likely further from a nucleus?

14.4 On a single plot, overlay the three X-ray spectra produced by the same conventional X-ray tube set at three different voltages.

14.5 Why is tungsten commonly used as a target in medical X-ray generating devices?

14.6 Tungsten, molybdenum, and copper are all used for X-ray production. For each anode material, calculate the electron energy needed for 100% efficiency of photon production. For the highest value, also determine the electron velocity. Does this seem reasonable for a clinical setting?

14.7 An electron beam measures 3.4 mm × 1.0 mm. If this beam strikes a target that is at a 17° angle relative to the beam direction, what is the cross-sectional area of the photon beam produced? Assume the longer dimension of the electron beam hits the target at 17°.

14.8 Typically, when an electron hits a target, only part of its energy goes into forming X-rays. What happens to the rest of the energy?

14.9 What energy X-ray is produced by a conventional X-ray tube operating at 350 kV and 150 mA with a tungsten (Z = 74) target?

14.10 Give two reasons why a ^{60}Co irradiator would be preferred for radiation therapy over a 400 kV orthovoltage machine.

14.11 Explain the general trend of greater PDD with X-ray energy observed in Figure 14.6 for the five curves. Why do the highest energy curves peak between 0 and 5 cm depth?

14.12 Give three reasons why megavoltage machines are favored for radiation therapy over orthovoltage units. Briefly describe each reason.

14.13 Carefully make drawings like Figures 14.8b and c, except change only the SDD. Which gives a larger penumbra? Repeat, this time move the patient a little closer to the source (reduce SSD). Which of these latter two drawings shows a smaller penumbra? You might want to use a computer-drawing program or graph paper to insure consistency between your drawings.

14.14 Briefly explain the main similarity and difference between the Bragg peak associated with charged particles interactions with matter and the fact that maximum dose is delivered inside a patient by high-energy photons.

14.15 Briefly explain how radiation therapy patients can be exposed to electron radiation during a photon treatment by a 4 MV linac machine, that is, the target and flattener are in place. Calculate the minimum possible LET (in air) of an electron produced by a 4 MV machine with SSD = 100 cm.

14.16 Give two unrelated reasons why the electrons produced in Question 14.15 are very likely to have a higher LET.

14.17 What is the purpose of a magnetron? A klystron?

14.18 Why are targets for higher energy X-ray machines thicker?

14.19 Sketch a diagram of a bent-beam accelerator. Label all major components.

14.20 Inorganic scintillator and semiconductor detectors are much more efficient at detecting high-energy photons. Why aren't they used in place of the ionization chamber in megavoltage machines?

14.21 What is an X-ray flattening filter? How does it differ from a scattering foil?

14.22 Chromatic also has meaning in music. Is there an analogy between this definition and the bending magnets described in this chapter? Explain your answer.

15 Dosimetry of Radiation Fields

Radiation therapy is all about zapping tumors with ionizing radiation (Section 9.2). Since tumor cells are rapidly dividing, they are much more susceptible to death by radiation than other cells. X-rays are typically used, but electron and proton beams are also used for certain types of tumors. Since it is difficult or impossible to directly measure dose inside a patient during treatment, radiation therapists need to calculate doses, not only to the tumor, but also to the surrounding (healthy) tissue. When done properly, dose to the tumor is maximized while dose to healthy tissue is minimized.

Measuring doses is generally done using a **phantom**, which is simply a human surrogate— something that behaves more or less the same way toward ionizing radiation but allows placement of a detector inside. Often it is simply a tub of water. Since humans are mostly water, and our remaining elemental makeup has a similar average atomic number (Z) and density, water is an excellent and inexpensive stand-in for human tissue. It can also be plastic (again, similar average Z to human tissue) where detectors can be placed below various thicknesses.

These measurements are usually expressed as a fraction of the amount of radiation reaching the detector relative to another value. Several different terms are used; we'll look at three different ways at looking at relative dose: percent depth dose (PDD), tissue air ratio (TAR), and tissue maximum ratio (TMR). Initially, we'll consider only X-rays for human therapy, and save electrons and protons for the end of the chapter.

15.1 PERCENT DEPTH DOSE

As we've already seen (Section 14.2) percent depth dose (PDD) is the percentage of the maximum dose remaining at a particular depth inside a human. This is done by making two measurements inside a phantom: the dose at the depth of interest (D_n) and at the depth where dose is maximized (D_{max}). Mathematically it is defined as:

$$\%D_n = \frac{D_n}{D_{max}} \times 100 \qquad (15.1)$$

where D_n is dose at depth n, D_{max} is the maximum dose, and $\%D_n$ is the percent depth dose. The fraction of D_n/D_{max} is known as the **depth dose**. As shown in Figure 15.1, dose measurements are made along the beam axis (center of the beam), directly below the radiation source.

Measuring PDD is represented schematically in Figure 15.1. The human/phantom is represented as a rounded rectangle, an approximation of a cross section of a human torso. Notice that the maximum dose (D_{max}) is shown as being at a distance of d_{max} below the skin. Remember that this is only true for the megavoltage machines (also known as linear accelerators, or linacs), as shown in Figure 14.6. For lower energy photons, D_{max} would be at the surface ($D_{max} = D_0$). Note also that the

DOI: 10.1201/9781003303787-15

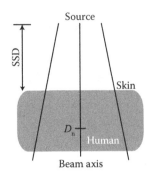

FIGURE 15.1   Percent depth dose measurements.

radiation source (where the X-rays are produced) is located at a distance of "SSD" from the patient's skin. As we saw in Chapter 14, SSD stands for "**source to surface** (or skin) **distance**. The importance of this distance will be discussed later, but it should be apparent that increasing this distance will decrease dose, according to the inverse square law (Section 13.3).

Example 15.1: A 6 MV linac[1] produces a beam with a PDD of 77.7% at 7.0 cm depth. If a 0.20 Gy dose is required at 7.0 cm, what is the maximum dose?

If we rearrange Equation 15.1 by solving for D_{max}, we can plug in the rest and solve.

$$D_{max} = \frac{D_n}{\%D_n} \times 100 = \frac{0.20 \text{ Gy}}{77.7} \times 100 = 0.26 \text{ Gy}$$

Notice that the maximum dose is higher than the dose delivered at 7.0 cm, as it should be. A 6 MV machine delivers its maximum dose at a depth of 1.5 cm—you can sort of see this in Figure 14.6.

We can calculate percent depth dose, and we have some understanding of how it is measured. What sorts of things affect PDD? First, let's consider beam energy. As the average energy of the photons in the beam increases, the maximum dose will be delivered deeper inside the patient. For example, an X-ray beam from a 20 MV linear accelerator has a d_{max} at 3 cm (compare to the 1.5 cm we saw for 6 MV). We know this from looking at Figure 14.6 and Table 14.1, and (hopefully) from reading some of the text around them. As a reminder, this is because higher-energy photons interact with low Z matter by scattering photons and electrons in the same direction as the beam. When we reach electron equilibrium, the dose from photons plus Compton electrons reaches a maximum (see Figure 14.7). How does this affect PDD below d_{max}? Since d_{max} is deeper, PDD values at a specific depth inside the patient will be higher for a higher-energy beam.

Second, let's consider **field size**. Field size is quite simply the cross-sectional area of the beam, typically measured at the end of the collimator (also known as the jaws). The beam widens as it travels down to the patient, but since distances between the source and patient can vary, it's less confusing to state the field size as the beam emerges from the collimator. The bottom line is that a wider beam means a larger field size, and interestingly, this will tend to increase percent depth dose. Remember that PDD is measured along the central beam axis, so a wider field will increase the probability that photons will be scattered to the detector from another part of the beam.

---

1   Remember that "6 MV" is not referring to the actual energy of the photons generated. It is a broad continuum with a maximum value slightly less than 6 *MeV*.

FIGURE 15.2 The effect of field size on PDD.

Imagine we first measure a dose from a beam that is narrower than our detector is wide (Figure 15.2a). As this beam travels through the air and the patient, some of its photons will be scattered out of the beam and miss the detector. Now, we increase the field size so that it is much larger than the area of our detector (Figure 15.2b). Now photons that would've missed the detector, had they not been scattered, have a chance of being scattered *into* the detector. Since there are many more photons that start out in a direction that would've missed the detector, than those that start out headed for the detector, the odds are that more photons will end up being scattered into the detector than get scattered away from it. The result is that the percent depth dose increases somewhat with field size.

Photons entering the detector directly from the source (not scattered) are called **primary (1°) radiation**, while those that are scattered into the detector are **secondary (2°) radiation**. The observed dose at any point within a wide beam is due to both primary and secondary radiation. Note that scattering can be due to X-ray interactions with the monitor, collimators, air, or patient.

This increase in PDD becomes less pronounced for very high-energy fields (> 15 MV) because higher-energy photons tend to scatter through smaller and smaller angles (scatter forward). At a certain point, these angles get too small for field size to affect PDD.

Example 15.2: Determine the percent contributions of primary and secondary radiation at 16 cm depth for a 10×10 cm² 10 MV field with an SSD of 100 cm. The percent depth dose at this depth is 58.8%.

We need one other piece of information to do this problem, the PDD for a "0" field at 16 cm depth for a 10 MV field at SSD =100 cm. In other words, keeping everything else constant, shrink the field size down to something so narrow, we'll only see primary radiation. For this problem, PDD is 54.4% for a zero field. Notice that this value is smaller than the PDD for the 10×10 cm² field. The difference is secondary radiation.

The percent contribution of primary radiation is:

$$\frac{54.4\%}{58.8\%} \times 100\% = 92.5\%$$

The percent contribution of secondary radiation is what's left over.

$$100\% - 92.5\% = 7.5\%$$

Okay, secondary radiation doesn't provide a huge contribution, but it's something.

So, how can we correct for changes in PDD when the size of the beam changes? Field size corrections (C_{fs}) will do this for us, but, unfortunately, they vary from one machine to another. Therefore, a field size correction specific to a particular instrument must be applied, and it must be determined for each individual machine. It is measured by comparing the dose at a certain depth (typically d_{max}) along the beam axis in any field, to the dose in the same place in a **standard field.** A standard field is usually 10×10 cm^2. Field size correction values range from ~0.90 to ~1.10 and depend on the collimator size (jaw settings), not SSD or photon energy. This is another reason why field size is measured at the jaws, not at the patient.

$$C_{fs} = \frac{D_n \text{ for field size (fs)}}{D_n \text{ for the standard field}}$$

Percent depth dose is also affected by **field symmetry**, which is simply the shape of the field. Figure 15.3 shows three different shapes superimposed on each other: a circle, a square, and a rectangle. All three have the same area but note that the square and especially the rectangle have parts of their areas that are further from the center than the circle. For objects of equal area, those with more of their area further from the center will be lower in symmetry. Therefore, the square is lower symmetry than the circle and the rectangle is lower symmetry than the square.

When symmetry is lowered, PDD generally goes down. Think about a simple change, from square to rectangle. If they are the same area, the rectangular field will scatter fewer photons into the beam axis. In the rectangle, more photons are further away from the beam axis, and the further away, the lower the probability of scattering all the way back to the beam axis. There are ways to convert PDD values from one simple shape to another (like rectangle to what's called an "equivalent square"), but we won't bother with that here. Modern-day radiation therapy often uses complex (and

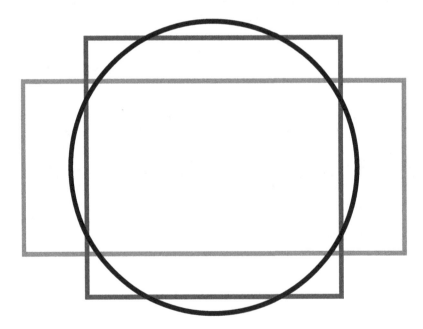

FIGURE 15.3 A circle, a square and a rectangle, all with equal areas, but different symmetry.

sometimes changing) shapes, and computers do dose calculations. We'll stick with square fields to keep things simple and use them to illustrate how these calculations are performed and allow us to focus on what other factors go into the math.

Percent depth dose is also dependent on the source to surface distance (SSD). Unfortunately, this dependence is a bit more complex than beam energy, or field size or symmetry. Ignoring the attenuation of air, increasing SSD will lower the dose rate at any specific point inside the patient according to the inverse square law.

$$\text{Dose rate} \propto \frac{1}{d^2} \quad \text{or} \quad \text{Dose Rate Change} = \frac{1}{\left(\dfrac{d_2}{d_1}\right)^2} \tag{15.2}$$

In these equations, d is distance, and d_1 is the original distance and d_2 is the new distance. We already know from the inverse square law (Section 13.3), that when the distance is doubled, the dose rate will be cut by one quarter. Let's try this out with Equation 15.2.

$$\text{Dose Rate Change} = \frac{1}{\left(\dfrac{2}{1}\right)^2} = \frac{1}{4}$$

Example 15.3: What happens to the surface dose rate if we increase SSD from 100 cm to 150 cm?

$$\text{Dose Rate Change} = \frac{1}{\left(\dfrac{150 \text{ cm}}{100 \text{ cm}}\right)^2} = 0.44$$

Meaning that the surface dose will be 0.44 times lower at 150 cm than at 100 cm. We could also say that the dose will be 44% lower at the new distance and that treatment time would need to be increased by a factor of $1/0.44 = 2.3$ to achieve the same dose at 100 cm.

Doses at all depths (including d_{max}) will decrease as SSD increases. As a result, we might intuitively expect PDD values below the surface to also decrease. As it turns out, the values of PDD *increase* as SSD increases. If we look again at Equation 15.1 we can see that as SSD increases, both D_n and D_{max} will decrease, but if D_{max} *decreases at a faster rate* than D_n, then $\%D_n$ will increase. This idea is mathematically demonstrated in the following example.

Example 15.4: Using the inverse square law, calculate the changes in D_{max} and D_n when SSD is increased from 100 cm to 150 cm. Assume $d_{max} = 0$ (i.e., D_{max} is at the surface), and D_n is at 10 cm depth. Also assume at least two significant figures in all values.

From Example 15.3, we already know that increasing SSD from 100 to 150 cm will decrease D_{max} (D_0) by a factor of 0.44. We can then calculate the change in D_{10} in the same way. Remember that we are now 10 cm below the surface, which means our original distance to the source is 100 cm + 10 cm = 110 cm, and our final distance is 150 cm + 10 cm = 160 cm.

$$\Delta D_n = \frac{1}{\left(\dfrac{160\ \text{cm}}{110\ \text{cm}}\right)^2} = 0.47$$

D_{max} decreases to 44% of its original value while D_{10} decreases to 47% of its original value. Both doses decrease, but D_{max} decreases to a greater extent than D_{10}. As you might guess, dose will always decrease at a slower rate for any point that is farther from the source.

Let's see how using a simpler example, shown in Figure 15.4. Initially, worried guy is 1 m from a radioactive source and happy guy is 2 m away. The source is then moved 2 m further away from the guys, making it 3 m from worried guy and 4 m from happy guy. The distance between worried guy and the source tripled, cutting his dose to 1/9 of the original. However, the distance doubled for happy guy meaning his dose decreased only to 1/4 of original. Worried guy is still getting a higher dose, but his dose is dropping at a faster rate than happy guy's. In general, whatever is closer to a source will see a more dramatic decrease in dose as the source moves away, because it also sees a more dramatic change in distance to the source.

Now, let's return to the work we did in Examples 15.3 and 15.4 and calculate percent depth doses at 10 cm depth for both 100 cm SSD and 150 cm SSD. We also need to know that $D_{max} = 100$ cGy and $D_{10} = 65$ cGy when SSD = 100 cm. Note that these values are specific to a particular instrument with certain settings and cannot be generalized to all X-ray machines. PDD at 100 cm SSD is just a matter of plugging in values to Equation 15.1.

$$\frac{65\ \text{cGy}}{100\ \text{cGy}} \times 100 = 65\%$$

At SSD = 150 cm, we'll need to first calculate the doses at the surface and at 10 cm depth. We can use the dose rate changes of 0.44 at the surface and 0.47 at 10 cm that were calculated in Examples 15.3 and 15.4.

$$100\ \text{cGy} \times 0.44 = 44\ \text{cGy}$$

$$65\ \text{cGy} \times 0.47 = 31\ \text{cGy}$$

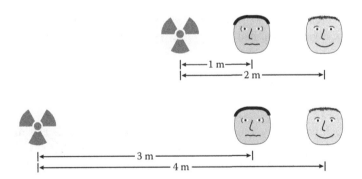

FIGURE 15.4 Effect of distance on dose.

Now we can determine PDD at 150 cm SSD using Equation 15.1.

$$\frac{31 \text{ cGy}}{44 \text{ cGy}} \times 100 = 69\%$$

PDD ($\%D_{10}$) *increases* as SSD increases! If you want to treat a tumor at any depth below d_{max}, it would seem that you would want to increase SSD (and therefore PDD) as much as possible in order to deliver more dose at the tumor (assuming it is below d_{max}). But wait, remember that PDD is a *relative* dose, i.e., the dose at the tumor relative to the dose someplace else. Dose rate (or photon flux) *decreases* with increasing SSD, which would mean longer treatment times to get the same dose delivered to the tumor. Deciding on an optimal SSD is therefore a compromise between PDD and dose rate.

We've already looked at how scattering affects PDD due to changes in field size and symmetry. When we did, we only considered scattering in the forward direction (same direction as the beam). What about backscatter? Remember that Compton scattering can occur through any angle up to 180°, and that backscatter is scattering in the direction opposite to the beam (Section 7.3.1). Therefore, a photon can be traveling through a patient, undergo backscattering, travel back the way it came, then interact again, depositing energy at less depth. We can determine the extent that backscatter contributes to dose through an experiment illustrated schematically in Figure 15.5. First, the dose in air is measured at a specific distance from the source. Second, a phantom is inserted into the beam such that D_{max} is observed at the same distance from the source and dose is measured again. The ratio of these two dose values is the backscatter factor (BSF).

$$\text{BSF} = \frac{D_{max}}{\text{Dose in air}}$$

If there's no backscatter, this ratio is 1.0. If there is some, it'll be greater than 1—in other words, backscattering only adds to dose. Generally speaking, it ranges from 1.0 to 1.5, and is only greater than 1 when the maximum photon energy is below 2 MeV. Above 2 MeV, Compton scattering is almost all in the forward direction (relatively low angle) and backscattering does not contribute significantly to dose. Therefore, this whole concept is pretty useless with most modern-day equipment. Despite its lack of utility with megavoltage machines, it is still applied, although it is called **peak scatter factor** (PSF), and it is almost always 1.0.

Now let's get a little interventionist. Sometimes it's desirable to place stuff in the beam. Why? The main reason would be to harden the beam. Remember (Section 7.4) that this means increasing

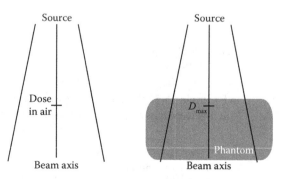

FIGURE 15.5 Determination of backscatter factor.

the average energy by filtering out more of the lower energy photons. This is easily, and commonly, accomplished by placing a plastic tray in the beam. There are also cases when having an obstruction like the patient's bed or the treatment platform in all or part of the beam is unavoidable. To account for this attenuation, the **equipment attenuation factor** (C_{attn}) is calculated from the ratio of dose without the obstruction to dose with the obstruction. These values are always less than 1 since the dose will always be lower with less beam intensity.

$$C_{attn} = \frac{\text{Dose with stuff in beam}}{\text{Dose without stuff in beam}}$$

We can now put everything we've learned in this section together and mathematically describe tumor dose based on depth dose and corrections to it.

$$\text{Tumor Dose} = N_{mu} \times C_{cal} \times \frac{D_n}{D_{max}} \times C_{fs} \times C_{attn} \qquad (15.3)$$

where N_{mu} is the number of **monitor units** (mu)—translated into English, a monitor unit is the amount of time it takes the machine to deliver certain dose. This "certain dose" is C_{cal}, also known as the **calibration factor**. It is usually set at 1.000 cGy/mu delivered at d_{max} at the calibrated field size (usually 10×10 cm^2) and the calibrated SSD (typically 80 or 100 cm). Note that multiplying the number of monitor units (N_{mu}) by the calibration factor (C_{cal}) is a measure of the (raw) dose, in cGy at d_{max}.

Normally we know what dose is needed, based on tumor imaging. What we need to calculate is the number of monitor units, i.e., the treatment time. Solving Equation 15.3 for N_{mu} gives Equation 15.4.

$$N_{mu} = \frac{\text{Tumor Dose}}{C_{cal} \times \frac{D_n}{D_{max}} \times C_{fs} \times C_{attn}} \qquad (15.4)$$

Example 15.5: Determine the treatment time (in monitor units) to deliver 200 cGy to a tumor 8.0 cm below the surface of a patient exposed to a 7×7 cm^2 6 MV field at 100 cm SSD. This accelerator is calibrated to deliver 0.998 cGy/mu, has a PDD of 74.6%, and a field correction factor of 0.978, under these conditions. There's nothing further attenuating the beam.

From the problem we can determine that depth dose is 0.746 (just divide PDD by 100) and assume $C_{attn} = 1.00$ (since there's nothing added to attenuate the beam).

$$N_{mu} = \frac{200 \text{ cGy}}{\left(0.998 \ \frac{\text{cGy}}{\text{mu}}\right) \times 0.746 \times 0.978 \times 1.00} = 275 \text{ mu}$$

What is the dose at d_{max} (D_{max}) in Example 15.5? Remember that we delivered 200 cGy at a depth of 8.0 cm with a PDD of 74.6%. With this wealth of knowledge, all we need to do is rearrange Equation 15.1, solving for D_{max}.

$$D_{max} = \frac{D_8}{\%D_8} \times 100 = \frac{200 \text{ cGy}}{74.6} \times 100 = 268 \text{ cGy}$$

As expected, the dose at d_{max} is greater than the dose at 8.0 cm (D_8).

Exit dose is the dose the patient receives where the photon beam leaves the body. With megavoltage machines the exit dose is often greater than the skin dose, where the beam enters the patient, and is therefore an important consideration in treatment planning. Assuming the patient is 30 cm thick ($\%D_{30} = 22.7\%$) where treatment is taking place in Example 15.5, we can again use a rearrangement of Equation 15.1 to calculate exit dose.

$$\text{Exit dose} = D_{30} = \frac{D_{max} \times \%D_{30}}{100} = \frac{268 \text{ cGy} \times 22.7}{100} = 60.8 \text{ cGy}$$

This is a fairly mild exit dose. It will be higher for skinnier patients (or body parts that aren't as thick) and for higher energy beams.

Example 15.6: What dose will be seen at d_{max} if 250 cGy is delivered at 9.5 cm by a 10×10 cm² 6 MV field at an SSD of 100 cm? Assume $\%D_9 = 71.3\%$ and $\%D_{10} = 67.7\%$ for this instrument configuration.

This problem provides a great deal of unnecessary information, but it does not give us a percent depth dose at 9.5 cm depth. We can make a pretty good guess by simple linear interpolation.

$$\%D_{9.5} = 67.7\% + \tfrac{1}{2}(71.3\% - 67.7\%) = 69.5\%$$

Now we can plug into Equation 15.1.

$$D_{max} = \frac{D_{9.5}}{\%D_{9.5}} \times 100 = \frac{250 \text{ cGy}}{69.5} \times 100 = 360 \text{ cGy}$$

You might think the linear interpolation in Example 15.6 is needlessly complex. Why not just average the two PDD values? You'd be right, but the process illustrated here is general for any linear interpolation. For example, what if we need to know the percent depth dose at 9.2 cm? Note that 9.2 cm is 1/5 of the way from 9.0 to 10.0 cm.

$$\%D_{9.2} = 67.7\% + \tfrac{1}{5}(71.3\% - 67.7\%) = 68.4\%$$

Before we finish our long look at percent depth dose, let's consider the effect on treatment time by changing the source to surface distance (SSD). Notice that SSD is not a part of Equation 15.4, rather it is incorporated into the depth dose (D_n/D_{max}) value. Remember that PDD changes if SSD is changed, so, if we've measured all of our PDD values for a particular machine at 100 cm SSD, we can't simply change SSD. We wouldn't know what depth dose value to plug into Equation 15.4.

Example 15.7: At a depth of 10.0 cm and an SSD of 150 cm, the $\%D_n$ is 71.3% for a 10×10 cm² beam from a 6 MV linac. What change in the number of monitor units (treatment time) is required to deliver the same dose as at 100 cm SSD, where the $\%D_n$ is 67.7% for the same field size?

Before we get into this, notice that PDD decreases with decreasing SSD, as expected. Now, as we try to wrap our brains around this problem, realize that there are two, competing effects. First, PDD is decreasing. A lower percentage of the dose delivered at depth means we'll need a longer treatment time. Second, the decrease in distance to the source means that dose rate will be going up, and treatment time will have to be shortened. Ouchies. Let's take them on one at a time. How much will we have to increase treatment time due to the change in PDD? It is simply the ratio of the two values. The larger value goes on top because treatment time will increase due to the change in PDD.

$$\frac{71.3\%}{67.7\%} = 1.05$$

Treatment time will have to be about 5% longer due to the decrease in PDD. Next, we should use the inverse square law to figure the change in treatment time due to the change in distance between the source and the tumor. Note that the distances are source to tumor, not source to surface. In other words, we'll need SSD plus 10 cm. Applying the inverse square law:

$$\frac{1}{\left(\dfrac{100 \text{ cm} + 10 \text{ cm}}{150 \text{ cm} + 10 \text{ cm}}\right)^2} = 2.12$$

What? Treatment time should be shortened by decreasing SSD, not doubled! Absolutely right. What we've calculated here is the effect on dose—remember *dose* follows the inverse square. If dose is roughly doubled, then treatment time will be approximately cut in half. Let's do it properly and invert 2.12.

$$\frac{1}{2.12} = 0.473$$

That's better. Treatment time will be cut to ~47.3% of its original value by the change in distance. We can combine the two conflicting effects by multiplying them together.

$$1.05 \times 0.473 = 0.498$$

This tells us that decreasing SSD from 150 cm to 100 cm will decrease treatment time by a factor of 0.498 (roughly half the original time).

The bottom line on depth dose is that it is sometimes an awkward measure of the fraction of radiation delivered to the target volume. The awkwardness can be seen in Example 15.7—dealing with changes in SSD can be more difficult than they need to be. We'll look (more briefly!) at two other ways of doing dose calculations (patient attenuation) that do not depend on SSD.

15.2 TISSUE–AIR RATIO

Percent depth dose is determined by measuring dose at two places inside the patient/phantom, and thus is a measurement of how much the patient attenuates the beam. It can also be considered as a *relative* dose, i.e., it is the dose in one place inside the patient relative to another. As we've seen, PDD changes (along with dose) with SSD in a manner that requires measuring PDD at every possible SSD. As we saw back in Section 9.2, it is desirable to rotate the source around the patient during treatment. To do so, and maintain the same SSD, the patient would have to be shaped like a cylinder with the tumor in the center of the cylinder.

To avoid the complexity of having to repeat all relative dose measurements at every conceivable SSD, it would be more convenient to measure patient attenuation another way. As shown in

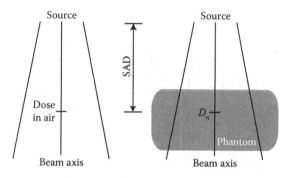

FIGURE 15.6 Determination of tissue–air ratio.

Figure 15.6, tissue–air ratio (TAR) is determined by first measuring the dose in air a specific distance from the radiation source. A second measurement is taken, at the same distance, but now with the patient in the beam. This ratio then reflects how much the patient (tissue) attenuates the beam relative to air.

If this looks a bit familiar, then you've been paying pretty close attention! If not, you should work on your active reading skills. Tissue–air ratio is determined a lot like backscatter factor (Figure 15.5), except we now measure anywhere along the beam axis inside the phantom/human, instead of just at d_{max}. The distance from the source to the point of measurement is the **source to axis distance** (SAD). Not the best acronym, but it could be worse. In radiation therapy we're most interested in the dose at the tumor, so we could instead define this distance as source–tumor distance (STD).

Mathematically, TAR is defined as the dose to the patient at SAD (D_n) divided by the dose in air at the same distance from the source.

$$\text{TAR} = \frac{\text{dose @ } SAD \text{ in phantom}}{\text{Dose in air @ SAD}}$$

Because dose changes with distance, so does the tissue–air ratio. Unlike PDD, we don't need to re-measure TAR at the new distance; all we need to do is apply the inverse square law. TAR is like PDD in that both increase with beam energy (less likely to interact, more dose delivered deeper) or field size (larger field means greater probability of scattering back to the center).

While it was commonly used in the early days of isocentric therapy, TAR is of relatively little importance today as it was developed when ^{60}Co therapy dominated—as an alternative to PDD. As a result, we won't do any math with it. For reasons we don't need to get into, it just didn't work well with linear accelerators. Another method of determining patient attenuation was necessary.

15.3 TISSUE MAXIMUM RATIO

The most commonly used patient attenuation factor today is the **tissue maximum ratio** (TMR). Like the others, it comes from two measurements (Figure 15.7), and like PDD, both are made inside the phantom. First, the dose at d_{max} (D_{max}) is measured at a specific distance from the radiation source (SAD). The second dose measurement is made with the patient/phantom moved up in the beam (closer to the source) so that depth of interest (typically the tumor depth) is now at the same distance from the source (SAD). Like TAR, the dose measurements are both made at the same distance from the source.

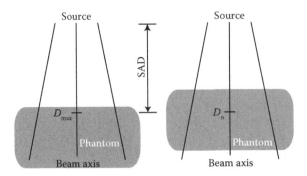

FIGURE 15.7   Determination of tissue maximum ratio.

Mathematically, TMR looks exactly like depth dose, except that the doses are measured at the same SAD instead of the same SSD. Tissue maximum ratio is the dose at some depth n inside the patient (D_n) divided by the dose at d_{max} (D_{max}).

$$\text{TMR} = \frac{\text{dose @ depth } n}{\text{Dose @ } d_{max}} = \frac{D_n}{D_{max}} \tag{15.5}$$

Like TAR, there is no dependence for TMR on SSD, since we are making measurements at the same distance from the source. TAR and TMR depend only on patient attenuation. Like TAR, if TMR is determined at one SAD, but treatment takes place at another, we'll need to use the inverse square law to perform treatment time calculations. For example, the treatment time equation (Equation 15.4) can be adapted to tissue maximum ratio by replacing depth dose with TMR and adding the inverse square law as an adjustment to TMR, as shown in Equation 15.6.

$$N_{mu} = \frac{\text{Tumor Dose}}{C_{cal} \times C_{fs} \times C_{attn} \times \text{TMR} \times \dfrac{1}{\left(\dfrac{\text{SAD}_{treat}}{\text{SAD}_{cal}}\right)^2}} \tag{15.6}$$

SAD_{treat} is the source to tumor distance during treatment, and SAD_{cal} is the source to dosimeter distance during calibration. Equation 15.6 can be rearranged to a more tractable form.

$$N_{mu} = \frac{\text{Tumor Dose}}{C_{cal} \times C_{fs} \times C_{attn} \times \text{TMR}} \times \left(\frac{\text{SAD}_{treat}}{\text{SAD}_{cal}}\right)^2$$

Example 15.8: Determine the treatment time (in monitor units) required to deliver a 250 cGy dose to a tumor 10.0 cm below the skin exposed to an 11×11 cm^2 6 MV field at 100 cm SSD. This accelerator is calibrated at SAD = 105 cm to deliver 1.007 cGy/mu under these conditions. A 0.6 cm Plexiglas®[2] tray is also placed in the beam (C_{attn} = 0.97). The 15×15 cm^2 field size correction

---

2   Plexiglas is poly(methyl methacrylate) and is a registered trademark of ELF Atochem, a subsidiary of Arkema in the U.S. It is a trademark of Evonik Röhm GmbH elsewhere.

factor for this machine is 1.031. The TMR values are 0.800 (for a 10×10 cm^2 field) and 0.810 (for a 12×12 cm^2 field).

The values of C_{fs} and TMR need to be interpolated. Remember that C_{fs} for a 10×10 cm^2 field is exactly 1.000.

$$C_{fs} = 1.000 + \tfrac{1}{5}(1.031 - 1.000) = 1.006$$

$$\text{TMR} = 0.800 + \tfrac{1}{2}(0.810 - 0.800) = 0.805$$

Okay, so the second interpolation could've been done in your head. At any rate, we're now ready to plug into Equation 15.6. Note that $\text{SAD}_{treat} = \text{SSD} + n = 110$ cm.

$$N_{mu} = \frac{250 \text{ cGy}}{1.007 \frac{\text{cGy}}{\text{mu}} \times 1.006 \times 0.97 \times 0.805} \times \left(\frac{110 \text{ cm}}{105 \text{ cm}}\right)^2 = 347 \text{ mu}$$

Finally, it should be noted that the doses calculated in the examples in this chapter are typically given over several treatments spread over several days to weeks. This is called **dose fractionation** and is done to allow the body time to recover from these relatively high doses.

15.4 ISODOSE CURVES

Up to this point, we've been looking at dosimetry in only one dimension—along the line defined by the beam axis (the center of the beam). We can easily go to two dimensions by looking at **isodose curves**: two-dimensional dose distribution lines representing equal absorbed dose (Figure 15.8). The numbers in Figure 15.8 indicate the percentage of the maximum dose. They are like the lines you sometimes see on weather maps indicating lines of equal temperature (isotherms) or equal pressure (isobars—not to be confused with nuclides with the same mass number or drinking establishments that look alike). Basically, you imagine looking at the patient's cross section and draw lines where you see specific dose values.

The isodose curves in Figure 15.8 have good flatness and symmetry (Section 14.2). Notice also that they spread further apart as the dose decreases, i.e., 50% is further from 60% than 80% is from 90%. The extent that they are spread depends on the energy of the beam. Isodose curves for lower-energy beams are more bunched together while those for higher-energy beams are more spread out.

FIGURE 15.8   Isodose curves.

FIGURE 15.9 Shaping dose with a wedge.

This is again a result of higher-energy beams having a lower probability of interacting with matter, depositing more dose deeper within the patient.

For sub-megavoltage machines, these curves will also show a wider penumbra since non-accelerator instruments generally have a larger source (focal spot). Penumbra would appear in Figure 15.8 as more angle in the parts of the curve that parallel the beam axis (more width in the curves if the beam axis were vertical). Sub-megavoltage isodose curves also show more bulging, i.e., the beam is wider, at greater depths. This is because lower-energy photons have a greater probability of scattering through larger angles.

If you look closely at the higher dose percentage curves in Figure 15.8, you'll notice they have horns (greater dose rates at the beam edge, Section 14.2). This comes from making the flattener a little too thick in the center and is quite deliberate. As the X-ray beam travels through the patient, more and more photons are scattered. Because there are no photons outside of the beam (beyond the edge), more photons are scattered into the center of the beam than are scattered into the edge. As a result, the isodose curves flatten out at greater depth. Since flatness is usually more desirable at depth (where treatment takes place), the flatteners are designed to initially overcompensate, attenuating more photons in the center of the beam.

How are isodose curves measured? If the phantom is just a box of water, it's pretty easy to move the detector around within it to map out the doses in two (or three) dimensions. You could also use an array of detectors inside a solid (or liquid) phantom and measure everything at once.

We can change the shape of the isodose curves by placing a **wedge** filter in the beam, as shown in Figure 15.9. The filter itself is simply a chunk of metal in the shape of a wedge—it attenuates the beam more dramatically where it is thicker, and less so where it is thin. This is pretty handy when the tumor is thicker in some parts than others. Similar sorts of dose shaping can also be done by adjusting the collimators during treatment such as is done with intensity-modulated radiation therapy (IMRT, Section 8.2).

In this section, we've made the jump from seeing dose in one dimension into two dimensions, but real doses are delivered in three dimensions. Instead of points or lines, we should really try to visualize isodose surfaces. It'd be a bit like a bunch of identical drinking glasses (or plastic food storage boxes) stacked one inside of another. The glasses (or plastic) represent the isodose surfaces. If viewed from the side, their outlines would look like Figure 15.8.

15.5 MOVING FIELDS

Radiation therapy photon sources generally move around a patient, maximizing tumor dose while minimizing dose elsewhere. As already mentioned, isocentric techniques would be much simpler if humans were perfect cylinders and tumors were always located along the central axis. Unfortunately, that isn't normally the case. Tumors locate where they please, and, in cross section, humans are a bit on the oval side.

Reality is more like Figure 15.10. The human is represented in cross section as an irregular oval, with the tumor located to the left and down a bit. Our problem is then to figure out what the correct

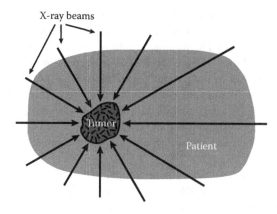

X-ray beams

Tumor

Patient

FIGURE 15.10   Isocentric therapy.

dose will be if it is to be delivered along two or more of the arrows drawn on Figure 15.10. We also have the option of rotating the beam continuously around all or part of the patient. The solution is to place the tumor at the isocenter of the instrument and measure how deep the tumor is at several angles around the patient, determine TMR (or TAR) for each angle, and average the results. Note that with the tumor at the isocenter, SAD remains constant, while SSD varies. Moving field calculations are greatly simplified by using tissue maximum ratios rather than percent depth dose values.

To minimize dose to healthy tissue, treatments are planned for a whole number of rotations for a specific therapy session. If this isn't possible, you can pick up where you left off at the next session.

15.6   PROTON AND ELECTRON BEAM DOSIMETRY

As mentioned in Section 9.2, electron beams are pretty easy to come by if you already have a linear accelerator X-ray therapy unit. Looking at Figures 14.13 and 14.14, you should realize that if the target and flattener are removed, an electron beam hits the patient instead of X-rays. The electron beam produced in these linear accelerators is very narrow and needs to be broadened and flattened for therapy. This is done by placing two scattering foils in the beam—the first spreads out the beam, and the second flattens it. These foils are thin pieces of metal—remember that electrons have a higher probability of interacting with matter than X-ray photons. If the foils are made too thick, they'd be no different than the target and they'd just make X-rays.

Because of the electrons' tendency to scatter, additional collimation (an **electron applicator** or **cone**) must be placed between the jaws (photon beam collimator) and the patient. This collimator is made from low Z material, which minimizes bremsstrahlung as the stray electrons slam into it. Because of its use, patients are typically treated with only one field (machine position relative to the patient) per day—isocentric treatment can still be done, albeit with separate fields on different days. To minimize air shielding, SSD values for electron therapy are kept low, typically less than 100 cm.

Blocking can further shape electron beams. Let's say the cone produces a 10×10 cm^2 square field at the surface of the patient, but you need a 5 cm diameter circular field. Take a 10×10 cm^2 sheet of lead and cut out a 5 cm diameter circle. This can then be placed directly on the patient's skin. A material called Cerrobend (an alloy of Bi, Pb, Sn, and Cd) is sometimes used instead of pure Pb. How thick does the lead need to be? The electrons hitting the metal will create X-rays, so it needs to be thick enough to absorb most of them. As it turns out, all you need to do is divide the electron beam energy by 2 MeV/mm—that gives the shielding thickness necessary in mm. For Cerrobend, multiply beam energy by 0.6 mm/MeV. Notice that these two different mathematical operations are quite similar in their results.

The beam can also be shaped at the patient's skin with a **bolus**. A bolus is made from material that will interact with ionizing radiation similar to human tissue and is typically pliable. There are two reasons to use a bolus: (1) The treatment volume is a little shallow for the electron beam energy available—adding a bolus effectively makes it deeper. In other words, adding a bolus brings the isodose curves closer to the patient's skin, delivering more dose closer to the surface. (2) The surface of the patient is uneven. Because electrons have a relatively high probability of interacting with matter, surface irregularities will create inhomogeneities in the doses delivered at depth. The bolus can help straighten out the isodose curves by providing a flat entry surface. Note that a bolus can also be used with high-energy X-ray beam therapy for the same reasons listed here.

Electrons are charged particles, and unlike photons, will deliver dose in the patient's body up until a point, then not much after that (Bragg peak, see Section 7.2). In other words, they have a definable range in matter. Range (R) in humans can be estimated as the energy of the electron beam (E) divided by 2.0 MeV/cm (Equation 7.10).

Some dose is delivered after this point, but this is largely due to X-rays created by the electron beam hitting things like the collimators and scattering foils as the beam emerges from the accelerator. This is often referred to as "X-ray contamination" and is inevitable when using electron beams. About 5% of the patient's total dose during electron beam therapy is due to X-rays.

What do isodose curves look like for electron beams? First, they are shallower, and more compressed than photon beams of "equivalent" energy. This is because electrons have a much higher probability of interacting with matter, and because electrons tend to deliver more of their dose at the Bragg peak. Electron isodose curves also "balloon" out at the sides because electrons tend to scatter through larger angles than high-energy X-rays. Percent depth dose (PDD) can be estimated for electron beams using simple (approximate!) formulas. Most electron beam therapy is performed at depths corresponding to PDD values of 90% and 80%. Ninety percent of the maximum dose will be delivered at a depth ($d_{90\%}$) equivalent to the energy of the electron beam (E) divided by 4.0 MeV/cm (Equation 15.7).

$$d_{90\%} = \frac{E}{4.0 \text{ MeV/cm}} \tag{15.7}$$

Eighty percent of the maximum dose is delivered at a depth corresponding to the beam energy divided by 3.0 MeV/cm (Equation 15.8).

$$d_{80\%} = \frac{E}{3.0 \text{ MeV/cm}} \tag{15.8}$$

Example 15.9: Electrons in a beam have an energy of 10 MeV. At what depth will the dose be 90% of the maximum?

$$d_{90\%} = \frac{10 \text{ MeV}}{4.0 \text{ MeV/cm}} = 2.5 \text{ cm}$$

What is the range, in tissue, of these electrons?

$$R_{e^-} = \frac{10 \text{ MeV}}{2.0 \text{ MeV/cm}} = 5.0 \text{ cm}$$

Like photons, increasing the energy of the electron beam will deliver more dose deeper, about one centimeter for every 2 MeV according to Equation 6.10. Even so, the depths where significant doses can be delivered remain fairly shallow for electron therapy, meaning it is only useful for relatively superficial treatments. Its most common applications today are for relatively shallow lesions with critical stuff behind them. An example would be treating the lymph nodes near the spinal column. The spinal column can't tolerate much of a radiation dose, making photon therapy less attractive.

Electron therapy is most commonly used to augment X-ray treatment of breast cancer. Photon therapy is generally performed tangential to the chest to minimize dose to the lungs. Because electron beams can't penetrate all the way to the lungs, they can be done *en face* (French for "straight out"). Since electron therapy is done as a complement to photon therapy when treating breast cancer, it is referred to as a "breast boost."

Electron therapy can also be used for skin, lip, head, and neck cancers. Total skin irradiation (TSI, six low-dose fields designed to irradiate all of the skin evenly) can be used in the treatment of some cases of skin cancer. During TSI, lead shielding is placed over the eyes and nail beds, which are more strongly affected by electron beam therapy. Also, a plastic sheet is placed in the beam to provide greater scattering.

Treatment times for electron beam therapy are calculated using a formula (Equation 15.9) similar to one used for X-rays (compare with Equation 15.4).

$$N_{mu} = \frac{\text{Tumor Dose}}{O_f \times \dfrac{D_n}{D_{max}} \times C_{fs}} \tag{15.9}$$

where O_f is the **output factor**, which is a calibrated dose rate for the machine. It is pretty much the same as C_{cal}, and usually has a value of 1.00 cGy/mu. The field size correction factor (C_{fs}) also includes any blocking.

Example 15.10: Using information from Example 15.9, how long will it take to deliver a 180 cGy dose at 2.5 cm if the output factor is 1.00 cGy/mu and the field size correction is 0.987?
The depth dose is 0.90 (from Example 15.9). Plug into Equation 15.9:

$$N_{mu} = \frac{180 \text{ cGy}}{1.00 \text{ cGy/mu} \times 0.90 \times 0.987} = 203 \text{ mu}$$

Like electrons, protons are charged particles, but they are much more massive. As a result, they have a much shorter range (Table 7.2), and don't scatter very much. Because they scatter less, they also have a much sharper Bragg peak than electrons, making them an excellent choice for radiation therapy, so long as the tumor is located at, or very near, the Bragg peak depth. That depth can be varied with the energy of the protons—something that can even be done during treatment! The peak also drops off very quickly shortly after delivering the bulk of the dose. No worries about an exit dose here—in fact, there's very little dose delivered beyond the peak (no pesky X-rays as with electrons). Because of the sharp Bragg peak, multi-field treatments with proton beams can yield very tight dose distributions—much better than photons.

It might be a good idea to remember how protons interact with matter (Section 7.2). Primarily, they interact with electrons—knocking them loose from their orbitals. These secondary electrons have a very short range and do their damage locally. Protons can also interact with nuclei, but typically this doesn't happen until they slow down quite a bit (in or near the Bragg peak). When protons crash into a nucleus, a nuclear reaction takes place, and fragments are scattered—again typically doing biological damage very close to where the nucleus was originally.

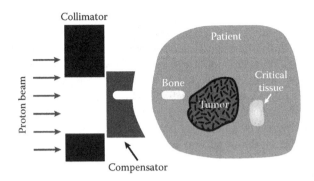

FIGURE 15.11   Sculpting a proton beam.

Protons can also be scattered by nuclei, through Coulomb repulsions. This causes the beam to widen as it penetrates the matter—creating penumbra. The amount of scattering depends on the energy of the proton beam. Higher-energy protons are more likely to scatter and give larger penumbra, lower-energy protons have smaller penumbra. We won't worry about why this is so.

Because protons have a much higher probability of interacting with matter than photons, tissue inhomogeneities (e.g., bones) are a bigger deal. It is important to have good anatomical data (computed tomography scans) before beginning treatment. With these data in hand, **compensators** (beam filters) are designed for each patient (Figure 15.11) to properly "sculpt" the beam.

The proton beam is first shaped by the collimators, which set the size of the beam to the approximate size of the tumor. The compensator then adjusts the proton flux according to its thickness—the thicker parts attenuate the proton beam to a greater degree. Notice how the right surface of the compensator generally mirrors that of the patient's left side. This way almost all protons in the beam travel the same distance through compensator and tissue. This places the Bragg peak at the same depth at every point in the beam. In other words, this makes for a relatively flat isodose curve on the backside of the tumor, protecting the critical tissue beyond. As you might guess, compensators are made from low Z materials to mimic proton interactions in tissues. A thicker compensator at the beam edges also minimizes penumbra by minimizing proton flux at the beam edges. Scattering from protons initially traveling closer to the center of the beam tends to even the dose inside the patient.

By taking a chunk out of the compensator for the part of the beam that passes through the bone, we can intensify the flux enough to even out the dose through that part of the tumor. Note that the removed chunk needs to be just the right size and perfectly aligned to properly dose the tumor and avoid excessive dose to healthy tissues beyond the tumor.

Like the photon and electron beams produced by linear accelerator beams, proton beams need to be broadened. Traditionally, this was done using scattering foils (just like broadening electron beams). These foils are thin pieces of high Z metals. This method is called **passive scattering**. Once broadened, a collimator trims the beam, and the compensator shapes its intensity.

Before scattering, the energy of the protons is varied using a **range-shifter** (also known as a range-modulator). By varying the beam energy, the depth of the Bragg peak can be moved forward and backward (along the beam axis—left and right on Figure 15.11) through the tumor. A problem with this set-up is that beam modulation is only one-dimensional. By shaping the compensator properly, we can get a very nice dose cutoff on the backside of the tumor but using a range-shifter means that the shape of the isodose curve on the nearside of the tumor will have the same shape as on the backside. If the tumor lacks this symmetry (very likely!) a significant dose will be delivered to healthy tissue on the nearside.

The solution to this problem is **active scattering**, as illustrated in Figure 15.12. Passing a beam of charged particles through a magnetic field will cause the beam to deflect. The degree of deflection

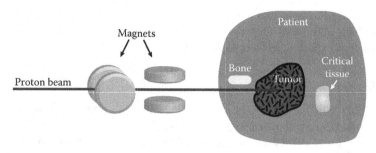

FIGURE 15.12 Active scattering of a proton beam.

will depend on the size and charge of the particle and the strength of the magnetic field. Two pairs of parallel magnets, offset 90° from each other, are used to move the Bragg peak in and out of the page (the magnets on the left) and up and down on the page (the magnets on the right). The strength of the magnetic fields is varied to allow continuous movement through the tumor in these two dimensions. Further upstream, the energy of the beam is dynamically varied, so the Bragg peak can be moved right and left inside of the tumor as pictured. This allows placement of the Bragg peak at any point, in three dimensions, within the tumor, and minimizes dose outside. If both sets of magnets are switched off, the proton beam will travel straight through. Notice also that no compensator, and generally no collimator, is needed with active scattering, just a lot of skill in changing proton energies and magnetic strengths in ways that provide an even dose throughout the tumor while minimizing dose to healthy tissues.

Proton therapy has proven especially effective in treating cancer in the eye, skull base sarcomas, and prostate cancer. Like electron therapy, it can sometimes be used effectively in conjunction with photon therapy.

It sounds like proton therapy is pretty good stuff, yet there are only a couple dozen treatment centers worldwide. Why isn't it more commonly used? First, it can only treat stuff that's fairly close to the surface (protons like to interact with matter). Also, as mentioned in Section 9.2, the equipment needed takes up a lot of space and requires some serious geeks to operate and maintain. The high-energy protons are generated in a cyclotron, which requires a fair bit of shielding. These protons have a lot of velocity, and it is hard to get the beam to turn. Really big magnets are necessary (don't take your credit card!). Because of the size of the cumbersome magnets, rotation of the gantry is limited to 180°, or less (take a look at Figure 9.4). Like linear accelerators, this gantry must have a great deal of precision in its movements. Typically beam variances at the isocenter must be less than 0.5 mm. Engineering this level of precision into an instrument this large does not come cheaply.

Because of the size requirements, a fair amount of space is required for a treatment facility. In addition to the cyclotron and its shielding, the beam lines and treatment rooms need to be heavily shielded. Constructing a proton therapy unit is roughly three times more expensive than a conventional photon therapy wing. Also, some radioactive contamination will develop wherever protons are used, from incidental nuclear reactions caused by the energetic protons. For example, a high-energy proton slamming into a nitrogen-14 atom (common in the air) creates oxygen-14 and a neutron.

$$^{14}_{7}\text{N} + ^{1}_{1}\text{H} \rightarrow ^{14}_{8}\text{O} + ^{1}\text{n}$$

Likewise, protons can react with ^{56}Fe (commonly used in construction) to create ^{56}Co and a neutron.

$$^{56}_{26}\text{Fe} + ^{1}_{1}\text{H} \rightarrow ^{56}_{27}\text{Co} + ^{1}\text{n}$$

The neutrons created in these two p,n reactions can go on to cause other nuclear reactions. Note also that the products in these two reactions are unstable, decaying via positron emission and electron capture, which is expected, since both are proton-rich.

QUESTIONS

15.1   The PDD is 70.7% at 9.0 cm for an X-ray beam from a 6 MV linac. What dose is required at d_{max} (1.5 cm) to deliver 250 cGy to a tumor at 9.0 cm?

15.2   The d_{max} for a 10 MV linear accelerator is only 2.5 cm. How can we treat tumors at greater depth without damaging the healthy tissue surrounding it?

15.3   Determine the percent contributions of primary and secondary radiation at 19 cm depth for a 20 MV, 15×15 cm^2 field at 100 cm SSD. Percent depth doses are 53.5% for a "0" field and 55.0% for a 15×15 cm^2 field. Repeat for a 6 MV machine. PDD values are now 39.2% and 43.9%, respectively. Can you explain the difference between the two linear accelerators?

15.4   Calculate the percent depth dose at 7.0 cm depth for an X-ray beam from a 10 MV linear accelerator at 125.0 cm SSD. For a 10 MV beam d_{max} = 2.5 cm and the PDD at 100.0 cm SSD = 84.0%.

15.5   A patient, 23 cm thick at the area of treatment, needs a 215 cGy dose delivered to a tumor 5.5 cm below his skin. Using a 20 MV machine at 100 cm SSD and a 12×12 cm^2 field, determine the dose at d_{max} and the exit dose. How deep is d_{max} for this machine? The PDD values in Table 15.1 apply only to the machine as configured in this question.

TABLE 15.1
Percent Depth Dose Values for Question 15.5

Depth (cm)	3.0	5.0	6.0	20.0	30.0
PDD	100.0	94.7	91.5	52.7	35.2

15.6   Using a PDD, determine the treatment time (in mu) to deliver 200 cGy to a tumor 8.0 cm below the surface of a patient exposed to a 7×7 cm^2 6 MV field at 150 cm SSD. This accelerator is calibrated to deliver 0.998 cGy/mu, has a field size correction factor of 0.978, an equipment attenuation factor of 0.95, and a percent depth dose of 74.6% at 8.0 cm depth and 100 cm SSD. The maximum dose is delivered at 1.5 cm.

15.7   What is the relationship between TAR and BSF? What is the relationship between TAR and TMR?

15.8   What happens to TAR and TMR values when SSD is decreased? Briefly explain.

15.9   Calculate the number of monitor units necessary to deliver 245 cGy to a tumor 6.5 cm below the patient's skin using a 15×15 cm^2 field on a 10 MV machine at 100 cm SAD. This accelerator is calibrated to deliver 1.02 cGy/mu at 100 cm SAD and has a TMR value of 0.932 under these conditions. The 15×15 cm^2 field size correction factor for this machine is 1.034.

15.10   Calculate the dose at d_{max} (at 2.5 cm) and the exit dose (at 15.0 cm, TMR = 0.741) for the previous question.

15.11   Determine the treatment time (in mu) required to deliver a 225 cGy dose to a tumor 7.3 cm below the skin exposed to an 7×7 cm^2 6 MV field at 110 cm SSD. This accelerator is calibrated at SAD = 100 cm to deliver 1.001 cGy/mu under these conditions. The

7 × 7 cm² field size correction factor for this machine is 0.978. The TMR values are 0.871 (at 7.0 cm depth) and 0.838 (at 8.0 cm). A 0.6 cm poly(methyl methacrylate) tray is also placed in the beam.

15.12 Define the following: secondary radiation, field symmetry, SAD, tissue-air ratio, and TMR.

15.13 Why are high Z metals used for blocking electron beams?

15.14 Why is X-ray production "inevitable" when using electron beams?

15.15 Give one reason why electron therapy would be preferred to photon therapy. Also give a reason why electron therapy might not be used.

15.16 Briefly explain the differences observed in the electron and photon isodose curves produced by the same MV machine. In terms of energy, what is the fundamental difference between these two beams?

15.17 What is the range, in tissue, of 12 MeV electrons? What treatment time (mu) will be required for a 170 cGy dose to be delivered 3.0 cm below the surface with a 12 MeV electron beam? The output factor is 1.00 cGy/mu and the field size correction is 0.985 for this beam.

15.18 Why doesn't C_{attn} appear in Equation 15.9?

15.19 What would be the effect of taking too big of a chunk out of the compensator to adjust proton flux for a bone between the skin and the tumor along the proton beam path? What if the chunk were properly sized, but not properly aligned?

15.20 Give one reason why proton therapy would be preferred to photon therapy. Also give a reason why proton therapy might not be used.

15.21 Why is it important to sculpt a proton beam for treatment?

Appendix A: Nuclide Data

Element	Z	N	Nuclide Symbol	Mass[1] (u)	Δ (MeV)[2]	Percent Abundance[3]	Half-life	Decay Mode[4]	Particle Energy[5] (MeV)	Gamma Energy[6] (keV)	Total Energy of Decay (MeV)
Neutron	0	1	^{1}n	1.008665	8.071	none	10.2 min	β⁻	0.782		0.782
Hydrogen	1	0	^{1}H	1.007825	7.289	99.9885	stable				
	1	1	^{2}H	2.014102	13.136	0.0115	stable				
	1	2	^{3}H	3.016049	14.950	none	12.32 a	β⁻	0.018591		0.018591
	1	3	^{4}H	4.0278	25.896	none	1.40E-22 s	n			2.91
Helium	2	1	^{3}He	3.016029	14.931	0.000134	stable				
	2	2	^{4}He	4.002603	2.425	99.999866	stable				
	2	3	^{5}He	5.0122	11.364	none	7.0E-22 s	n			
	2	4	^{6}He	6.018888	17.594	none	807 ms	β⁻	3.510		3.508
Lithium	3	3	^{6}Li	6.015123	14.087	7.59	stable				
	3	4	^{7}Li	7.016005	14.908	92.41	stable				
	3	5	^{8}Li	8.022487	20.946	none	0.840 s	β⁻	13		16.0052
Beryllium	4	3	^{7}Be	7.016929	15.769	none	53.3 d	EC		478	0.8619
	4	4	^{8}Be	8.005305	4.942	none	7.0E-17 s	α	0.0461		
	4	5	^{9}Be	9.012182	11.348	100	stable				
	4	6	^{10}Be	10.013534	12.607	none	1.5E+06 a	β⁻	0.556		0.556
Boron	5	4	^{9}B	9.013329	12.416	none	8E-19 s	αα	0.0461		
	5	5	^{10}B	10.012937	12.051	19.9	stable				
	5	6	^{11}B	11.009305	8.668	80.1	stable				
	5	7	^{12}B	12.014352	13.369	none	20.20 ms	β⁻	13.37	4439	13.369
	5	8	^{13}B	13.017780	16.562	none	17.4 ms	β⁻	13.4	3684	13.437
	5	9	^{14}B	14.02540	23.660	none	13 ms	β⁻	14	6092	20.64
Carbon	6	4	^{10}C	10.016853	15.699	none	19.308 s	β⁺	1.87	718	3.6480
	6	5	^{11}C	11.011434	10.651	none	20.36 min	β⁺ EC	0.960		1.982
	6	6	^{12}C	12.000000	0.000	98.93	stable				

Element	Z	N	Nuclide	Atomic mass	Mass excess	Abundance (%)	Half-life	Decay mode	E (MeV)	E_γ (keV)	Q (MeV)
Nitrogen	6	7	^{13}C	13.003355	3.125	1.07	stable				
	6	8	^{14}C	14.003242	3.020	none	5715 a	β-	0.157		0.156476
	7	6	^{13}N	13.005739	5.345	none	9.97 min	β+	1.190		2.2205
	7	7	^{14}N	14.003074	2.863	99.636	stable				
	7	8	^{15}N	15.000109	0.101	0.364	stable				
	7	9	^{16}N	16.006101	5.683	none	7.13 s	β-	4.27 / 10.44	6129 / 7115	10.421
	7	12	^{19}N	19.01702	15.854	none	0.27 s	β- + n	4.06	96	12.5
Oxygen	8	4	^{12}O	12.03440	32.043	none	1E-21 s	2p			1.77
	8	7	^{15}O	15.003065	2.855	none	2.037 min	β+ / EC	1.72		2.7542
	8	8	^{16}O	15.994915	-4.737	99.757	stable				
	8	9	^{17}O	16.999132	-0.809	0.038	stable				
	8	10	^{18}O	17.999161	-0.782	0.205	stable				
	8	11	^{19}O	19.003579	3.334	none	26.9 s	β-	3.3 / 4.60	197 / 1357	4.822
	8	12	^{20}O	20.004076	3.797	none	13.5 s	β-	2.75	1057	3.815
Fluorine	9	8	^{17}F	17.002095	1.952	none	1.08 min	β+	1.74		2.7605
	9	9	^{18}F	18.000938	0.873	none	1.8293 h	β+ / EC	0.635		1.655
	9	10	^{19}F	18.998403	-1.487	100	stable				
	9	11	^{20}F	19.999981	-0.017	none	11.1 s	β-	5.39	1634	7.0245
	9	16	^{25}F	25.01210	11.271	none	0.07 s	β-		1703	13.4
Neon	10	8	^{18}Ne	18.005697	5.307	none	1.667 s	β+	3.42	1041	4.443
	10	9	^{19}Ne	19.001880	1.751	none	17.22 s	β+ / EC	2.24	110 / 1357	3.2388
	10	10	^{20}Ne	19.992440	-7.042	90.48	stable				
	10	11	^{21}Ne	20.993847	-5.732	0.27	stable				
	10	12	^{22}Ne	21.991385	-8.025	9.25	stable				
	10	13	^{23}Ne	22.994467	-5.154	none	37.1 s	β-	4.38 / 3.95	440	4.3758

(continued)

Element	Z	N	Nuclide Symbol	Mass¹ (u)	Δ (MeV)²	Percent Abundance³	Half-life	Decay Mode⁴	Particle Energy⁵ (MeV)	Gamma Energy⁶ (keV)	Total Energy of Decay (MeV)
	10	15	^{25}Ne	24.99779	−2.059	none	0.61 s	β⁻	7.2 6.3	90 980	7.25
Sodium	11	11	^{22}Na	21.994437	−5.182	none	2.604 a	β⁺ EC	0.546	1275	2.8423
	11	12	^{23}Na	22.989769	−9.530	100	stable				
	11	13	^{24}Na	23.990963	−8.418	none	14.97 h	β⁻	1.391	1369 2754	5.5155
	11	14	^{25}Na	24.989954	−9.358	none	59.3 s	β⁻	3.8	975 585 390 1612	3.835
Magnesium	12	11	^{23}Mg	22.994125	−5.473	none	11.32 s	β⁺	3.09	440	4.056
	12	12	^{24}Mg	23.985042	−13.934	78.99	stable				
	12	13	^{25}Mg	24.985837	−13.193	10.00	stable				
	12	14	^{26}Mg	25.982593	−16.215	11.01	stable				
	12	15	^{27}Mg	26.984341	−14.587	none	9.45 min	β⁻	1.75 1.59	844 1014	2.6100
	12	16	^{28}Mg	27.983877	−15.018	none	21.0 h	β⁻	0.459	31 1342	1.832
Aluminum	13	12	^{25}Al	24.990429	−8.916	none	7.17 s	β⁺	3.26	1612	4.2767
	13	13	^{26}Al	25.986892	−12.210	none	7.1E+05 a	β⁺ EC	1.17	1809	4.0043
	13	14	^{27}Al	26.981539	−17.197	100	stable				
	13	15	^{28}Al	27.981910	−16.851	none	2.25 min	β⁻	2.86	1779	4.6424
Silicon	14	11	^{25}Si	25.00411	3.828	none	220 ms	β⁺		452 493 1612 945	12.74
	14	13	^{27}Si	26.986705	−12.384	none	4.14 s	β⁺	3.85	2210	4.8124
	14	14	^{28}Si	27.976927	−21.493	92.223	stable				
	14	15	^{29}Si	28.976495	−21.895	4.685	stable				

	14	16	^{30}Si	29.973770	−24.433	3.092	stable				
	14	17	^{31}Si	30.97536	−22.952	none	2.62 h	β⁻	1.48	1266	1.4919
	14	18	^{32}Si	31.974148	−24.081	none	1.6E+02 a	β⁻	0.221		0.2243
Phosphorus	15	15	^{30}P	29.978314	−20.201	none	2.50 min	β⁺, EC	3.24	2235	4.2324
	15	16	^{31}P	30.973762	−24.441	100	stable				1.7105
	15	17	^{32}P	31.973907	−24.305	none	14.28 d	β⁻	1.709		
	15	18	^{33}P	32.971725	−26.338	none	25.3 d	β⁻	0.249		0.249
Sulfur	16	15	^{31}S	30.979554	−19.045	none	2.56 s	β⁺	4.39	1266	5.396
	16	16	^{32}S	31.972071	−26.016	94.99	stable				
	16	17	^{33}S	32.971459	−26.586	0.75	stable				
	16	18	^{34}S	33.967867	−29.932	4.25	stable				
	16	19	^{35}S	34.969032	−28.846	none	87.2 d	β⁻	0.1674		0.1672
	16	20	^{36}S	35.967081	−30.664	0.01	stable				
	16	21	^{37}S	36.971126	−26.896	none	5.05 min	β⁻	1.76	3104	4.8652
Chlorine	17	17	^{34m}Cl	33.973919	−24.294	none	32.2 min	IT, β⁺	2.5, 1.3	2128, 1176, 3304, 146	5.638, 0.146
	17	17	^{34}Cl	33.973762	−24.441	none	1.527 s	β⁺	4.47		5.4920
	17	18	^{35}Cl	34.968853	−29.014	75.76	stable				
	17	19	^{36}Cl	35.968307	−29.522	none	3.01E+05 a	β⁻, EC	0.709, 0.12		0.7097, 1.1422
	17	20	^{37}Cl	36.965903	−31.762	24.24	stable				
	17	21	^{38}Cl	37.968011	−29.798	none	37.2 min	β⁻	4.91	2168, 1642	4.9165
Argon	18	17	^{35}Ar	34.975257	−23.048	none	1.775 s	β⁺	4.943, 1.11	1219, 1763, 2694	5.966
	18	18	^{36}Ar	35.967545	−30.232	0.3365	stable				
	18	19	^{37}Ar	36.966776	−30.948	none	35.0 d	EC			0.8139
	18	20	^{38}Ar	37.962732	−34.715	0.0632	stable				
	18	21	^{39}Ar	38.964313	−33.242	none	269 a	β⁻	0.565		0.565
	18	22	^{40}Ar	39.962383	−35.040	99.6003	stable				

(continued)

Element	Z	N	Nuclide Symbol	Mass¹ (u)	Δ (MeV)²	Percent Abundance³	Half-life	Decay Mode⁴	Particle Energy⁵ (MeV)	Gamma Energy⁶ (keV)	Total Energy of Decay (MeV)
	18	23	^{41}Ar	40.964501	−33.067	none	1.83 h	β⁻	1.198	1294	2.4916
	18	24	^{42}Ar	41.963045	−34.423	none	32.9 a	β⁻	0.60		0.599
Potassium	19	18	^{37}K	36.973377	−24.799	none	1.23 s	β⁺	5.13	2796	6.1475
	19	19	^{38}K	37.969080	−28.802	none	7.63 min	β⁺	2.68	2168	5.9139
	19	20	^{39}K	38.963707	−33.807	93.2581	stable				
	19	21	^{40}K	39.963999	−33.535	0.0117	1.25E+09 a	β⁻ / EC / β⁺	1.33	1461	1.3111 / 1.5047
	19	22	^{41}K	40.961826	−35.559	6.7302	stable				
	19	23	^{42}K	41.962403	−35.021	none	12.36 h	β⁻	3.52	1525	3.5255
	19	24	^{43}K	42.960716	−36.593	none	22.3 h	β⁻	0.83	373 / 618	1.82
Calcium	20	20	^{40}Ca	39.962591	−34.846	96.941	stable				
	20	21	^{41}Ca	40.962278	−35.138	none	1.03E+05 a	EC			0.4213
	20	22	^{42}Ca	41.958618	−38.547	0.647	stable				
	20	23	^{43}Ca	42.958767	−38.409	0.135	stable				
	20	24	^{44}Ca	43.955482	−41.468	2.086	stable				
	20	25	^{45}Ca	44.956186	−40.812	none	162.7 d	β⁻	0.258	12	0.256
	20	26	^{46}Ca	45.953693	−43.135	0.004	stable				
	20	27	^{47}Ca	46.954547	−42.339	none	4.536 d	β⁻	0.694 / 1.990	1297	1.992
	20	28	^{48}Ca	47.952534	−44.214	0.187	4E+19 a	β⁻β⁻			4.272
	20	29	^{49}Ca	48.955673	−41.290	none	8.72 min	β⁻	2.18	3084	5.263
Scandium	21	23	^{44}Sc	43.959403	−37.816	none	3.93 h	β⁺ / EC	1.47	1157	3.652
	21	24	^{45}Sc	44.955912	−41.068	100	stable				
	21	25	^{46}Sc	45.955170	−41.759	none	83.81 d	β⁻	0.357	1121 / 889	2.366
	21	26	^{47}Sc	46.952408	−44.332	none	3.349 d	β⁻	0.439 / 0.600	159	0.600
Titanium	22	22	^{44}Ti	43.9596902	−37.548	none	59.2 a	EC		78 / 68	0.268

Element	Z	N	Nuclide	Atomic mass	Mass excess	% Abundance	Half-life	Decay			
	22	23	45Ti	44.958124	−39.007	none	3.078 h	β+ EC	1.04	720 1408	2.062
	22	24	46Ti	45.952632	−44.123	8.25	stable				
	22	25	47Ti	46.951763	−44.932	7.44	stable				
	22	26	48Ti	47.947946	−48.488	73.72	stable				
	22	27	49Ti	48.947870	−48.559	5.41	stable				
	22	28	50Ti	49.944791	−51.427	5.81	stable				
	22	29	51Ti	50.946616	−49.727	none	5.76 min	β−	2.14	320	2.474
	23	25	48V	47.952255	−44.474	none	15.98 d	β+ EC	0.694	984 1312	4.012
Vanadium	23	26	49V	48.948517	−47.956	none	331 d	EC			0.602
	23	27	50V	49.947158	−49.222	0.250	1.4E+17 a	EC β−	1553.8	783	2.205 1.0379
	23	28	51V	50.943960	−52.201	99.750	stable				
	23	29	52V	51.944780	−51.437	none	3.76 min	β−	2.47	1434	3.976
	24	24	48Cr	47.954036	−42.815	none	21.6 h	EC β+		308 112	1.66
Chromium	24	25	49Cr	48.951341	−45.326	none	42.3 min	β+	1.39 1.45	91 153 62	2.626
	24	26	50Cr	49.946044	−50.260	4.345	stable				
	24	27	51Cr	50.944772	−51.445	none	27.702 d	EC		320	0.7526
	24	28	52Cr	51.940507	−55.417	83.789	stable				
	24	29	53Cr	52.940649	−55.285	9.501	stable				
	24	30	54Cr	53.938880	−56.933	2.365	stable				
	24	31	55Cr	54.940844	−55.103	none	3.497 min	β−	2.49	1528	2.6031
Manganese	25	28	53Mn	52.941295	−54.683	none	3.7E+06 a	EC			0.5968
	25	29	54Mn	53.940363	−55.552	none	312.1 d	EC β− β+		835	1.377 0.697 1.377

(continued)

Element	Z	N	Nuclide Symbol	Mass¹ (u)	Δ (MeV)²	Percent Abundance³	Half-life	Decay Mode⁴	Particle Energy⁵ (MeV)	Gamma Energy⁶ (keV)	Total Energy of Decay (MeV)
	25	30	^{55}Mn	54.938045	−57.711	100	stable				
	25	31	^{56}Mn	55.938909	−56.906	none	2.578 h	β⁻	2.84, 1.04	847, 1811, 2113	3.6956
Iron	26	26	^{52}Fe	51.94812	−48.326	none	8.28 h	β⁺ / EC	0.80	169, 169	2.37
	26	27	^{53}Fe	52.945312	−50.942	none	8.51 min	β⁺ / EC	2.8, 2.4	378	3.743
	26	28	^{54}Fe	53.939611	−56.252	5.845	stable				
	26	29	^{55}Fe	54.938298	−57.475	none	2.75 a	EC		126	0.2312
	26	30	^{56}Fe	55.934937	−60.606	91.754	stable				
	26	31	^{57}Fe	56.935394	−60.180	2.119	stable				
	26	32	^{58}Fe	57.933276	−62.153	0.282	stable				
	26	33	^{59}Fe	58.934880	−60.659	none	44.50 d	β⁻	0.466, 0.274	1099, 1292	1.565
	26	40	^{60}Fe	59.934077	−5650.371	none	2.62E+06 a	β⁻	0.14	59	0.237
Cobalt	27	30	^{57}Co	56.936296	−59.340	none	271.8 d	EC		122, 137, 14	0.836
	27	31	^{58}Co	57.935758	−59.841	none	70.88 d	EC / β⁺	0.474	811	2.308
	27	32	^{59}Co	58.933195	−62.228	100	stable				
	27	33	^{60m}Co	59.933885	−61.586	none	10.47 min	IT / β⁻	1.6	59, 1333	0.059, 2.883
	27	33	^{60}Co	59.933822	−61.644	none	5.271 a	β⁻	0.318	1333, 1173	2.8231
Nickel	28	27	^{55}Ni	54.95134	−45.326	none	202 ms	β⁺	7.66		8.69
	28	28	^{56}Ni	55.94214	−53.896	none	5.9 d	EC		158, 812	2.14
	28	29	^{57}Ni	56.939800	−56.076	none	35.6 h	EC / β⁺	0.85	1378	3.262

Z	N	Element	Nuclide	Atomic mass	Mass excess	Abundance	Half-life	Decay mode	Particle energy (MeV)	γ (keV)	Q (MeV)
28	30		^{58}Ni	57.935343	−60.228	68.0769	stable				
28	31		^{59}Ni	58.934352	−61.151	none	7.6E+04 a	EC, β+			1.0728
28	32		^{60}Ni	59.930786	−64.472	26.2231	stable				
28	33		^{61}Ni	60.931056	−64.221	1.1399	stable				
28	34		^{62}Ni	61.928345	−66.746	3.6345	stable				
28	35		^{63}Ni	62.929673	−65.509	none	101 a	β−	0.0669	none	0.06698
28	36		^{64}Ni	63.927966	−67.099	0.9256	stable				
28	37		^{65}Ni	64.930088	−65.123	none	2.517 h	β−	2.14, 0.65	1482, 1116	2.138
29	32	Copper	^{61}Cu	60.933462	−61.980	none	3.35 h	β+, EC	1.21	283, 656	2.237
29	33		^{62}Cu	61.932587	−62.795	none	9.74 min	β+, EC	2.93	1173, 876	3.948
29	34		^{63}Cu	62.929597	−65.580	69.15	stable				
29	35		^{64}Cu	63.929768	−65.421	none	12.701 h	EC, β−, β+	0.578, 0.651	1346	1.675, 0.579, 1.6750
29	36		^{65}Cu	64.927789	−67.264	30.85	stable				
29	37		^{66}Cu	65.928873	−66.254	none	5.10 min	β−	2.63	1039	2.641
30	32	Zinc	^{62}Zn	61.93433	−61.171	none	9.22 h	β+, EC	0.66	597, 41, 548, 508	1.63
30	33		^{63}Zn	62.933216	−62.209	none	38.5 min	β+, EC	2.32	670, 962	3.367
30	34		^{64}Zn	63.929142	−66.004	48.268	stable				
30	35		^{65}Zn	64.929245	−65.908	none	244.0 d	EC, β+	0.325	1116	1.3521
30	36		^{66}Zn	65.926033	−68.900	27.975	stable				
30	37		^{67}Zn	66.927127	−67.881	4.102	stable				

Element	Z	N	Nuclide Symbol	Mass[1] (u)	Δ (MeV)[2]	Percent Abundance[3]	Half-life	Decay Mode[4]	Particle Energy[5] (MeV)	Gamma Energy[6] (keV)	Total Energy of Decay (MeV)
	30	38	⁶⁸Zn	67.924844	−70.007	19.024	stable				
	30	38	⁶⁹ᵐZn	68.927025	863.518	none	13.76 h	IT		439	0.439
								β⁻		574	1.345
	30	39	⁶⁹Zn	68.926554	−68.415	none	56 min	β⁻	0.90	319	0.910
	30	40	⁷⁰Zn	69.925319	−69.565	0.631	stable				
	30	42	⁷²Zn	71.926861	−68.129	none	46.5 h	β⁻	0.30	145	0.46
Gallium	31	36	⁶⁷Ga	66.928205	−66.877	none	3.2611 d	EC		93	1.001
										185	
										300	
	31	37	⁶⁸Ga	67.927984	−67.082	none	1.130 h	β⁺	1.899	1077	2.921
								EC			
	31	38	⁶⁹Ga	68.925574	−69.327	60.108	stable				
	31	39	⁷⁰Ga	69.926028	−68.904	none	21.14 min	β⁻	1.65	1039	1.653
										176	
								EC			0.655
	31	40	⁷¹Ga	70.924701	−70.141	39.892	stable				
	31	41	⁷²Ga	71.926370	−68.586	none	14.10 h	β⁻	0.97	834	3.997
									0.74	2202	
										630	
	31	42	⁷³Ga	72.925170	−69.704	none	4.87 h	β⁻	1.2	297	1.598
										326	
										53	
										13	
Germanium	32	32	⁶⁴Ge	63.9416	−54.399	none	1.06 min	β⁺	3.0	427	4.48
								EC		667	
										128	
	32	36	⁶⁸Ge	67.928097	−66.977	none	270.8 d	EC			0.11
	32	37	⁶⁹Ge	68.927972	−67.094	none	1.63 d	EC		1107	2.227
								β⁺	1.21	574	
										872	
	32	38	⁷⁰Ge	69.924247	−70.563	20.38	stable				

Element	Z	N	Isotope	Atomic mass	Mass excess	Abundance	Half-life	Decay	Energy	γ (keV)	Q
	32	39	^{71}Ge	70.924954	−69.905	none	11.4 d	EC			0.2325
	32	40	^{72}Ge	71.922076	−72.586	27.31	stable				
	32	41	^{73}Ge	72.923459	−71.297	7.76	stable				
	32	42	^{74}Ge	73.921178	−73.422	36.72	stable				
	32	43	^{75}Ge	74.922860	−71.855	none	1.380 h	β−	1.19	265	1.176
	32	44	^{76}Ge	75.921403	−73.213	7.83	1.3E+21 a	β−β−			2.04
	32	45	^{77}Ge	76.923549	−71.214	none	11.22 h	β−	2.20	264	2.703
									1.38	211	
										216	
Arsenic	33	40	^{73}As	72.923825	−70.957	none	80.3 d	EC		53	0.341
										13	
	33	41	^{74}As	73.923929	−70.860	none	17.78 d	EC	0.941	596	2.562
								β+	1.350	635	1.353
								β−	0.717		
	33	42	^{75}As	74.921596	−73.033	100	stable				
	33	43	^{76}As	75.922394	−72.290	none	26.3 h	β−	2.97	559	2.963
									2.41		
Selenium	34	38	^{72}Se	71.92711	−67.897	none	8.5 d	EC		46	0.34
	34	39	^{73}Se	72.92677	−68.213	none	7.1 h	β+	1.32	361	2.74
								EC		67	
	34	40	^{74}Se	73.922476	−72.213	0.89	stable				
	34	41	^{75}Se	74.922524	−72.168	none	119.78 d	EC		265	0.863
										136	
										280	
	34	42	^{76}Se	75.919214	−75.252	9.37	stable				
	34	43	^{77}Se	76.919914	−74.600	7.63	stable				
	34	44	^{78}Se	77.917309	−77.026	23.77	stable				
	34	45	^{79}Se	78.918500	−75.917	none	3.5E+05 a	β−	0.16		0.151
	34	46	^{80}Se	79.916521	−77.760	49.61	stable				
	34	47	^{81}Se	80.917993	−76.389	none	18.5 min	β−	1.58	276	1.585
										290	
	34	48	^{82}Se	81.916699	−77.594	8.73	9E+19 a	β−β−			2.995

(continued)

Element	Z	N	Nuclide Symbol	Mass[1] (u)	Δ (MeV)[2]	Percent Abundance[3]	Half-life	Decay Mode[4]	Particle Energy[5] (MeV)	Gamma Energy[6] (keV)	Total Energy of Decay (MeV)
	34	49	83Se	82.919119	−75.340	none	22.3 min	β⁻	1.0	357	3.67
									1.8	510	
										225	
Bromine	35	42	77mBr	76.921494	−73.128	none	4.3 min	IT		106	0.106
	35	42	77Br	76.921380	−73.234	none	2.376 d	EC		239	1.365
										521	
	35	43	78Br	77.921146	−73.452	none	6.45 min	β⁺	0.34		3.574
								β⁺	2.5	614	
								EC			
								β⁻			0.727
	35	44	79Br	78.918337	−76.069	50.69	stable				
	35	45	80Br	79.918530	−75.889	none	17.66 min	β⁻	2.00	617	2.003
								EC		666	1.8705
								β⁺	0.85		
	35	46	81Br	80.916291	−77.974	49.31	stable				
	35	47	82Br	81.916805	−77.496	none	1.471 d	β⁻	0.444	777	3.093
										554	
										619	
	35	51	86Br	85.918806	−75.632	none	55.1 s	β⁻	3.3	1565	7.63
									7.4	2751	
Krypton	36	41	77Kr	76.924668	−70.171	none	1.24 h	β⁺	1.88		3.065
									1.70		
								EC		130	
										147	
	36	42	78Kr	77.920365	−74.180	0.355	1.5E+21 a	ECEC			2.85
	36	43	79Kr	78.920083	−74.442	none	1.455 d	EC	0.60	2	1.626
										261	
										398	
										606	
	36	44	80Kr	79.916379	−77.892	2.286	stable				
	36	45	81Kr	80.916592	−77.694	none	2.3E+05 a	EC		276	0.2808
	36	46	82Kr	81.913484	−80.589	11.593	stable				

Element	Z	N	Nuclide	Atomic mass	Mass excess (MeV)	Abundance (%)	Half-life	Decay mode	Eβ (MeV)	Eγ (keV)	Decay energy (MeV)
	36	47	^{83}Kr	82.914136	−79.982	11.500	stable				
	36	48	^{84}Kr	83.911507	−82.431	56.987	stable				
	36	49	^{85}Kr	84.912527	−81.481	none	10.76 a	β⁻	0.687	514	0.687
	36	50	^{86}Kr	85.910611	−83.266	17.279	stable				
	36	51	^{87}Kr	86.913354	−80.710	none	1.27 h	β⁻	3.5, 3.9	403, 2555	3.8884
	36	56	^{92}Kr	91.92615	−68.791	none	1.84 s	β⁻	6.0	142, 1219	5.99
Rubidium	37	44	^{81}Rb	80.918994	−75.457	none	4.57 h	EC, β⁺	1.05	190, 446	2.24
	37	45	^{82}Rb	81.918208	−76.189	none	1.260 min	β⁺, EC	3.4	777	4.401
	37	46	^{83}Rb	82.915112	−79.073	none	86.2 d	EC		520, 530, 553	0.91
	37	47	^{84}Rb	83.914385	−79.750	none	33.2 d	EC, β⁺	1.66, 0.78, 0.894	882	2.681, 0.894
	37	48	^{85}Rb	84.911790	−82.167	72.17	stable				
	37	49	^{86}Rb	85.911167	−82.747	none	18.65 d	β⁻, EC	1.775	1077	1.777, 0.5186
	37	50	^{87}Rb	86.909181	−84.598	27.83	4.9E+10 a	β⁻			
	37	51	^{88}Rb	87.911319	−82.606	none	17.7 min	β⁻	5.31	1836, 898	5.313
	37	55	^{92}Rb	91.919725	−74.776	none	4.48 s	β⁻	8.10	815, 2821, 570	8.10
	37	56	^{93}Rb	92.922033	−72.626	none	5.85 s	β⁻	7.46	433, 986, 213	7.47
	37	57	^{94}Rb	93.926407	−68.551	none	2.71 s	β⁻	9.5	837, 1578	10.29
Strontium	38	44	^{82}Sr	81.918401	−76.009	none	25.36 d	EC			0.18

(continued)

Element	Z	N	Nuclide Symbol	Mass[1] (u)	Δ (MeV)[2]	Percent Abundance[3]	Half-life	Decay Mode[4]	Particle Energy[5] (MeV)	Gamma Energy[6] (keV)	Total Energy of Decay (MeV)
	38	45	^{83}Sr	82.917555	−76.797	none	1.350 d	EC β⁺	 1.23	763 382	2.28
	38	46	^{84}Sr	83.913425	−80.644	0.56	stable				
	38	47	^{85}Sr	84.912933	−81.102	none	64.84 d	EC		514	1.065
	38	48	^{86}Sr	85.909260	−84.524	9.86	stable				
	38	49	^{87}Sr	86.908877	−84.881	7.00	stable				
	38	50	^{88}Sr	87.905612	−87.922	82.58	stable				
	38	51	^{89}Sr	88.907453	−86.207	none	50.61 d	β⁻	1.488	909	1.493
	38	52	^{90}Sr	89.907738	−85.941	none	28.8 a	β⁻	0.546		0.546
	38	54	^{92}Sr	91.911030	−82.875	none	2.61 h	β⁻	0.54	1384	1.95
	38	56	^{94}Sr	93.915360	−78.842	none	1.25 min	β⁻	2.08	1428	3.51
	38	62	^{100}Sr	99.9354	−60.175	none	201 ms	β⁻		964 899 66	7.1
Yttrium	39	49	^{88}Y	87.909503	−84.297	none	106.63 d	EC β⁺	 0.76	1836 898	3.623
	39	50	^{89m}Y			none	15.7 s	IT		909	0.9091
	39	50	^{89}Y	88.905848	−87.702	100	stable				
	39	51	^{90}Y	89.907151	−86.488	none	2.669 d	β⁻	2.281	2186	2.280
	39	53	^{92}Y	91.90895	−84.813	none	3.54 h	β⁻	3.64	935 1405	3.64
	39	56	^{94}Y	93.911594	−1013.844	none	18.7 min	β⁻	4.92	919 1139 551	4.92
	39	63	^{102}Y	101.93356	−61.888	none	0.30 s	β⁻		152 1211	9.9
Zirconium	40	49	^{89}Zr	88.908889	−84.869	none	3.27 d	EC β⁺	 0.90	909	2.833
	40	50	^{90}Zr	89.904704	−88.768	51.45	stable				
	40	51	^{91}Zr	90.905646	−87.890	11.22	stable				

Element	Z	N	Isotope	Atomic mass	Mass excess	Abundance (%)	Half-life	Decay mode	Decay energy	γ energy	Q
	40	52	^{92}Zr	91.905041	−88.454	17.15	stable				
	40	53	^{93}Zr	92.906476	−87.117	none	1.5E+06 a	β⁻	0.060	31	0.091
	40	54	^{94}Zr	93.906315	−87.267	17.38	stable				1.124
	40	55	^{95}Zr	94.908043	−85.657	none	64.02 d	β⁻	0.368, 0.400	757, 724	
	40	56	^{96}Zr	95.908273	−85.443	2.80	2.0E+19 a	β⁻β⁻			3.35
	40	57	^{97}Zr	96.910951	−82.949	none	16.75 h	β⁻	1.92	743, 469	2.659
	40	59	^{99}Zr	98.91651	−77.770	none	2.2 s	β⁻	3.54	546, 594	4.56
	40	62	^{102}Zr	101.92298	−71.744	none	2.9 s	β⁻	3.50	600, 535, 65	4.61
Niobium	41	51	^{92}Nb	91.907193	−86.449	none	3.5E+07 a	EC		561, 935	2.006
	41	52	^{93}Nb	92.906378	−87.208	100	stable				
	41	53	^{94}Nb	93.907284	−86.364	none	2.0E+04 a	β⁻	0.473	871, 703	2.045
	41	58	^{99}Nb	98.91162	−82.325	none	15.0 s	β⁻	3.5	138, 98	3.64
	41	63	^{104}Nb	103.9224	−72.284	none	4.9 s	β⁻		192	8.1
Molybdenum	42	49	^{91}Mo	90.91175	−82.204	none	15.5 min	β⁺, EC	3.44	1637, 1581	4.43
	42	50	^{92}Mo	91.906811	−86.805	14.77	stable				
	42	51	^{93}Mo	92.906812	−86.804	none	3.5E+03 a	EC		31	0.405
	42	52	^{94}Mo	93.905088	−88.410	9.23	stable				
	42	53	^{95}Mo	94.905842	−87.708	15.90	stable				
	42	54	^{96}Mo	95.904679	−88.791	16.68	stable				
	42	55	^{97}Mo	96.906021	−87.541	9.56	stable				
	42	56	^{98}Mo	97.905408	−88.112	24.19	stable				
	42	57	^{99}Mo	98.907712	−85.966	none	2.7476 d	β⁻	1.214	740, 181	1.357
	42	58	^{100}Mo	99.90748	−86.182	9.67	7.0E+18 a	β⁻β⁻		141	3.034

(continued)

Element	Z	N	Nuclide Symbol	Mass¹ (u)	Δ (MeV)²	Percent Abundance³	Half-life	Decay Mode⁴	Particle Energy⁵ (MeV)	Gamma Energy⁶ (keV)	Total Energy of Decay (MeV)
	42	59	^{101}Mo	100.910347	−83.511	none	14.61 min	β⁻	0.7 2.23	591 192 1013 506	2.82
	42	60	^{102}Mo	101.91030	−83.555	none	11.3 min	β⁻	1.2	212 148 224	1.01
	42	61	^{103}Mo	102.91320	−80.854	none	1.13 min	β⁻	3.7	83 424 46	3.8
	42	62	^{104}Mo	103.91376	−80.332	none	1.00 min	β⁻	2.02	69 70	2.16
Technetium	43	53	^{96}Tc	95.907871	−85.818	none	4.3 d	EC		778 850 813	2.97
	43	54	^{97m}Tc	96.90647	−87.123	none	91 d	IT EC		97	0.097 0.417
	43	54	^{97}Tc	96.906365	−87.220	none	2.6E+06 a	EC			0.320
	43	55	^{98}Tc	97.907216	−86.428	none	4.2E+06 a	β⁻	0.40	745 652	1.80
	43	56	^{99m}Tc	98.906409	−87.179	none	6.008 h	IT β⁻	0.435	140.5 322	0.143 0.437
	43	56	^{99}Tc	98.906255	−87.323	none	2.13E+05 a	β⁻	0.294	90	0.294
	43	61	^{104}Tc	103.91144	−82.493	none	18.2 min	β⁻	5.3	358	5.60
Ruthenium	44	51	^{95}Ru	94.91041	−83.453	none	1.64 h	EC β⁺	1.20 0.91	336 1097 627	2.57
	44	52	^{96}Ru	95.90760	−86.070	5.54	stable				
	44	53	^{97}Ru	96.90756	−86.107	none	2.89 d	EC		216 325	1.11
	44	54	^{98}Ru	97.90529	−88.222	1.87	stable				

Element	Z	N	Isotope	Mass	Mass excess	Abundance	Half-life	Decay	E (MeV)	E$_\gamma$ (keV)	Q (MeV)
	44	55	^{99}Ru	98.905939	−87.617	12.76	stable				
	44	56	^{100}Ru	99.904219	−89.219	12.60	stable				
	44	57	^{101}Ru	100.905582	−87.950	17.06	stable				
	44	58	^{102}Ru	101.904349	−89.098	31.55	stable				
	44	59	^{103}Ru	102.906324	−87.259	none	39.27 d	β⁻	0.223	497	0.763
	44	60	^{104}Ru	103.905433	−88.089	18.62	stable				
	44	61	^{105}Ru	104.907750	−85.930	none	4.44 h	β⁻	1.187, 1.11, 1.134	724, 469, 676	1.918
Rhodium	45	54	^{99}Rh	98.908132	−85.574	none	16 d	EC, β⁺	0.68, 0.54, 1.10	528, 353, 90	2.04
	45	57	^{102}Rh	101.906843	−86.775	none	207 d	EC, β⁺, β⁻	1.30, 0.82, 1.15	475	2.323, 1.150
	45	58	^{103}Rh	102.905504	−88.022	100	stable				
	45	59	^{104}Rh	103.906655	−86.950	none	42.3 s	β⁻, EC	2.44	557	2.440, 1.139
	45	60	^{105}Rh	104.905692	−87.847	none	35.4 h	β⁻	0.566, 0.248	556, 319	0.567
Palladium	46	53	^{99}Pd	98.91177	−82.186	none	21.4 min	β⁺, EC	2.18	136	3.39
	46	55	^{101}Pd	100.90829	−85.427	none	8.4 h	EC, β⁺	0.776	296, 591	1.980
	46	56	^{102}Pd	101.905609	−87.925	1.02	stable				
	46	57	^{103}Pd	102.906087	−87.479	none	16.99 d	EC		40, 358	0.543
	46	58	^{104}Pd	103.904036	−89.390	11.14	stable				
	46	59	^{105}Pd	104.905085	−88.413	22.33	stable				
	46	60	^{106}Pd	105.903486	−89.902	27.33	stable				
	46	61	^{107}Pd	106.905129	−88.372	none	6.5E+06 a	β⁻	0.04		0.034
	46	62	^{108}Pd	107.903892	−89.524	26.46	stable				

(continued)

Element	Z	N	Nuclide Symbol	Mass[1] (u)	Δ (MeV)[2]	Percent Abundance[3]	Half-life	Decay Mode[4]	Particle Energy[5] (MeV)	Gamma Energy[6] (keV)	Total Energy of Decay (MeV)
	46	63	^{109}Pd	108.905954	−87.603	none	13.46 h	β⁻	1.028	88	1.116
	46	64	^{110}Pd	109.90515	−88.352	11.72	stable				
	46	65	^{111}Pd	110.90764	−86.033	none	23.4 min	β⁻	2.1	580 / 70	2.22
	46	72	^{118}Pd	117.9190	−75.451	none	2.1 s	β⁻		125 / 224	4.1
Silver	47	52	^{99}Ag	98.9176	−76.755	none	2.1 min	β⁺ / EC		265	5.4
	47	57	^{104}Ag	103.908628	−85.112	none	1.2 h	EC / β⁺	0.99	556 / 768	4.279
	47	59	^{106}Ag	105.906666	−86.940	none	24.0 min	β⁺ / EC / β⁻	1.96	512	2.965 / 0.20
	47	60	^{107}Ag	106.905097	−88.402	51.839	stable				
	47	61	^{108}Ag	107.905954	−87.603	none	2.39 min	β⁻ / EC / β⁺	1.65 / 0.88	633 / 434 / 619	1.65 / 1.92
	47	62	^{109}Ag	108.904752	−88.723	48.161	stable				
	47	63	^{110}Ag	109.906111	−87.457	none	24.6 s	β⁻ / EC	2.891	658	2.891 / 0.89
Cadmium	48	56	^{104}Cd	103.90985	−83.974	none	58 min	EC / β⁺	0.29	84 / 709	1.14
	48	57	^{105}Cd	104.90947	−84.328	none	55.5 min	EC / β⁺	1.69	962 / 347 / 1303	2.738
	48	58	^{106}Cd	105.90646	−87.132	1.25	stable				
	48	59	^{107}Cd	106.906614	−86.988	none	6.52 h	EC		93 / 829	1.417
	48	60	^{108}Cd	107.90418	−89.256	0.89	stable	β⁺	0.302		

Z	Element	Nuclide	N	Atomic mass	Mass excess	Abundance	Half-life	Decay mode	E (MeV)	γ (keV)	Q (MeV)
48		^{109}Cd	61	108.904986	−88.505	none	462 d	EC		88	0.214
48		^{110}Cd	62	109.903002	−90.353	12.49	stable				
48		^{111}Cd	63	110.904178	−89.258	12.80	stable				
48		^{112}Cd	64	111.902758	−90.580	24.13	stable				0.32
48		^{113}Cd	65	112.904402	−89.049	12.22	8.0E+15 a	β⁻	1.11		
48		^{114}Cd	66	113.903359	−90.021	28.73	stable		0.593		1.446
48		^{115}Cd	67	114.905431	−88.090	none	2.228 d	β⁻		336 528	
48		^{116}Cd	68	115.904756	−88.719	7.49	3.0E+19 a	β⁻β⁻	0.67	315	2.81
48		^{117}Cd	69	116.907218	−86.426	none	2.49 h	β⁻	2.2	273 1303	2.52
49	Indium	^{104}In	55	103.9183	−76.103	none	1.81 min	β⁺ EC	4.30	658 834	7.9
49		^{111}In	62	110.905111	−88.389	none	2.8049 d	EC		245 171	0.862
49		^{112}In	63	111.905533	−87.995	none	14.4 min	EC β⁺	1.56	617 607	2.584
49		^{113m}In	64	112.904479	−88.977	none	1.658 h	β⁻ IT	0.66	392	0.665 0.392
49		^{113}In	64	112.904058	−89.369	4.29	stable				
49		^{114}In	65	113.904917	−88.569	none	1.198 min	β⁻ EC β⁺	1.984	1300	1.989 1.449
49		^{115}In	66	114.903878	−89.537	95.71	4.4E+14 a	β⁻	0.40	558 576	0.499
49		^{116}In	67	115.905260	−88.250	none	14.1 s	β⁻	0.49 3.3	498	3.278
50	Tin	^{111}Sn	61	110.907735	−85.944	none	35 min	EC β⁺	1.5	1294 463	0.469 2.45
50		^{112}Sn	62	111.904818	−88.661	0.97	stable			1153 1915 762 1611	

(continued)

Element	Z	N	Nuclide Symbol	Mass[1] (u)	Δ (MeV)[2]	Percent Abundance[3]	Half-life	Decay Mode[4]	Particle Energy[5] (MeV)	Gamma Energy[6] (keV)	Total Energy of Decay (MeV)
	50	63	^{113}Sn	112.905173	−88.331	none	115.1 d	EC		392	1.037
	50	64	^{114}Sn	113.902779	−90.561	0.66	stable				
	50	65	^{115}Sn	114.903342	−90.036	0.34	stable				
	50	66	^{116}Sn	115.901741	−91.528	14.54	stable				
	50	67	^{117}Sn	116.902952	−90.400	7.68	stable				
	50	68	^{118}Sn	117.901603	−91.656	24.22	stable				
	50	69	^{119}Sn	118.903308	−90.068	8.59	stable				
	50	70	^{120}Sn	119.902195	−91.105	32.58	stable				
	50	71	^{121}Sn	120.904237	−89.203	none	1.128 d	β⁻	0.383		0.391
	50	72	^{122}Sn	121.903439	−89.946	4.63	stable				
	50	73	^{123}Sn	122.905722	−87.819	none	129.2 d	β⁻	1.42	1089	1.404
	50	74	^{124}Sn	123.905274	−88.237	5.79	stable				
	50	75	^{125}Sn	124.907785	−85.898	none	9.64 d	β⁻	2.35	1067 1089 822 916	2.357
	50	78	^{128}Sn	127.91054	−83.331	none	59.1 min	β⁻	0.63	482 75	1.27
	50	80	^{130}Sn	129.91385	−80.248	none	3.72 min	β⁻	1.1	193 780 70	2.15
	50	82	^{132}Sn	131.91774	−76.625	none	39.7 s	β⁻	1.76	86 341 247 890 993	3.12
	50	85	^{135}Sn	134.9347	−60.827	none	0.52 s	β⁻		282 732 923	8.9
Antimony	51	69	^{120}Sb	119.905074	−88.423	none	15.89 min	EC β⁺	1.72	1171	2.68
	51	70	^{121}Sb	120.903816	−89.595	57.21	stable				

Z	N	Nuclide	Atomic mass	Abundance (%)	Half-life	Decay mode	Decay energy	γ energy	
51	71	122Sb	121.905175	none	2.72 d	β⁻	1.414, 1.980	564	1.984
51	71					EC, β⁺	0.570	1141	1.616
51	72	123Sb	122.904214	42.79	stable				
51	73	124Sb	123.905938	none	60.20 d	β⁻	0.61, 2.301	603, 1691, 723	2.904
51	75	126Sb	125.90725	none	12.4 d	β⁻	1.9	666, 695, 415	3.67
51	79	130Sb	129.91155	none	39.5 min	β⁻	2.9	839, 794, 331, 182	5.06
Tellurium									
52	67	119Te	118.906408	none	16.0 h	EC, β⁺	0.627	644	2.293
52	68	120Te	119.90402	0.09	stable				
52	69	121Te	120.90493	none	19.1 d	EC		573, 508	1.04
52	70	122Te	121.903044	2.55	stable				
52	71	123Te	122.904270	0.89	9E+16 a	EC			0.052
52	72	124Te	123.902818	4.74	stable				
52	73	125mTe	124.904587	none	58 d	IT		36, 109	0.145
52	73	125Te	124.904431	7.07	stable				
52	74	126Te	125.903312	18.84	stable				
52	75	127Te	126.905217	none	9.4 h	β⁻	0.69	418, 360	0.702
52	76	128Te	127.904463	31.74	4E+24 a	β⁻β⁻			0.87
52	77	129Te	128.906596	none	33.6 d	β⁻	1.45	460, 28	1.500
52	78	130Te	129.906224	34.08	6E+20 a	β⁻β⁻			
52	79	131Te	130.908522	none	25.0 min	β⁻	2.1	150, 452	2.235

(continued)

Element	Z	N	Nuclide Symbol	Mass¹ (u)	Δ (MeV)²	Percent Abundance³	Half-life	Decay Mode⁴	Particle Energy⁵ (MeV)	Gamma Energy⁶ (keV)	Total Energy of Decay (MeV)
	52	82	¹³⁴Te	133.91154	−82.400	none	42.0 min	β⁻	0.6 0.7	767 211 278 80	1.51
Iodine	52	86	¹³⁸Te	137.9292	−65.950	none	1.15 s	β⁻			6.4
	53	70	¹²³I	122.905598	−87.935	none	13.221 h	EC		159	1.229
	53	72	¹²⁵I	124.904624	−88.842	none	59.4 d	EC		35.5	0.1858
	53	73	¹²⁶I	125.905620	−87.914	none	12.89 d	EC β⁻ β⁺	0.87 1.13	666 389	2.154 1.258 2.154
	53	74	¹²⁷I	126.904473	−88.983	100	stable				
	53	75	¹²⁸I	127.905805	−87.742	none	25.00 min	β⁻ EC β⁺	2.13	443 743 743	2.122 1.254
	53	76	¹²⁹I	128.904988	−88.503	none	1.57E+07 a	β⁻	0.15	40	0.194
	53	77	¹³⁰I	129.906674	−86.933	none	12.36 h	β⁻	1.04 0.62	536 669 740	2.949
	53	78	¹³¹I	130.906124	−87.445	none	8.023 d	β⁻	0.606	365	0.971
	53	84	¹³⁷I	136.91787	−76.504	none	24.5 s	β⁻		1218 601 1303 1220	5.88
Xenon	54	66	¹²⁰Xe	119.91215	−81.832	none	46 min	EC β⁺		25 73 178	1.62
	54	69	¹²³Xe	122.90847	−85.260	none	2.00 h	EC β⁺	1.51	149 178 330	2.69
	54	70	¹²⁴Xe	123.905893	−87.660	0.0952	1.80E+22 a	ECEC			2.86

54	71	^{125m}Xe	124.906670	none	57 s	IT		140 112	0.253
54	71	^{125}Xe	124.906398	none	17.1 h	EC β+	0.47	188	1.644
54	72	^{126}Xe	125.90427	0.0890	stable				
54	73	^{127}Xe	126.905180	none	36.4 d	EC		203 172	0.662
54	74	^{128}Xe	127.903531	1.9102	stable				
54	75	^{129}Xe	128.904779	26.4006	stable				
54	76	^{130}Xe	129.903508	4.0710	stable				
54	77	^{131}Xe	130.905082	21.2324	stable				
54	78	^{132}Xe	131.904153	26.9086	stable				
54	79	^{133}Xe	132.905906	none	5.243 d	β-	0.346	81	0.427
54	80	^{134}Xe	133.905394	10.4357	stable				
54	81	^{135}Xe	134.90721	none	9.10 h	β-	0.91	250	1.165
54	82	^{136}Xe	135.90722	8.8573	stable				
54	83	^{137}Xe	136.911563	none	3.82 min	β-	4.1 3.6	456	4.17
Cesium									
55	75	^{130}Cs	129.906706	none	29.21 min	EC β+ β-	1.96 0.36	536	2.98
55	77	^{132}Cs	131.906430	none	6.48 d	EC β+ β-	0.40	668 465	0.36 2.125
55	78	^{133}Cs	132.905452	100	stable				
55	79	^{134}Cs	133.906713	none	2.065 a	β-	0.658 0.089	605 796	1.279 2.0587
55	80	^{135}Cs	134.905972	none	2.3E+06 a	EC β-	0.21	819 1048 341	1.233 0.269 2.548
55	81	^{136}Cs	135.907306	none	13.10 d	β-	0.341		
55	82	^{137}Cs	136.907084	none	30.07 a	β-	0.514	662	1.1756

(continued)

Element	Z	N	Nuclide Symbol	Mass¹ (u)	Δ (MeV)²	Percent Abundance³	Half-life	Decay Mode⁴	Particle Energy⁵ (MeV)	Gamma Energy⁶ (keV)	Total Energy of Decay (MeV)
	55	83	¹³⁸Cs	137.91101	−82.894	none	32.2 min	β⁻	2.9	1436 463 1010	5.37
	55	85	¹⁴⁰Cs	139.917277	−77.056	none	1.06 min	β⁻	5.7 6.21	602	6.22
Barium	56	73	¹²⁹Ba	128.90867	−85.073	none	2.2 h	β⁺	1.42	214 221 129	2.44
	56	74	¹³⁰Ba	129.906321	−87.261	0.106	2E+21 a	ECEC			2.62
	56	75	¹³¹Ba	130.906931	−86.693	none	11.53 d	EC β⁺		496 124 216	1.38
	56	76	¹³²Ba	131.905061	−88.435	0.101	3E+21 a	ECEC		356 81 303	0.846
	56	77	¹³³Ba	132.906002	−87.559	none	10.54 a	EC			0.517
	56	78	¹³⁴Ba	133.904508	−88.950	2.417	stable				
	56	79	¹³⁵Ba	134.905689	−87.851	6.592	stable				
	56	80	¹³⁶Ba	135.904576	−88.887	7.854	stable				
	56	81	¹³⁷ᵐBa	136.906638	−86.966	none	2.552 min	IT		662	0.662
	56	81	¹³⁷Ba	136.905827	−87.721	11.232	stable				
	56	82	¹³⁸Ba	137.905247	−88.262	71.698	stable				
	56	83	¹³⁹Ba	138.908835	−84.920	none	1.396 h	β⁻	2.27 2.14	166	2.318
	56	84	¹⁴⁰Ba	139.910600	−83.276	none	12.75 d	β⁻	1.0 0.48 1.02	537 30	1.05
	56	85	¹⁴¹Ba	140.914406	−79.730	none	18.3 min	β⁻	2.59 2.73	190 304 277 344	3.21

Element	Z	N	Isotope	Atomic mass	Mass excess	Abundance	Half-life	Decay	β energy	γ energy	
Lanthanum	57	73	130La	129.9123	−81.692	none	8.7 min	β+ EC	3.7 4.3	357 551 908 545	5.63
	57	80	137La	136.90647	−87.123	none	6E+04 a	EC			0.62
	57	81	138La	137.907112	−86.525	0.090	1.05E+11 a	EC	0.25	1436 789	1.737 1.04
	57	82	139La	138.906353	−87.232	99.910	stable				
	57	83	140La	139.909473	−84.325	none	1.678 d	β−	1.35 1.24 1.67	1596 487 816 329	3.762
	57	84	141La	140.910957	−82.943	none	3.90 h	β−	2.43	1355 131	2.502
Cerium	58	72	130Ce	129.9147	−79.456	none	23 min	EC			2.21
	58	75	133mCe	132.9116	−82.344	none	5.4 h	β+ EC	1.3	477 510 58	2.937
	58	77	135Ce	134.90915	−84.626	none	17.7 h	EC β+	0.8	266 300 607	2.026
	58	78	136Ce	135.90717	−86.471	0.185	stable				
	58	79	137Ce	136.90778	−85.902	none	9.0 h	EC β+		447	1.222
	58	80	138Ce	137.90599	−87.570	0.251	stable				
	58	81	139Ce	138.906647	−86.958	none	137.6 d	EC		166	0.28
	58	82	140Ce	139.905439	−88.083	88.450	stable				
	58	83	141Ce	140.908271	−85.445	none	32.50 d	β−	0.436 0.581	145	0.581
	58	84	142Ce	141.909244	−84.539	11.114	stable				
	58	85	143Ce	142.912381	−81.617	none	1.377 d	β−	1.110 1.404	293 57	1.461

(continued)

Element	Z	N	Nuclide Symbol	Mass[1] (u)	Δ (MeV)[2]	Percent Abundance[3]	Half-life	Decay Mode[4]	Particle Energy[5] (MeV)	Gamma Energy[6] (keV)	Total Energy of Decay (MeV)
Praseodymium	59	81	^{140}Pr	139.909071	−84.700	none	3.39 min	EC β+	2.37	1596 307	3.39
	59	82	^{141}Pr	140.907653	−86.021	100	stable				
	59	83	^{142}Pr	141.910040	−83.797	none	19.12 h	β− EC	2.162	1576 641	2.162 0.746
Neodymium	59	84	^{143}Pr	142.910812	−83.078	none	13.57 d	β−	0.933	742	0.934
	60	81	^{141}Nd	140.909605	−84.202	none	2.49 h	β+	0.802		1.823
	60	82	^{142}Nd	141.907723	−85.955	27.2	stable				
	60	83	^{143}Nd	142.909814	−84.008	12.2	stable				
	60	84	^{144}Nd	143.910087	−83.753	23.8	2.38E+15 a	α	1.83		1.905
	60	85	^{145}Nd	144.912574	−81.437	8.3	stable				
	60	86	^{146}Nd	145.913117	−80.931	17.2	stable				
	60	87	^{147}Nd	146.916096	−78.156	none	10.98 d	β−	0.805	91 531	0.896
	60	88	^{148}Nd	147.916893	−77.414	5.7	stable				
	60	89	^{149}Nd	148.920144	−74.385	none	1.73 h	β−	1.42 1.13 1.03	211 114 270	1.690
	60	90	^{150}Nd	149.920891	−73.690	5.6	8E+18 a	β−β−			3.37
	60	91	^{151}Nd	150.923825	−70.957	none	12.4 min	β−	1.2	117 256 1181	2.442
Promethium	61	84	^{145}Pm	144.912744	−81.278	none	17.7 a	EC		73 67	0.163
	61	85	^{146}Pm	145.914692	−79.464	none	5.53 a	EC		454 736	1.471
								β−	0.795	747 633	1.542

Element	Z	N	Nuclide	Atomic mass	Mass excess	Abundance (%)	Half-life	Decay	E (MeV)	γ (keV)	Q (MeV)
	61	86	^{147}Pm	146.915134	−79.052	none	2.6234 a	β⁻	0.224	121	0.2241
	61	87	^{148}Pm	147.917468	−76.878	none	5.37 d	β⁻	2.47	1465	2.47
									1.02	550	
	61	88	^{149}Pm	148.918330	−76.075	none	2.212 d	β⁻	1.072	286	1.071
	61	89	^{150}Pm	149.92098	−73.607	none	2.68 h	β⁻	2.3	334	3.45
									1.8	1325	
									1.6	1166	
										832	
	61	90	^{151}Pm	150.921203	−73.399	none	1.183 d	β⁻	0.84	340	1.187
										168	
										275	
Samarium	62	81	^{143}Sm	142.914624	−79.527	none	8.83 min	EC	2.42	1057	3.443
								β⁺		1515	
	62	82	^{144}Sm	143.911999	−81.972	3.07	stable				
	62	83	^{145}Sm	144.913406	−80.662	none	340 d	EC		61	0.616
	62	84	^{146}Sm	145.913041	−81.002	none	1.03E+08 a	α	2.455		2.31
	62	85	^{147}Sm	146.914898	−79.272	14.99	1.17E+11 a	α	2.23		1.99
	62	86	^{148}Sm	147.914823	−79.342	11.24	7E+15 a	α	1.96		
	62	87	^{149}Sm	148.917185	−77.142	13.82	stable				
	62	88	^{150}Sm	149.917276	−77.057	7.38	stable				
	62	89	^{151}Sm	150.919928	−74.587	none	90 a	β⁻	0.076	22	0.077
	62	90	^{152}Sm	151.919732	−74.769	26.75	stable				
	62	91	^{153}Sm	152.922094	−72.569	none	1.928 d	β⁻	0.69	103	0.808
									0.64	70	
	62	92	^{154}Sm	153.922209	−72.462	22.75	stable				
	62	93	^{155}Sm	154.924636	−70.201	none	22.2 min	β⁻	1.52	104	1.627
	63	87	^{150}Eu	149.919698	−74.801	none	36 a	EC		334	2.26
										439	
										584	
Europium	63	88	^{151}Eu	150.919850	−74.659	47.81	stable				
	63	89	^{152m}Eu	151.921898	−72.752	none	1.60 h	IT		90	0.148
	63	89	^{152m}Eu	151.921788	−72.854	none	9.29 h	β⁻	1.86	842	1.865
								EC		963	1.920

(continued)

Element	Z	N	Nuclide Symbol	Mass[1] (u)	Δ (MeV)[2]	Percent Abundance[3]	Half-life	Decay Mode[4]	Particle Energy[5] (MeV)	Gamma Energy[6] (keV)	Total Energy of Decay (MeV)
	63	89	^{152}Eu	151.921740	−72.899	none	13.54 a	EC β+ β−	0.727 0.699	122 1408 344	1.874 1.820
	63	90	^{153}Eu	152.921230	−73.374	52.19	stable				
	63	91	^{154}Eu	153.922975	−71.748	none	8.60 a	β− EC	0.58 0.27	123 1275 82 185	1.969 0.717
Gadolinium	64	87	^{151}Gd	150.920344	−74.199	none	124 d	EC		154 243	0.464
	64	88	^{152}Gd	151.919791	−74.714	0.20	1.1E+14 a	α α	2.60 2.14		2.205
	64	89	^{153}Gd	152.921746	−72.893	none	241.6 d	EC		97 103	0.484
	64	90	^{154}Gd	153.920866	−73.713	2.18	stable				
	64	91	^{155}Gd	154.922622	−72.077	14.80	stable				
	64	92	^{156}Gd	155.922123	−72.542	20.47	stable				
	64	93	^{157}Gd	156.923960	−70.831	15.65	stable				
	64	94	^{158}Gd	157.924104	−70.697	24.84	stable				
	64	95	^{159}Gd	158.926385	−68.572	none	18.5 h	β−	0.96	364 58	0.971
	64	96	^{160}Gd	159.927054	−67.949	21.86	stable				
	64	97	^{161}Gd	160.929666	−65.516	none	3.66 min	β−	1.56	361 315 102	1.955
Terbium	65	93	^{158}Tb	157.925410	−69.480	none	1.8E+02 a	EC β−	0.83	944 962 99	1.220 0.935
	65	94	^{159}Tb	158.925347	−69.539	100	stable				
	65	95	^{160}Tb	159.927164	−67.846	none	72.3 d	β−	0.57 0.86	879 299 966	1.835

Element	Z	N	Nuclide	Atomic mass	Mass excess	Abundance	Half-life	Decay	(MeV)	(keV)	
Dysprosium	65	103	^{168}Tb	167.9434	−52.723	none	8 s	β−			6
	66	88	^{154}Dy	153.92442	−70.402	none	3E+06 a	α	2.87		2.095
	66	89	^{155}Dy	154.92575	−69.163	none	9.9 h	EC, β+	0.845	227	
	66	90	^{156}Dy	155.92428	−70.533	0.056	stable				
	66	91	^{157}Dy	156.925461	−69.433	none	8.1 h	EC		326	1.34
	66	92	^{158}Dy	157.924409	−70.413	0.095	stable				
	66	93	^{159}Dy	158.925736	−69.176	none	144.4 d	EC		58	0.366
	66	94	^{160}Dy	159.925198	−69.678	2.329	stable				
	66	95	^{161}Dy	160.926933	−68.061	18.889	stable				
	66	96	^{162}Dy	161.926798	−68.187	25.475	stable				
	66	97	^{163}Dy	162.928731	−66.387	24.896	stable				
	66	98	^{164}Dy	163.929175	−65.973	28.260	stable				
	66	99	^{165}Dy	164.931700	−63.621	none	2.33 h	β−	1.29	95, 362	1.287
	66	102	^{168}Dy	167.9371	−58.591	none	8.7 min	β−	0.870	192	1.5
Holmium	67	96	^{163}Ho	162.928730	−66.388	none	4.57E+03 a	EC			0.00256
	67	97	^{164}Ho	163.930231	−64.989	none	29 min	EC, β−	0.96, 0.88	73, 91	0.986, 0.962
	67	98	^{165}Ho	164.930322	−64.905	100	stable				
	67	99	^{166}Ho	165.932281	−63.080	none	1.118 d	β−	1.855	81, 1379	1.855
	67	101	^{168}Ho	167.9355	−60.081	none	2.99 min	β−	1.773, 1.90	741	2.93, 6.9
Erbium	68	78	^{146}Er	145.9521	−44.619	none	2 s	β+			
	68	93	^{161}Er	160.93000	−65.205	none	3.21 h	β+, EC	0.82	827, 211	1.99
	68	94	^{162}Er	161.928778	−66.343	0.139	stable				
	68	95	^{163}Er	162.930029	−65.178	none	1.25 h	EC, β+	0.19	1114, 436, 440	1.210
	68	96	^{164}Er	163.929200	−65.950	1.601	stable				
	68	97	^{165}Er	164.930723	−64.531	none	10.36 h	EC			0.376

(continued)

Element	Z	N	Nuclide Symbol	Mass¹ (u)	Δ (MeV)²	Percent Abundance³	Half-life	Decay Mode⁴	Particle Energy⁵ (MeV)	Gamma Energy⁶ (keV)	Total Energy of Decay (MeV)
Erbium	68	98	^{166}Er	165.930293	−64.932	33.503	stable				
	68	99	^{167}Er	166.932048	−63.297	22.869	stable				
	68	100	^{168}Er	167.932370	−62.997	26.978	stable				
	68	101	^{169}Er	168.934588	−60.931	none	9.39 d	β⁻	0.34	8	0.351
	68	102	^{170}Er	169.935464	−60.115	14.910	stable				
	68	103	^{171}Er	170.938026	−57.728	none	7.516 h	β⁻	1.065	308, 296, 112	1.491
	68	104	^{172}Er	171.939352	−56.493	none	2.05 d	β⁻	0.36, 0.28	610, 407	0.891
Thulium	69	78	^{147}Tm	146.9610	−36.328	none	0.58 s	EC, p	1.051	81	10.7, 1.061
	69	99	^{168}Tm	167.934170	−61.320	none	93.1 d	EC, β⁺, β⁻		198, 816, 448, 184	1.679, 0.257
	69	100	^{169}Tm	168.934213	−61.280	100	stable				
	69	101	^{170}Tm	169.935798	−59.804	none	128.6 d	β⁻, EC	0.968, 0.883	84, 79	0.968, 0.314
Ytterbium	70	97	^{167}Yb	166.934947	−60.596	none	17.5 min	EC, β⁺	0.64	113, 106, 176	1.954
	70	98	^{168}Yb	167.933897	−61.575	0.13	stable				
	70	99	^{169}Yb	168.935187	−60.373	none	32.03 d	EC		63, 198	0.910
	70	100	^{170}Yb	169.934762	−60.769	3.04	stable				
	70	101	^{171}Yb	170.936326	−59.312	14.28	stable				
	70	102	^{172}Yb	171.936381	−59.261	21.83	stable				
	70	103	^{173}Yb	172.938211	−57.556	16.13	stable				

	70	104	^{174}Yb	173.938862	−56.950	31.83	stable				
	70	105	^{175}Yb	174.941273	−54.704	none	4.185 d	β^-	0.466	396 283 114	0.470
	70	106	^{176}Yb	175.942572	−53.494	12.76	stable				
	70	107	^{177}Yb	176.945257	−50.993	none	1.911 h	β^-	1.40	150 1080	1.400
Lutetium	71	97	^{168}Lu	167.9387	−57.101	none	5.5 min	EC		1484 42	4.51
	71	101	^{172m}Lu	171.939127	−56.703	none	3.7 min	IT			0.0419
	71	103	^{174}Lu	173.940334	−55.579	none	3.3 a	EC		1242 77	1.374
	71	104	^{175}Lu	174.940772	−55.171	97.41	stable	β^+	0.38		
	71	105	^{176}Lu	175.942686	−53.388	2.59	3.76E+10 a	β^-	0.57	307 202	1.190
	71	106	^{177}Lu	176.943755	−52.392	none	6.65 d	β^-	0.497	208 113	0.501
Hafnium	72	96	^{168}Hf	167.9406	−55.331	none	26.0 min	EC		57	1.7
	72	100	^{172}Hf	171.93946	−56.393	none	1.87 a	β^+, EC	0.236	24 126 67 82	0.34
	72	101	^{173}Hf	172.9407	−55.238	none	23.6 h	EC		124 297	1.47
	72	102	^{174}Hf	173.940046	−55.847	0.16	2.0E+15 a	α	2.50		
	72	103	^{175}Hf	174.941503	−54.490	none	70 d	EC		343	0.687
	72	104	^{176}Hf	175.941409	−54.577	5.26	stable				
	72	105	^{177}Hf	176.943221	−52.889	18.60	stable				
	72	106	^{178}Hf	177.943699	−52.444	27.28	stable				
	72	107	^{179}Hf	178.945816	−50.472	13.62	stable				
	72	108	^{180}Hf	179.946550	−49.788	35.08	stable				
	72	109	^{181}Hf	180.949099	−47.414	none	42.4 d	β^-	0.405	482 133 346	1.030

(continued)

Element	Z	N	Nuclide Symbol	Mass¹ (u)	Δ (MeV)²	Percent Abundance³	Half-life	Decay Mode⁴	Particle Energy⁵ (MeV)	Gamma Energy⁶ (keV)	Total Energy of Decay (MeV)
	72	110	¹⁸²Hf	181.950553	−46.060	none	8.9E+06 a	β⁻		270	0.37
Tantalum	73	95	¹⁶⁸Ta	167.9480	−48.438	none	2.0 min	β⁺	5.82	124	6.97
								EC			
	73	106	¹⁷⁹Ta	178.945934	−50.362	none	1.82 a	EC			0.1056
	73	107	¹⁸⁰mTa	179.947546	−48.861	0.012	7.1E+15 a				
	73	107	¹⁸⁰Ta	179.947465	−48.936	none	8.15 h	β⁻	0.71	103	0.708
									0.61		
								EC		93	0.852
	73	108	¹⁸¹Ta	180.947996	−48.441	99.988	stable				
	73	109	¹⁸²Ta	181.950152	−46.433	none	114.43 d	β⁻	0.522	68	1.814
									0.25	1121	
	73	111	¹⁸⁴Ta	183.95401	−42.839	none	8.7 h	β⁻	1.19	414	2.87
										253	
										921	
Tungsten	74	105	¹⁷⁹W	178.94707	−49.304	none	37.8 min	EC		31	1.06
	74	106	¹⁸⁰W	179.946704	−49.645	0.12	stable				
	74	107	¹⁸¹W	180.948198	−48.253	none	121.2 d	EC		6	0.188
	74	108	¹⁸²W	181.948204	−48.248	26.50	stable				
	74	109	¹⁸³W	182.950223	−46.367	14.31	stable				
	74	110	¹⁸⁴W	183.950931	−45.707	30.64	stable				
	74	111	¹⁸⁵W	184.953421	−43.388	none	74.8 d	β⁻	0.433	125	0.432
	74	112	¹⁸⁶W	185.954364	−42.510	28.43	stable				
	74	113	¹⁸⁷W	186.957158	−39.907	none	23.9 h	β⁻	0.622	686	1.311
									1.312	480	
	74	114	¹⁸⁸W	187.958487	−38.669	none	69.78 d	β⁻	0.349	291	0.349
										227	
										64	
Rhenium	75	109	¹⁸⁴Re	183.952524	−44.224	none	35 d	EC		903	1.481
										792	
	75	110	¹⁸⁵Re	184.952955	−43.822	37.40	stable				
	75	111	¹⁸⁶Re	185.954987	−41.929	none	3.718 d	β⁻	1.071	137	1.069
									0.933		
	75	112	¹⁸⁷Re	186.955753	−41.216	62.60	4.12E+10 a	EC		123	0.579
								β⁻	0.002470		0.002469

Osmium	75	113	188Re	187.958112	−39.018	none	17.004 h	β⁻	2.118 / 1.962	155	2.1203
	76	107	183Os	182.9531	−43.687	none	13 h	EC / β⁺		382 / 114 / 168	2.1
	76	108	184Os	183.952489	−44.256	0.02	stable				
	76	109	185Os	184.954043	−42.809	none	93.6 d	EC		646 / 875 / 880 / 717	1.0128
	76	110	186Os	185.953838	−43.000	1.59	2E+15 a	α	2.758		2.82
	76	111	187Os	186.955750	−41.219	1.96	stable				
	76	112	188Os	187.955838	−41.137	13.24	stable				
	76	113	189Os	188.958147	−38.986	16.15	stable				
	76	114	190Os	189.958447	−38.706	26.26	stable				
	76	115	191Os	190.960928	−36.395	none	15.4 d	β⁻	0.143	129	0.313
	76	116	192Os	191.961481	−35.880	40.78	stable				
	76	117	193Os	192.964148	−33.396	none	30.11 h	β⁻	1.13	139 / 461 / 73	1.141
Iridium	77	113	190Ir	189.9606	−36.701	none	11.8 d	EC		187 / 605 / 519	1.955
	77	114	191Ir	190.960594	−36.706	37.3	stable				
	77	115	192Ir	191.962602	−34.836	none	73.83 d	β⁻	0.672 / 0.535	317 / 468 / 485	1.460 / 1.047
	77	116	193Ir	192.962926	−34.534	62.7	stable				
	77	117	194Ir	193.965076	−32.531	none	19.3 h	β⁻	2.251	329	2.234
Platinum	78	111	189Pt	188.96083	−36.487	none	10.9 h	EC / β⁺	0.89	721 / 608 / 94 / 569 / 244	1.97

(continued)

Element	Z	N	Nuclide Symbol	Mass[1] (u)	Δ (MeV)[2]	Percent Abundance[3]	Half-life	Decay Mode[4]	Particle Energy[5] (MeV)	Gamma Energy[6] (keV)	Total Energy of Decay (MeV)
	78	112	^{190}Pt	189.95993	−37.325	0.014	6.5E+11 a	α	3.18		3.249
	78	113	^{191}Pt	190.961685	−35.690	none	2.88 d	EC		539	1.008
										410	
										360	
	78	114	^{192}Pt	191.961038	−36.293	0.782	stable				
	78	115	^{193}Pt	192.962985	−34.479	none	50 a	EC			0.0568
	78	116	^{194}Pt	193.962680	−34.763	32.967	stable				
	78	117	^{195}Pt	194.964791	−32.797	33.832	stable				
	78	118	^{196}Pt	195.964952	−32.647	25.242	stable				
	78	119	^{197}Pt	196.967323	−30.438	none	19.96 h	β⁻	0.642	77	0.719
									0.719	191	
	78	120	^{198}Pt	197.967893	−29.907	7.163	stable				
	78	121	^{199}Pt	198.970576	−27.408	none	30.8 min	β⁻	1.69	543	1.703
										494	
										317	
Gold	79	115	^{194}Au	193.965419	−32.212	none	38 h	EC	1.49	329	2.50
								β⁺			0.227
	79	116	^{195}Au	194.965018	−32.586	none	186.12 d	EC		99	0.227
	79	117	^{196}Au	195.966551	−31.158	none	6.167 d	EC		356	1.507
										333	
								β⁻	0.259	426	0.687
								β⁺			1.507
	79	118	^{197}Au	196.966569	−31.141	100	stable				
	79	119	^{198}Au	197.968225	−29.598	none	2.6949 d	β⁻	0.961	412	1.372
Mercury	80	114	^{194}Hg	193.965449	−32.184	none	447 a	EC		780	0.028
	80	115	^{195}Hg	194.96664	−31.075	none	10.53 h	EC		61	1.57
	80	116	^{196}Hg	195.965833	−31.826	0.15	stable				
	80	117	^{197}Hg	196.967195	−30.558	none	2.672 d	EC		77	0.600
	80	118	^{198}Hg	197.966769	−30.954	9.97	stable				
	80	119	^{199}Hg	198.968280	−29.547	16.87	stable				
	80	120	^{200}Hg	199.968326	−29.504	23.10	stable				
	80	121	^{201}Hg	200.970302	−27.664	13.18	stable				

Element	Z	N	Nuclide	Atomic mass	Mass excess	Abundance (%)	Half-life	Decay	E (MeV)	E_γ (keV)	Q (MeV)
	80	122	^{202}Hg	201.970643	−27.346	29.86	stable				
	80	123	^{203}Hg	202.972857	−25.284	none	46.61 d	β⁻	0.213	279	0.492
	80	124	^{204}Hg	203.973494	−24.690	6.87	stable				
	80	125	^{205}Hg	204.976056	−22.304	none	5.2 min	β⁻	1.6	204, 305	1.533
	80	126	^{206}Hg	205.97750	−20.959	none	8.2 min	β⁻	0.94	167	1.31
Thallium	81	120	^{201}Tl	200.97080	−27.200	none	3.043 d	EC		135	0.48
	81	121	^{202}Tl	201.97209	−25.998	none	12.23 d	EC, β⁺		440	1.36
	81	122	^{203}Tl	202.972344	−25.761	29.52	stable				
	81	123	^{204}Tl	203.973849	−24.359	none	3.78 a	β⁻, EC	0.7634		0.7638, 0.344
Lead	81	124	^{205}Tl	204.974428	−23.820	70.48	stable				
	81	125	^{206}Tl	205.976095	−22.267	none	4.20 min	β⁻	1.528	803	1.532
	82	121	^{203}Pb	202.973376	−24.800	none	2.16 d	EC		279	0.97
	82	122	^{204}Pb	203.973044	−25.109	1.4	stable				
	82	123	^{205}Pb	204.974467	−23.784	none	1.5E+07 a	EC			0.0505
	82	124	^{206}Pb	205.974465	−23.786	24.1	stable				
	82	125	^{207}Pb	206.975897	−22.452	22.1	stable				
	82	126	^{208}Pb	207.976652	−21.749	52.4	stable				
	82	127	^{209}Pb	208.981075	−17.629	none	3.25 h	β⁻	0.645		0.644
	82	128	^{210}Pb	209.984173	−14.743	none	22.3 a	β⁻, α	0.017, 0.061, 3.72	47	0.0635
	82	129	^{211}Pb	210.988732	−10.497	none	36.1 min	β⁻	1.38	405, 832, 427	1.37
	82	130	^{212}Pb	211.991888	−7.557	none	10.64 h	β⁻	0.335, 0.569	239, 300	0.570
	82	131	^{213}Pb	212.996500	−3.260	none	10.2 min	β⁻	0.67	352	2.05
	82	132	^{214}Pb	213.999798	−0.188	none	27 min	β⁻	0.73	295, 242	1.02

(continued)

Element	Z	N	Nuclide Symbol	Mass[1] (u)	Δ (MeV)[2]	Percent Abundance[3]	Half-life	Decay Mode[4]	Particle Energy[5] (MeV)	Gamma Energy[6] (keV)	Total Energy of Decay (MeV)
Bismuth	83	125	208Bi	207.979726	−18.885	none	3.68E+05 a	EC		2614	2.878
	83	126	209Bi	208.980399	−18.258	100	2.0E+19 a	α	3.08		3.14
	83	127	210Bi	209.984105	−14.806	none	5.01 d	β⁻	1.162	305	1.161
								α	4.648	266	
									4.687		
Polonium	84	124	208Po	207.981246	−17.469	none	2.898 a	α	5.115	292	
								EC		570	
										602	
	84	125	209Po	208.982430	−16.366	none	102 a	α	4.880	261	
								EC		263	
										896	
	84	126	210Po	209.982874	−15.953	none	138.38 d	α	5.3044	803	
	84	128	212Po	211.988868	−10.369	none	0.298 μs	α	8.7844		
	84	130	214Po	213.995186	−4.484	none	163.7 μs	α	7.6869	799	
	84	134	218Po	218.008966	8.352	none	3.10 min	α	6.0024	510	
								β⁻			
Astatine	85	124	209At	208.986159	−12.893	none	5.4 h	EC		545	3.49
										782	
										790	
	85	125	210At	209.987131	−11.987	none	8.1 h	α	5.647	1181	3.98
								EC		245	
										1483	
									5.524	83	
								α	5.442	106	
									5.361		
	85	126	211At	210.987496	−11.647	none	7.21 h	EC	5.868	687	0.785
								α		670	
Radon	86	133	219Rn	219.009480	8.831	none	3.96 s	α	6.8193	271	
									6.553	402	
									6.425		

	86	135	²²¹Rn	221.01554	14.475	none	25 min	β⁻	0.83	186	1.19
								α	6.037	254	
										265	
	86	136	²²²Rn	222.017578	16.374	none	3.8235 d	α	5.4895	510	5.590
	86	137	²²³Rn	223.0218	20.307	none	23 min	β⁻		592	1.9
										635	
										416	
										654	
	86	138	²²⁴Rn	224.0241	22.449	none	1.8 h	β⁻		261	0.8
										266	
Francium	87	135	²²²Fr	222.01754	16.338	none	14.3 min	β⁻	1.72	206	2.03
	87	136	²²³Fr	223.019731	18.379	none	21.8 min	β⁻	1.92	111	1.149
									1.15	50	
										80	
										235	
								α	5.291	151	
									5.314	59	
										145	
	87	137	²²⁴Fr	224.02324	21.648	none	3.0 min	β⁻	2.6	216	2.83
									1.8	132	
										837	
Radium	88	135	²²³Ra	223.018502	17.235	none	11.435 d	α	5.7164	269	
									5.607	154	
										324	
	88	136	²²⁴Ra	224.020212	18.827	none	3.63 d	α	5.6855	241	0.356
									5.449		
	88	137	²²⁵Ra	225.023605	21.988	none	14.9 d	β⁻	0.32	40	
								α	5.006		
									4.976		
	88	138	²²⁶Ra	226.025410	23.669	none	1599 a	α	4.7844	186	
									4.602		
	88	139	²²⁷Ra	227.029171	27.173	none	42 min	β⁻	1.31	27	1.328
									1.03	300	
										303	
										284	

(continued)

Element	Z	N	Nuclide Symbol	Mass[1] (u)	Δ (MeV)[2]	Percent Abundance[3]	Half-life	Decay Mode[4]	Particle Energy[5] (MeV)	Gamma Energy[6] (keV)	Total Energy of Decay (MeV)
	88	140	^{228}Ra	228.031064	28.936	none	5.76 a	β-	0.039 0.015 0.026	14	0.046
Actinium	89	136	^{225}Ac	225.02323	21.639	none	10.0 d	α	5.829 5.793 5.731	100	
	89	137	^{226}Ac	226.026090	24.303	none	1.224 d	β- EC	0.89 1.11	230 158 254 186	1.113 0.641
	89	138	^{227}Ac	227.027747	25.846	none	21.772 a	α β- α	5.4 0.045 4.9534 4.941	15 100	0.045
	89	139	^{228}Ac	228.031015	28.890	none	6.15 h	β-	1.158 1.731	911 969 338	2.124
Thorium	90	135	^{225}Th	225.02395	22.309	none	8.72 min	α α	4.27 6.479 6.441 6.501	322 246 359	0.67
	90	139	^{229}Th	229.031762	29.586	none	7.4E+03 a	α	4.845 4.901 4.814	194 86 211 32	
	90	140	^{230}Th	230.033134	30.864	none	7.56E+04 a	α	4.688 4.621	68 110 620	
	90	141	^{231}Th	231.036297	33.810	none	1.063 d	β-	0.305	26 84	0.392
	90	142	^{232}Th	232.038055	35.448	100	1.40E+10 a	α SF	4.012 3.974	64 141	

Z	N	Nuclide	Element	Atomic mass	Δ	Abundance	Half-life	Decay mode	Energy (MeV)	Intensity	γ (MeV)
90	143	²³³Th		233.041577	38.729	none	21.83 min	β⁻	1.245	87; 29; 459	1.243
90	144	²³⁴Th		234.043596	40.609	none	24.10 d	β⁻	0.198	63; 92; 93	0.273
91	139	²³⁰Pa	Protactinium	230.034533	32.167	none	17.4 d	EC; β⁻; α	0.51; 4.766; 5.345	952; 315	1.311; 0.560
91	140	²³¹Pa		231.035884	33.426	none	3.28E+04 a	α	5.013; 4.950; 5.029	27; 300	1.34
91	141	²³²Pa		232.038582	35.939	none	1.32 d	β⁻	0.31; 0.29	969; 894	
91	142	²³³Pa		233.040240	37.483	none	26.967 d	EC; β⁻	0.256; 0.15	312	0.50; 0.570
91	143	²³⁴Pa		234.043302	40.336	none	6.69 h	β⁻	0.48; 0.65	131; 34; 1938	2.195
92	140	²³²U	Uranium	232.037156	34.611	none	69.8 a	α	5.3203; 5.2635	58	
92	141	²³³U		233.039635	36.920	none	1.592E+05 a	α	4.824	43	
92	142	²³⁴U		234.040952	38.147	0.0054	2.46E+05 a	SF; α	4.783; 4.776	97; 55	
92	143	²³⁵U		235.043930	40.921	0.7204	7.04E+08 a	α	4.725; 4.398	53; 121	
92	144	²³⁶U		236.045568	42.446	none	2.342E+07 a	SF; α	4.366; 4.494; 4.445	186; 144; 49; 113	

(continued)

Element	Z	N	Nuclide Symbol	Mass¹ (u)	Δ (MeV)²	Percent Abundance³	Half-life	Decay Mode⁴	Particle Energy⁵ (MeV)	Gamma Energy⁶ (keV)	Total Energy of Decay (MeV)
	92	145	^{237}U	237.048724	45.386	none	6.752 d	β⁻	0.24 0.25	60 208	0.519
	92	146	^{238}U	238.050788	47.309	99.2742	4.468E+ 09 a	α SF	4.197 4.147	50	
	92	147	^{239}U	239.054288	50.569	none	23.47 min	β⁻	1.21 1.28	75 44	1.261
	92	148	^{240}U	240.056586	52.710	none	14.1 h	β⁻	0.36	44	0.40
Neptunium	93	144	^{237}Np	237.048173	44.873	none	2.14E+06 a	α	4.788 4.771	29 87	
	93	145	^{238}Np	238.050941	47.451	none	2.103 d	β⁻	0.263 1.248	985 1029	1.2915
	93	146	^{239}Np	239.502931	468.477	none	2.356 d	β⁻	0.438 0.341	106 278 228	0.723
Plutonium	94	144	^{238}Pu	238.049560	46.165	none	87.7 a	α SF	5.4992 5.4565	44 100	
	94	145	^{239}Pu	239.052163	48.590	none	2.410E+ 04 a	α SF	5.156 5.144 5.105	52 30 1057	5.24
	94	146	^{240}Pu	240.053814	50.127	none	6.56E+03 a	α SF	5.1685 5.1241	45 104	
	94	147	^{241}Pu	241.056845	52.951	none	14.29 a	β⁻ α	0.0208 4.897 4.853	149 104	0.0208
	94	148	^{242}Pu	242.058743	54.719	none	3.75E+05 a	α SF	4.901 4.856	45	

Element	Z	N	Nuclide	Atomic mass	Mass excess	Abundance	Half-life	Decay	Energy	Intensity	(γ MeV)
	94	149	^{243}Pu	243.061997	57.750	none	4.956 h	β⁻	0.58, 0.49	84	0.579
	94	150	^{244}Pu	244.06420	59.802	none	8.11E+07 a	α, SF	4.589, 4.546		
Americium	95	145	^{240}Am	240.05529	51.502	none	2.12 d	SF, EC		988, 889	1.38
	95	146	^{241}Am	241.056829	52.936	none	432.7 a	α, α	5.378, 5.4857, 5.4430	60, 26, 955	
	95	147	^{242}Am	242.059543	55.464	none	16.02 h	SF, β⁻	0.63, 0.67	42	0.6645
	95	148	^{243}Am	243.061381	57.176	none	7.37E+03 a	EC, α	5.276, 5.234	45, 75, 31, 662	0.751
	95	149	^{244m}Am	244.064373	59.963	none	26 min	SF, β⁻	1.50	43, 744, 154, 898	1.516
	95	149	^{244m}Am	244.064279	59.876	none	10.1 h	β⁻	0.387		1.427
Curium	96	145	^{241}Cm	241.057647	53.698	none	32.8 d	EC, α	5.939, 5.929, 5.884	472, 146	0.767
	96	146	^{242}Cm	242.058836	54.805	none	162.8 d	α, SF	6.1127, 6.0694	44	
	96	147	^{243}Cm	243.061389	57.183	none	29.1 a	α, EC, SF	5.785, 5.742	278, 228	

(continued)

Element	Z	N	Nuclide Symbol	Mass[1] (u)	Δ (MeV)[2]	Percent Abundance[3]	Half-life	Decay Mode[4]	Particle Energy[5] (MeV)	Gamma Energy[6] (keV)	Total Energy of Decay (MeV)
	96	148	^{244}Cm	244.062753	58.454	none	18.1 a	α	5.8048	43	
									5.7627		
								SF			
	96	149	^{245}Cm	245.065491	61.004	none	8.5E+03 a	α	5.362	175	
								SF		133	
	96	150	^{246}Cm	246.067224	62.619	none	4.77E+03 a	α	5.386	45	
									5.342		
								SF			
	96	151	^{247}Cm	247.070354	65.534	none	1.56E+07 a	SF		403	
								α	4.870		
									5.267		
Berkelium	97	149	^{246}Bk	246.06867	63.966	none	1.80 d	EC		799	
	97	150	^{247}Bk	247.07031	65.493	none	1.4E+03 a	α	5.532	84	1.4
									5.711	268	
	97	151	^{248}Bk	248.07308	68.074	none	23.7 h	β^-	0.86	551	0.87
								EC			0.72
	97	152	^{249}Bk	249.074980	69.843	none	330 d	β^-	0.124	328	0.124
								α	5.417	308	
Californium	98	153	^{251}Cf	251.079587	74.135	none	9.0E+02 a	SF		177	
								α	5.677	227	
									5.852		
									6.014		
	98	154	^{252}Cf	252.08163	76.038	none	2.646 a	α	6.1181	43	
								SF	6.0756	100	
	98	155	^{253}Cf	253.085127	79.295	none	17.8 d	β^-	0.27		0.29
								α	5.98		
									5.92		
Einsteinium	99	150	^{249}Es	249.07641	71.175	none	1.70 h	EC		380	1.45
								α	6.77	813	

Z	N	Nuclide	Atomic mass	Abundance	Half-life	Decay mode	MeV	keV	
99	151	²⁵⁰Es	250.0787	none	8.6 h	EC		829	2.1
								303	
								350	
99	152	²⁵¹Es	251.079984	none	1.38 d	EC		178	0.38
								153	
						α	6.492		
							6.462		
99	153	²⁵²Es	252.0830	none	1.29 a	α	6.632	52	1.26
						α	6.562	64	
								418	
						EC		785	
								139	
99	154	²⁵³Es	253.084825	none	20.47 d	α	6.6327	42	
							6.5916	389	
99	155	²⁵⁴Es	254.088022	none	276 d	SF		64	
						α	6.429		
					40 d	β⁻			
99	156	²⁵⁵Es	255.09027	none		α	6.300	33	0.29
							6.260		
						SF			
100	151	²⁵¹Fm	251.081567	none	5.3 h	SF		881	1.47
						EC		453	
						α	6.833	451	
								425	
								480	
								358	
100	152	²⁵²Fm	252.08247	none	1.058 d	α	7.039	96	
							6.998	42	
						SF			
100	153	²⁵³Fm	253.085185	none	3.0 d	EC	6.943	272	0.336
						α	6.676		
100	154	²⁵⁴Fm	254.086854	none	3.240 h	α	7.192	99	
							7.150	43	
						SF			

Fermium

(continued)

Element	Z	N	Nuclide Symbol	Mass[1] (u)	Δ (MeV)[2]	Percent Abundance[3]	Half-life	Decay Mode[4]	Particle Energy[5] (MeV)	Gamma Energy[6] (keV)	Total Energy of Decay (MeV)
	100	155	^{255}Fm	255.08996	83.797	none	20.1 h	α SF	7.022 6.963	81 58	
	100	156	^{256}Fm	256.09177	85.483	none	2.63 h	SF			
	100	157	^{257}Fm	257.09510	88.585	none	100.5 d	α	6.92 6.519	241 179	
Mendelevium	101	154	^{255}Md	255.09108	84.840	none	27 min	SF EC		170 231 453 406	1.04
	101	155	^{256}Md	256.0941	87.654	none	1.29 h	α EC	7.327	634 644 409 379	2.1
	101	156	^{257}Md	257.095541	88.996	none	5.5 h	EC α	7.206 7.142	371	0.41
	101	157	^{258}Md	258.098431	91.688	none	51.5 d	α α	7.074 6.718 6.763	368	
	101	158	^{259}Md	259.1005	93.615	none	1.64 h	SF			
Nobelium	102	150	^{252}No	252.08898	82.884	none	2.45 s	SF α	8.42 8.37		
	102	151	^{253}No	253.0907	84.487	none	1.6 min	α	8.01 8.04 8.06	222 280 152	
	102	152	^{254}No	254.09096	84.729	none	50 s	EC α EC SF	8.09		3.2 1.2

102	153	^{255}No	255.09324	86.853	none	3.1 min	α	8.12, 7.93, 8.08	187	2.01
							EC			
102	154	^{256}No	256.09428	87.821	none	2.9 s	α	8.45		
							SF			
102	155	^{257}No	257.09685	90.215	none	25 s	α	8.22, 8.27, 8.32	77, 124, 101	
102	156	^{258}No	258.0982	91.473	none	1.2 ms	SF			
102	157	^{259}No	259.1010	94.081	none	58 min	α	7.520, 7.551, 7.581		0.5
Lawrencium										
103	152	^{255}Lr	255.0967	90.075	none	21 s	EC			
							α	8.37, 8.42		
103	153	^{256}Lr	256.0986	91.845	none	27 s	EC			3.2
							α	8.43, 8.39, 8.52		
103	154	^{257}Lr	257.0996	92.777	none	0.65 s	EC			4.0
							SF			
							α	8.86, 8.80		
103	155	^{258}Lr	258.1018	94.826	none	4.1 s	EC			2.5
							α	8.60, 8.62, 8.56		
103	156	^{259}Lr	259.1029	95.851	none	6.2 s	EC			
							α	8.44		
							SF			
103	157	^{260}Lr	260.1055	98.273	none	3 min	α	8.03		
							EC			
103	158	^{261}Lr	261.1069	99.577	none	40 min	α			2.7
							EC			
103	159	^{262}Lr	262.1097	102.185	none	3.6 h	SF			2.2
							EC			

(continued)

Element	Z	N	Nuclide Symbol	Mass[1] (u)	Δ (MeV)[2]	Percent Abundance[3]	Half-life	Decay Mode[4]	Particle Energy[5] (MeV)	Gamma Energy[6] (keV)	Total Energy of Decay (MeV)
Rutherfordium	104	155	^{259}Rf	259.1056	98.366	none	2.5 s	α	8.77		
									8.86		
	104	156	^{260}Rf	260.1064	99.111	none	20 ms	SF			
								EC			
	104	157	^{261}Rf	261.10877	101.319	none	4.2 s	SF			
	104	158	^{262}Rf	262.1099	102.371	none	2.3 s	SF			
	104	159	^{263}Rf	263.1125	104.793	none	17 min	SF			
Dubnium	105	150	^{255}Db	255.1074	100.042	none	1.6 s	α			
								SF			
	105	151	^{256}Db	256.1081	100.695	none	1.6 s	α	9.01		
								SF			
	105	152	^{257}Db	257.1077	100.322	none	1.5 s	α	8.97		
									9.07		
								EC			6.5
	105	153	^{258}Db	258.1092	101.719	none	4.2 s	SF			
								EC			
								α	9.17		
									9.08		
	105	154	^{259}Db	259.1096	102.092	none	0.5 s	EC			5.3
								α	9.47		
								SF			
	105	155	^{260}Db	260.1113	103.675	none	1.5 s	α	9.05		
									9.08		
									9.13		
	105	156	^{261}Db	261.1121	104.420	none	1.8 s	SF			
								α	8.93		
								SF			
	105	157	^{262}Db	262.1141	106.283	none	34 s	α	8.45		
									8.53		
									8.67		
								SF			

(continued)

Element	Z	N	Isotope	Mass		Mass excess	Half-life	Decay	Energy
	105	158	^{263}Db	263.1150	none	107.122	27 s	SF / α / EC	8.35
Seaborgium	106	152	^{258}Sg	258.1132	none	105.445	2.9 ms	SF	
	106	153	^{259}Sg	259.1145	none	106.656	0.5 s	α	9.62
	106	154	^{260}Sg	260.11442	none	106.582	4 ms	SF / α	9.77, 9.72
	106	155	^{261}Sg	261.1161	none	108.146	0.23 s	α	9.56
	106	156	^{262}Sg	262.1164	none	108.426	12 ms	SF	
	106	157	^{263}Sg	263.1183	none	110.196	1.0 s	SF / α	9.06, 9.25
	106	158	^{264}Sg	264.1189	none	110.755	0.05 s	SF	
	106	159	^{265}Sg	265.1211	none	112.804	8 s	α	8.84, 8.94
	106	160	^{266}Sg	266.1221	none	113.735	21 s	SF / α	8.77, 8.52
	106	165	^{271}Sg	271.133	none	123.889	3 min	SF / α	8.5
Bohrium	107	153	^{260}Bh	260.122	none	113.642	35 ms	α / SF	10.2
	107	157	^{264}Bh	264.1246	none	116.064	1.0 s	α	9.49, 9.62
	107	159	^{266}Bh	266.1269	none	118.207	2 s	α	9.2
	107	165	^{272}Bh	272.138	none	128.546	9.8 s	α	9.02
Hassium	108	157	^{264}Hs	264.12839	none	−811.899	0.08 ms	α / SF	10.43
	108	162	^{270}Hs	270.1347	none	125.472	3.6 s	SF / α	9.2

Element	Z	N	Nuclide Symbol	Mass[1] (u)	Δ (MeV)[2]	Percent Abundance[3]	Half-life	Decay Mode[4]	Particle Energy[5] (MeV)	Gamma Energy[6] (keV)	Total Energy of Decay (MeV)
	108	167	^{275}Hs	275.146	135.998	none	0.2 s	α	9.4		
Meitnerium	109	157	^{266}Mt	266.1373	127.894	none	2.6 ms	α	11.05		
	109	158	^{267}Mt	267.1375	128.080	none					
	109	159	^{268}Mt	268.1387	129.198	none	0.02 s	α	10.10		
									10.24		
	109	165	^{274}Mt	274.147	136.930	none	440 ms	α	10		
	109	167	^{276}Mt	276.151	140.656	none	0.72 s	α	9.7		
Darmstadtium	110	157	^{267}Ds	267.1443	134.415	none	3 µs	α	11.60		
	110	169	^{279}Ds	279.159	148.108	none	0.19 s	SF			
								α	9.70		
	110	171	^{281}Ds	281.162	150.902	none	9.6 s	SF			
Roentgenium	111	161	^{272}Rg	272.1536	143.077	none	2 ms	α	10.99		
	111	167	^{278}Rg	278.162	150.902	none	4.2 ms	α	10.7		
	111	169	^{280}Rg	280.164	152.765	none	3.6 s	α	9.7		
Copernicium	112	165	^{277}Cn	277.1639	152.672	none	0.69 ms	α	11.45		
									11.31		
	112	170	^{282}Cn	282.170	158.354	none	0.7 ms	SF			
	112	173	^{285}Cn	285.174	162.080	none	34 s	α	9.2		
Nihonium	113	170	^{283}Nh	283.176	163.943	none	100 ms	α	10.1		
	113	171	^{284}Nh	284.178	165.806	none	0.48 s	α	10.0		
Flerovium	114	174	^{288}Fl	288.186	173.258	none	0.8 s	α	9.9		
	114	175	^{289}Fl	289.187	174.189	none	2.7 s	α	9.8		
Moscovium	115	172	^{287}Mc	287.191	177.915	none	32 ms	α	10.6		
	115	173	^{288}Mc	288.192	178.847	none	87 ms	α	10.5		

	Z	N	Symbol	Atomic Mass[1]	Mass Excess[2]	Abundance[3]	Half-life	Decay[4]	Energy[5,6]
Livermorium	116	174	^{290}Lv	290.199	185.367	none	8 ms	α	10.9
	116	175	^{291}Lv	291.200	186.299	none	18 ms	α	10.7
	116	176	^{292}Lv	292.200	186.299	none	18 ms	α	10.7
Tennessine	117	176	^{293}Ts	293.208	193.751	none	18 ms	α	11.0
	117	177	^{294}Ts	294.210	195.614	none	290 ms	α	10.8
Oganesson	118	176	^{294}Og	294.214	199.340	none	1.4 ms	α	11.7

1 The atomic mass of a nuclide includes the mass of Z electrons.
2 Mass Excess.
3 The percent abundance of a nuclide is its atomic percent present in nature.
4 Not all decay modes are listed for every nuclide, some, relatively minor, modes may be omitted.
5 Only the most commonly observed energies are listed. Values for β⁻ and β⁺ are maximum energy values (E_{max}).
6 Only the most commonly observed energies, associated with each decay mode, are listed. Gamma energies listed after EC/positron could be associated with one or both.

Appendix B: Table of Chi-Squared

Degrees of Freedom	Probability of observing an outcome at least as large as the chi-squared value						
	0.99	0.95	0.90	0.50	0.10	0.05	0.01
2	0.020	0.103	0.211	1.386	4.605	5.991	9.210
3	0.115	0.352	0.584	2.366	6.251	7.815	11.35
4	0.297	0.711	1.064	3.357	7.779	9.488	13.28
5	0.554	1.145	1.610	4.351	9.236	11.07	15.09
6	0.872	1.635	2.204	5.348	10.65	12.59	16.81
7	1.239	2.167	2.833	6.346	12.02	14.07	18.48
8	1.646	2.733	3.490	7.344	13.36	15.51	20.09
9	2.088	3.325	4.168	8.343	14.68	16.92	21.67
10	2.558	3.940	4.865	9.342	15.99	18.31	23.21
11	3.053	4.575	5.578	10.34	17.28	19.68	24.73
12	3.571	5.226	6.304	11.34	18.55	21.03	26.22
13	4.107	5.892	7.042	12.34	19.81	22.36	27.69
14	4.660	6.571	7.790	13.34	21.06	23.69	29.14
15	5.229	7.261	8.547	14.34	22.31	25.00	30.58
16	5.812	7.962	9.312	15.34	23.54	26.30	32.00
17	6.408	8.672	10.09	16.34	24.77	27.59	33.41
18	7.015	9.390	10.87	17.34	25.99	28.87	34.81
19	7.633	10.12	11.65	18.34	27.20	30.14	36.19
20	8.260	10.85	12.44	19.34	28.41	31.41	37.57
21	8.897	11.59	13.24	20.34	29.62	32.67	38.93
22	9.542	12.34	14.04	21.34	30.81	33.92	40.29
23	10.20	13.09	14.85	22.34	32.01	35.17	41.64
24	10.86	13.85	15.66	23.34	33.20	36.42	42.98
25	11.52	14.61	16.47	24.34	34.38	37.65	44.31
26	12.20	15.38	17.29	25.34	35.56	38.89	45.64
27	12.88	16.15	18.11	26.34	36.74	40.11	46.96
28	13.57	16.93	18.94	27.34	37.92	41.34	48.28
29	14.26	17.71	19.77	28.34	39.09	42.56	49.59
30	14.95	18.49	20.60	29.34	40.26	43.77	50.89
40	22.16	26.51	29.05	39.34	51.81	55.76	63.69
50	29.71	34.76	37.69	49.33	63.17	67.51	76.15
60	37.49	43.19	46.46	59.33	74.40	79.08	88.38
70	45.44	51.74	55.33	69.33	85.53	90.53	100.4
80	53.54	60.39	64.28	79.33	96.58	101.88	112.3
90	61.75	69.13	73.29	89.33	107.6	113.1	124.1
100	70.07	77.93	82.36	99.33	118.5	124.3	135.8

Appendix C: Useful Constants, Conversion Factors, SI Prefixes, and Formulas

CONSTANTS

Speed of light	$c = 2.998 \times 10^8$ m/s
Planck's constant	$h = 6.626 \times 10^{-34}$ J·s
Avogadro's number	6.022×10^{23} things/mole
W-quantity in air	33.85 eV/IP
Proton mass	1.673×10^{-27} kg = 1.007276 u
Neutron mass	1.675×10^{-27} kg = 1.008665 u
Electron mass	9.109×10^{-31} kg = 5.486×10^{-4} u
Radius constant	$r_0 = 1.4$ fm
Boltzmann's constant	1.3806×10^{-23} J/K = 8.63×10^{-5} eV/K

CONVERSION FACTORS

1 u = 1.66054×10^{-24} g	1 Ci = 3.7×10^{10} dps	1 μCi = 2.22×10^6 dpm
1 Bq = 1 dps	5.259×10^5 min = 1 a = 365.24 d	1 J = 6.242×10^{18} eV
$1\,J = 1\,\dfrac{kg \cdot m^2}{s^2}$	$1R = \dfrac{2.58 \times 10^{-4} C}{kg}$	$1\dfrac{J}{kg} = 1Gy$
$1\,rad = \dfrac{100\,erg}{g} = \dfrac{10^{-2}J}{kg} = 1cGy$	931.5 MeV = 1 u	1 rem = 10^{-2} Sv = 1 cSv = 10 mSv

SI PREFIXES

Prefix	Symbol	Factor
tera	T	10^{12}
giga	G	10^9
mega	M	10^6
kilo	k	10^3
centi	c	10^{-2}

(continued)

Prefix	Symbol	Factor
milli	m	10^{-3}
micro	μ	10^{-6}
nano	n	10^{-9}
pico	p	10^{-12}
femto	f	10^{-15}

FORMULAS

$c = \lambda v$ $\qquad\qquad$ $E = hv$ $\qquad\qquad$ $A = Z + N$

$\ln\dfrac{A_1}{A_2} = \lambda t$ $\qquad\qquad$ $A_2 = A_1 e^{-\lambda t}$ $\qquad\qquad$ $\lambda = \dfrac{\ln 2}{t_{1/2}}$

$\ln\dfrac{N_1}{N_2} = \lambda t$ $\qquad\qquad$ $N_2 = N_1 e^{-\lambda t}$ $\qquad\qquad$ $A = \lambda N$

$\%\text{ efficiency} = \dfrac{\text{cpm}}{\text{dpm}} \times 100\%$ $\qquad$ $\text{specific activity} = \dfrac{\text{activity}}{\text{mass}}$ $\qquad$ $N_{2D} = \left(e^{\lambda t} - 1\right)N_{2P} + N_{1D}$

Secular equilibrium: $N_B = \dfrac{\lambda_A}{\lambda_B} N_A (1 - e^{-\lambda_B t})$ $\qquad$ $A_B = A_A (1 - e^{-\lambda_B t})$

at equilibrium: $\quad A_B = A_A$ $\qquad$ $\dfrac{N_B}{N_A} = \dfrac{\lambda_A}{\lambda_B} = \dfrac{t_{1/2(B)}}{t_{1/2(A)}}$ $\quad$ or $\quad N_A \lambda_A = N_B \lambda_B$

Transient equilibrium: $\quad \dfrac{N_B}{N_A} = \dfrac{t_{1/2(B)}}{t_{1/2(A)} - t_{1/2(B)}}$ $\qquad$ $\dfrac{A_A}{A_B} = 1 - \dfrac{t_{1/2(B)}}{t_{1/2(A)}}$ $\qquad$ $t_{max} = \left[\dfrac{1.44 t_{1/2(A)} t_{1/2(B)}}{\left(t_{1/2(A)} - t_{1/2(B)}\right)}\right] \times \ln\dfrac{t_{1/2(A)}}{t_{1/2(B)}}$

$\sigma \approx \sqrt{x}$ or $\sigma \approx \sqrt{\bar{x}}$ $\qquad$ $RSD = \dfrac{\sigma}{x} \approx \dfrac{\sqrt{x}}{x} = x^{-\frac{1}{2}}$ or $RSD = \dfrac{\sigma}{\bar{x}} \approx \dfrac{\sqrt{\bar{x}}}{\bar{x}} = \bar{x}^{-\frac{1}{2}}$ $\qquad$ $\sigma_R = \dfrac{\sigma}{t} \approx \sqrt{\dfrac{R}{t}}$

$\chi^2 = \dfrac{1}{\bar{x}} \sum (\bar{x} - x)^2$ $\qquad$ $\sigma_{R_n} = \sqrt{\sigma_{R_g}^2 + \sigma_{R_b}^2} = \sqrt{\dfrac{R_g}{t_g} + \dfrac{R_b}{t_b}}$ $\qquad$ $\dfrac{t_b}{t_g} = \dfrac{t_b}{t_t - t_b} = \sqrt{\dfrac{R_b}{R_g}}$

$$E = mc^2$$

$K_\alpha = K_T \left(\dfrac{M}{M + m}\right)$ $\qquad\qquad$ $K_F = K_T \left(\dfrac{m}{M + m}\right)$ $\qquad\qquad$ $\alpha_T = \dfrac{I_e}{I_\gamma}$

$r = r_0 \sqrt[3]{A}$ $\qquad$ $E_B = a_v A - a_s A^{2/3} - a_c \left(\dfrac{Z^2}{A^{1/3}}\right) - a_a \dfrac{(N - Z)^2}{A} + \delta$ $\qquad$ $\delta = \pm a_\delta A^{-1/2}$ $\qquad$ $Z_A = \dfrac{A}{2}\left(\dfrac{81}{80 + 0.6A^{2/3}}\right)$

$KE = \tfrac{1}{2} mv^2$ $\qquad\qquad$ $KE = \left(\dfrac{1}{\sqrt{1 - \dfrac{v^2}{c^2}}} - 1\right)mc^2$ $\qquad\qquad$ $v = c\sqrt{1 - \left(\dfrac{mc^2}{KE + mc^2}\right)^2}$

$LET = SI \times W$ $\qquad\qquad$ $\text{Range} = R = \dfrac{E}{LET}$ $\qquad\qquad$ $R_{\alpha\text{-air}} \cong 0.31 E_\alpha^{3/2}$

$$RSP = \frac{R_{air}}{R_{abs}}$$

$$SI = \frac{4500 \frac{IP}{m}}{\frac{v^2}{c^2}}$$

$$R_{e^-} = \frac{E}{2.0 \text{ MeV/cm}}$$

$$E_{SC} = \frac{E_0}{1 + \left(\frac{E_0}{0.511}\right) \times (1 - \cos\theta)}$$

$$I = I_0 e^{-\mu x}$$

$$\ln\frac{I}{I_0} = -\mu x$$

$$HVL = \frac{\ln 2}{\mu}$$

$$TVL = \frac{\ln 10}{\mu}$$

$$\mu_m = \frac{\mu}{\rho}$$

$$\mu_a = \frac{\mu \times \text{atomic mass}}{\rho \times \text{Avogadro's \#}}$$

$$\mu_{eff} = \frac{\ln\frac{I_0}{I}}{x}$$

$$\mu_{eff} = \frac{\ln 2}{HVL}$$

$$E_{ED} = \frac{E_0^2}{E_0 + 0.2555}$$

$$E_{BS} = \frac{E_0}{1 + (3.91 \times E_0)}$$

$$E_0 = E_{ED} + E_{BS}$$

$$E_{tr} = -Q \times \left(\frac{A_A + A_X}{A_A}\right)$$

$$E_{ecb} \approx 1.11 \times \left(\frac{A_A + A_X}{A_A}\right) \times \left(\frac{Z_A Z_X}{A_A^{1/3} + A_X^{1/3}}\right)$$

$$E_{KC} \approx E_{KX} \times \left(\frac{A_X}{A_C}\right)$$

$$N_B = \sigma\Phi N_A t$$

$$N_B = \frac{\sigma\Phi N_A}{\lambda}\left(1 - e^{-\lambda t}\right)$$

$$A_0 = \sigma\Phi N_A \left(1 - e^{-\lambda t}\right)$$

$$A_{sat} = \sigma\Phi N_A$$

$$A_0 = \sigma\Phi N_A \lambda t$$

$$K_T = 0.80 \times \frac{Z_1 Z_2}{\sqrt[3]{A_1} + \sqrt[3]{A_2}}$$

$$K_1 = K_T \left(\frac{A_2}{A_1 + A_2}\right)$$

$$K_2 = K_T \left(\frac{A_1}{A_1 + A_2}\right)$$

$$\frac{rate_A}{rate_B} = \sqrt{\frac{MW_B}{MW_A}}$$

$$H = \text{dose} \times Q$$

$$f = \frac{\text{dose in medium}}{\text{dose in air}}$$

$$X = \frac{\Gamma \cdot A \cdot t}{d^2}$$

$$X = \frac{\Gamma \cdot A \cdot t}{d^2} B$$

$$B = \frac{I}{I_0}$$

$$f = 3.5 \times 10^{-4} \, ZE$$

$$\%D_n = \frac{D_n}{D_{max}} \times 100$$

$$\text{Dose Rate Change} = \frac{1}{\left(\frac{d_2}{d_1}\right)^2}$$

$$N_{mu} = \frac{\text{Tumor Dose}}{C_{cal} \times \frac{D_n}{D_{max}} \times C_{fs} \times C_{attn}}$$

$$N_{mu} = \frac{\text{Tumor Dose}}{C_{cal} \times C_{fs} \times C_{attn} \times TMR} \times \left(\frac{SAD_{treat}}{SAD_{cal}}\right)^2$$

$$d_{90\%} = \frac{E}{4.0 \text{ MeV/cm}}$$

$$d_{80\%} = \frac{E}{3.0 \text{ MeV/cm}}$$

$$N_{mu} = \frac{\text{Tumor Dose}}{O_f \times \frac{D_n}{D_{max}} \times C_{fs}}$$

Appendix D: Periodic Table of the Elements

Periodic Table of the Elements

1A																	8A
1 **H** 1.008	2A											3A	4A	5A	6A	7A	2 **He** 4.003
3 **Li** 7.0	4 **Be** 9.012											5 **B** 10.8	6 **C** 12.01	7 **N** 14.01	8 **O** 16.00	9 **F** 19.00	10 **Ne** 20.18
11 **Na** 22.99	12 **Mg** 24.30	3B	4B	5B	6B	7B		8B		1B	2B	13 **Al** 26.98	14 **Si** 28.08	15 **P** 30.97	16 **S** 32.1	17 **Cl** 35.45	18 **Ar** 39.95
19 **K** 39.10	20 **Ca** 40.08	21 **Sc** 44.96	22 **Ti** 47.87	23 **V** 50.94	24 **Cr** 52.00	25 **Mn** 54.94	26 **Fe** 55.84	27 **Co** 58.93	28 **Ni** 58.69	29 **Cu** 63.55	30 **Zn** 65.38	31 **Ga** 69.72	32 **Ge** 72.63	33 **As** 74.92	34 **Se** 78.97	35 **Br** 79.90	36 **Kr** 83.80
37 **Rb** 85.47	38 **Sr** 87.62	39 **Y** 88.90	40 **Zr** 91.22	41 **Nb** 92.91	42 **Mo** 95.95	43 **Tc** (99)	44 **Ru** 101.1	45 **Rh** 102.9	46 **Pd** 106.4	47 **Ag** 107.9	48 **Cd** 112.4	49 **In** 114.8	50 **Sn** 118.7	51 **Sb** 121.8	52 **Te** 127.6	53 **I** 126.9	54 **Xe** 131.3
55 **Cs** 132.9	56 **Ba** 137.3	71 **Lu** 175.0	72 **Hf** 178.5	73 **Ta** 180.9	74 **W** 183.8	75 **Re** 186.2	76 **Os** 190.2	77 **Ir** 192.2	78 **Pt** 195.1	79 **Au** 197.0	80 **Hg** 200.6	81 **Tl** 204.4	82 **Pb** 207.2	83 **Bi** 209.0	84 **Po** (209)	85 **At** (210)	86 **Rn** (222)
87 **Fr** (223)	88 **Ra** (226)	103 **Lr** (261)	104 **Rf** (263)	105 **Db** (268)	106 **Sg** (269)	107 **Bh** (270)	108 **Hs** (269)	109 **Mt** (278)	110 **Ds** (280)	111 **Rg** (281)	112 **Cn** (283)	113 **Nh** (285)	114 **Fl** (288)	115 **Mc** (289)	116 **Lv** (291)	117 **Ts** (294)	118 **Og** (294)

57 **La** 138.9	58 **Ce** 140.1	59 **Pr** 140.9	60 **Nd** 144.2	61 **Pm** (145)	62 **Sm** 150.4	63 **Eu** 152.0	64 **Gd** 157.2	65 **Tb** 158.9	66 **Dy** 162.5	67 **Ho** 164.9	68 **Er** 167.3	69 **Tm** 168.9	70 **Yb** 173.0
89 **Ac** (227)	90 **Th** 232.0	91 **Pa** 231.0	92 **U** 238.0	93 **Np** (237)	94 **Pu** (244)	95 **Am** (243)	96 **Cm** (247)	97 **Bk** (247)	98 **Cf** (251)	99 **Es** (252)	100 **Fm** (257)	101 **Md** (258)	102 **No** (259)

Bibliography

Acton, James M., and Mark Hibbs. 2012. *Why Fukushima was preventable*. The Carnegie Papers. Washington, DC: Carnegie Endowment.

Alley, William M., and Rosemarie Alley. 2013. *Too hot to touch*. New York: Cambridge University Press.

Allison, Wade. 2009. *Radiation and reason: The impact of science on a culture of fear*. York, UK: York.

American Nuclear Society. 2012. *Fukushima Daiichi: ANS committee report*. www.ans.org. Accessed on 5/25/12.

Baum, Edward M., Mary C. Ernesti, Harold D. Knox, Thomas R. Miller, and Aaron M. Watson. 2010. *Nuclides and isotopes: Chart of the nuclides*. 17th ed. Schenectady, NY: Bechtel Marine Propulsion Corporation.

Bertulani, Carlos A. and Helio Schechter. 2002. *Introduction to nuclear physics*. New York: Nova Science

Chandra, Ramesh 2011. *Nuclear medicine physics: The basics*. 7th ed. Philadelphia: Lippincott, Williams & Wilkins.

Chase, Grafton D. and Joseph L. Rabinowitz. 1967. *Principles of radioisotope methodology*. 3rd ed. Minneapolis: Burgess.

Cherry, Simon R., James A. Sorenson, and Michael E. Phelps. 2012. *Physics in nuclear medicine*. 4th ed. Philadelphia: Saunders.

Choppin, Gregory, Jan-Olov Liljenzin, Jan Rydberg, and Christian Ekberg. 2013. *Radiochemistry and nuclear chemistry*. 4th ed. Oxford: Academic Press.

Clark, Claudia. 1997. *Radium girls: Women and industrial health reform 1910–1935*. Chapel Hill, NC: University of North Carolina Press.

Cravins, Gwyneth. 2007. *Power to save the world: The truth about nuclear energy*. New York: Knopf.

De Sanctis, Enzo, Stefano Monti, and Marco Ripani. 2016. *Energy from nuclear fission: An introduction*. Switzerland: Springer.

Dunlap, Richard A. 2004. *An introduction to the physics of nuclei and particles*. Belmont, CA: Brooks/Cole.

Ehmann, William D. and Diane E. Vance. 1991. *Radiochemistry and nuclear methods of analysis*. New York: Wiley.

Friedlander, Gerhart, Joseph W. Kennedy, and Julian M. Miller. 1981. *Nuclear and radiochemistry*. 3rd ed. New York: Wiley.

Ferguson, Charles D. 2011. *Nuclear energy: What everyone needs to know*. New York: Oxford.

Gale, Robert Peter, and Eric Lax. 2013. *Radiation: What it is, what you need to know*. New York: Random House.

Gordon, Glen E., and Charles D. Coryell. 1967. Models for nuclear structure of spherical nuclei. *J. Chem. Educ.* 44(11), 636–646.

Graham, Donald T., Paul Cloke, and Martin Vesper. 2011. *Principles and applications of radiological physics*. 6th ed. Edinburgh: Churchill Livingstone.

Hála, Jirí and James D. Navratil. 2003. *Radioactivity, ionizing radiation and nuclear energy*. 2nd ed. Berkova, Czech Republic: Konvoj.

Heath, R. L. 1997. *Scintillation spectrometry: Gamma-ray spectrum catalogue*. Vol 1. 2nd ed. Idaho National Laboratory. https://gammaray.inl.gov/SitePages/Home.aspx. Accessed on 5/27/20.

Hendee, William R., and E. Russell Ritenour. 2002. *Medical imaging physics*. 4th ed. New York: Wiley.

Hendee, William R., Geoffrey S. Ibbott, and Eric G. Hendee. 2005. *Radiation therapy physics*. 3rd ed. New York: Wiley.

Hoppe, Michael E., and Taly G. Schmidt. 2012. Estimation of organ and effective dose due to Compton backscatter security scans. *Med. Phys.* 39(6), 3396–3403.

Khan, Faiz M. 2014. *The physics of radiation therapy*. 5th ed. Baltimore: Lippincott, Williams & Wilkins.

Knoll, Glenn F. 2010. *Radiation detection and measurement*. 4th ed. New York: Wiley.

Kondo, Sohei. 1993. *Health effects of low-level radiation*. Osaka: Kinki University Press.

Kröner, Markus, Christof H. Weber, Stefan Wirth, Klaus-Jürgen Pfeifer, and Maximilian F . Eriser, and Marcus Terrill. 2007. Advances in digital radiography: physical principles and system overview. *RadioGraphics* 27:675–686.

Lawrence Berkeley Laboratory. 2009. *The ABC's of Nuclear Science*. www.lbl.gov/abc/. Berkeley, CA: The Regents of the University of California. Accessed on 8/4/17.

Lieser, Karl H. 2001. *Nuclear and radiochemistry: Fundamentals and applications.* 2nd ed. Berlin: Wiley-VCH.

Loveland, Walter, David J. Morrissey, and Glenn T. Seaborg. 2017. *Modern nuclear chemistry.* 2nd ed. New York: Wiley.

Malley, Marjorie C. 2011. *Radioactivity: A history of a mysterious science.* New York: Oxford.

Mann, Robert. 2010. *An introduction to particle physics and the standard model.* Boca Raton: CRC Press.

Mayer, Maria Goeppert, J. Hans D. Jensen. 1955. *Elementary theory of nuclear shell structure.* New York: Wiley.

Moody, Kenton J., Ian D. Hutcheon, and Patrick M. Grant. 2014. *Nuclear forensic analysis.* 2nd ed. Boca Raton: Taylor & Francis.

Muller, Richard A. 2008. *Physics for future presidents: The science behind the headlines.* New York: W. W. Norton.

Nero, Anthony V. 1979. *A guidebook to nuclear reactors.* Berkeley, CA: University of California Press.

Prekeges, Jennifer 2012. *Nuclear medicine instrumentation.* 2nd ed. Boston: Jones and Bartlett.

Saha, Gopal B. 2010. *Fundamentals of nuclear pharmacy.* 6th ed. New York: Springer.

Saha, Gopal B. 2012. *Physics and radiobiology of nuclear medicine.* 4th ed. New York: Springer.

Semat, Henry and John R. Albright. 1972. *Introduction to atomic and nuclear physics.* 5th ed. New York: Holt, Rinehart and Winston.

Stanton, Robert and Donna Stinson. 1996. *Applied physics for radiation oncology.* Madison, WI: Medical Physics.

Tammemagi, Hans and David Jackson. 2009. *Half-lives: A guide to nuclear technology in Canada.* Don Mills, Ontario: Oxford.

Walker, J. Samuel. 2004. *Three Mile Island: A nuclear crisis in historical perspective.* Berkeley, CA: University of California Press

Weart, Spencer R. 2012. *The rise of nuclear fear.* Cambridge, MA: Harvard University Press.

Webb, Steve, ed. 1988. *The physics of medical imaging.* Bristol, UK: Adam Hilger.

Wolfson, Richard. 1991. *Nuclear choices: A citizen's guide to nuclear technology.* Cambridge, MA: MIT Press.

Index

Printed in the United States
by Baker & Taylor Publisher Services